Automation in Textile Machinery

Automation in Textile Machinery

Instrumentation and Control System Design Principles

L. Ashok Kumar

M. Senthilkumar

CRC Press
Taylor & Francis Group
Boca Raton London New York

CRC Press is an imprint of the
Taylor & Francis Group, an **informa** business

CRC Press
Taylor & Francis Group
6000 Broken Sound Parkway NW, Suite 300
Boca Raton, FL 33487-2742

First issued in paperback 2020

© 2018 by Taylor & Francis Group, LLC
CRC Press is an imprint of Taylor & Francis Group, an Informa business

No claim to original U.S. Government works

ISBN-13: 978-0-367-57187-0 (pbk)
ISBN-13: 978-1-4987-8193-0 (hbk)

**Visit the Taylor & Francis Web site at
http://www.taylorandfrancis.com**

**and the CRC Press Web site at
http://www.crcpress.com**

Contents

Organization of the Book

Chapter 1 provides the reader a deep knowledge about electrical and electronic terminologies and about the basic working principles of the important components. The control systems principles and their types are explained. Also, different types of controllers used in control systems are detailed.

Chapter 2 helps users understand the classification of different types of instruments. The working and methods of measurement of electrical and non-electrical parameters is explained. The concept of working of sensors and transducers and their classifications have been elaborated in detail.

Chapter 3 explains the working of programmable logic controller (PLC) and gives a clear view of programming methods of PLC. The types of inputs and outputs used in PLC are explained in detail. The user will understand the procedure to program a PLC for an industrial concept.

Chapter 4 deals with the process involved in the blowroom sequence, and the instruments and measuring devices used in the blowroom process are explained. Different control systems used in the process are explained to give a clear picture of a working blowroom sequence.

Chapter 5 explains the working of carding machine and the working principle of different sensors and transducers used in the carding machine. The importance of automation in the carding process is explained.

Chapter 6 deals with the working of draw frame and speed frame machines. The detailed explanation of sensors, transducers, and measuring devices used in these machines is given. The importance of the monitoring various parameters used in the machines is explained.

Chapter 7 describes the control systems used in the ring and rotor spinning machine. The working of different instruments and measuring devices is explained in detail.

Chapter 8 deals with various control systems, and automation processes are discussed in detail. The sensors and transducers used in the machine are explained clearly to help the user understand the working of the cone winding machine.

Chapter 9 describes the working principle of the warping and sizing machines. The control system concepts and the importance of measuring devices are explained in detail.

Chapter 10 deals with the control system concepts used in the weaving machine. The necessary measuring device used in the machinery is explained with its working principle. The interface of the weaving machine with monitoring systems is also discussed in detail.

Chapter 11 explains the working concept of the knitting machine and the control process used in the knitting machine. The computer interface of the knitting machine is explained in detail.

Chapter 12 discusses the importance of textile testing instruments. The working concepts of textile testing instruments are discussed in detail. The user will learn the concept of using sensors and transducers in measuring textile parameters.

Chapter 13 explains the concept of automation used in chemical processing. The importance of measuring the textile parameters is discussed in detail. The different types of plant manager systems in chemical processing are explained.

Chapter 14 deals with the machinery used in the garment industry. The working and electronic components used in the machinery are described in detail. The automation involved in the garment machinery is explained.

Chapter 15 discusses the importance of CAD/CAM for the textile industry. The design procedure and CAM usage in garment industries are explained in detail.

About This Book

Automation or automatic control is the use of various control systems for operating equipment such as machinery, processes in textile industries. Automation referred to as "the creation and application of technology to monitor and control the production and delivery of products and services." A control and instrumentation knowledge is essential for textile and electrical engineers to work in various sectors in textile manufacturing industries. These engineers are responsible for designing, developing, installing, managing, and/or maintaining equipment that is used to monitor and control engineering systems, machinery, and processes. *Automation in Textile Machinery: Instrumentation and Control System Design Principles*— is a comprehensive analysis of the current trends and technologies in automation and control systems used in textile engineering. Instrumentation is the use of measuring instruments to monitor and control a process. It is the art and science of measurement and control of process variables within a production, laboratory, or manufacturing area. In this text, we dissect the important components of an integrated control system in spinning, weaving, knitting, chemical processing, and garment industries and then determine if and how the components converge to provide manageable and reliable systems throughout the chain from fiber to the ultimate customer. Although the implementation of advanced process control strategies is not foreseen in the immediate future, it is apparent that the textile industry is slowly moving toward modular machines and systems. The dedicated systems still prevalent today are gradually being replaced by standard units, distributed automation concepts, and an increasing connectivity of the production floor with planning and scheduling systems.

We hope that this text will provide a guideline to engineers, researchers, scientists, and industrialists, as well as students of various disciplines such as EEE, ECE, robotics, instrumentation and control engineering, and textile engineering, and be a useful source of information in automation, instrumentation, and control systems in textile machinery.

Salient Features

The salient features of this book include:

- Fundamentals of electrical and electronic engineering in textiles
- Application of sensors, transducers, and control systems in textiles
- Instrumentation and control systems in spinning, weaving, knitting, and wet processing in the garment industry

Preface

The textile industry is one of the oldest industries in the world. In most sectors of textile manufacturing, automation is one major key to quality improvement and cost competitiveness. Early modernization and technical developments in textiles concentrated on the automation of individual machines and their processes. The textile industry has made many strides thanks to the advent of automation. The term *automation* is defined as the use of equipment and machinery to help make production easier and more efficient. Textiles such as cloth, yarn, cotton, and other fabrics have been made easier to produce thanks to automation.

Automation made it possible for the same tasks to be performed but with fewer hours of labor for employees. For example, inventions such as Eli Whitney's cotton gin made it possible to separate the seeds from cotton without using manual labor. Similar inventions of automation were created with the purpose of making textile-related jobs easier to perform and with less human labor.

Automation in the textile industry has provided safer working conditions for employees. The textile industry is known for transforming various cloths and fibers into fabrics. This process often includes dyeing and spinning, which are textile processes that can be relatively dangerous to an individual. Automation has created equipment to handle the bulk of these processes, making working conditions safer for all in the textile industry.

Automation in spinning has taken place in various processes such as picking and ginning, which were completely manual in the past. The high volume instrument (HVI) system has made it possible to carry out the cotton fiber test in seconds; this process used to take hours. HVI tests have improved the accuracy in measuring the cotton's staple lengths, color grade, micronaire, strength, elongation, and uniformity index. Cotton mixing has been automated so that uniformity can be achieved in the yarn. Blowroom performance has been improved by using a sequence of different machines, arranged in series and connected by transport ducts, for opening, cleaning, and blending. Automation is recently being used to separate the contamination of any color, size, and nature in the fiber. Machines using ultraviolet, optic, and acoustic technologies are being used for the detection and elimination of contaminant of any color, size, and nature, thus improving the overall quality of the final yarn produced. Automation has been achieved in spinning by the invention of machines such as ring spinning, air-jet spinning, rotor spinning, vortex spinning, and so on. Improvements in ring spinning machines have taken place through drive systems, drafting systems, and use of robotics. Yarn fault detection has been automated to improve production and to achieve uniform yarn quality. Yarn knots have now been replaced with the joints using splicing techniques such as air splicing, wet splicing, hot air splicing, and moist air splicing, which minimizes the defects in the final fabric.

All this automation in the spinning process has reduced the need for skilled manpower. Weaving machines have improved greatly in last three decades, resulting in improved quality and production. Major developments such as automatic shuttle and shuttleless looms have taken the industry to a new level. Shuttleless machines have made possible the efficient production of fault-free cloth. Developments in shuttleless machines have spanned three basic picking principles: rapier, projectile, and air-jet and water-jet.

Microprocessors are now integrated with weaving machines that monitor, control, regulate, and optimize all key aspects of the machines. Automation has been achieved with the help of microelectronics to control warp tension, picking of multi-filling colors, break detection, data collection, and so on, which has helped in improving the production as a lesser labor cost. Garmenting has undergone many advancements in the recent past.

Weaving and knitting machine builders have been leading the way in utilizing computer technology in textile manufacturing for many years with their use of CAD, bidirectional communication, and artificial intelligence. With the availability of electronic dobby and jacquard heads, automatic pick finding, needle selection, and so on, these technologies are easily integrated into computer networks of any production machines. Bidirectional communication systems can be used to control many functions on a weaving machine.

The automatic control of dyeing machines dates well back into the 1960s, and each succeeding year has shown miniaturization and enhancement in the management of information on a timelier basis. Automation started with the introduction of a system that controlled a set temperature by switching heaters on or off. A short time later, these were replaced by systems that controlled the dyeing cycle according to a time/temperature sequence. The processes of dye and auxiliary chemical addition as well as loading and unloading of textile materials were also automated to result in automated dye-house management. Now, the jiggers have been fully computerized with total automated control over the process. In the pad-batch dyeing system, the most outstanding development is the special dye dispensing system, online color monitoring, and dye pickup control.

Recent advances in imaging technology have resulted in high-quality image acquisition and advances in computer technology that allow image processing to be performed quickly and cheaply. This has given rise not only to a number of developments for laboratory quality testing equipment for fibers, yarns, and fabrics but also to development of online equipment for continuous monitoring of quality in textiles such as the fiber contamination eliminator, intelligent yarn grader, and automatic fabric inspection.

Research and development is being done in the textile machinery to achieve further automation and enhancements. The emphasis is on further improving quality and production and at the same time bringing down costs. Advancements are taking place to reduce the space and power requirements for various textile machinery, increasing their speed and efficiency. Big data and the Internet of things is also going to play a big role in future textile machinery to analyze machine behavior and proactively make decisions to improve the quality and productivity of the machines.

Automation technologies have helped the textile industry to increase output multiple times at a cheaper cost. Automation products and solutions are available now not only for the individual process or machine, but for the entire production line. Some of the key benefits achieved through automation are improved production at cheaper cost, better quality, safety for humans and machines, predictable production and inventory, energy savings, lower impact on the environment, better machine uptimes, self-diagnostics and predictive maintenance, efficient packaging and transport, and improved customer satisfaction.

Acknowledgments

The authors are always thankful to the Almighty for perseverance and achievements. The authors owe their gratitude to Shri L. Gopalakrishnan, Managing Trustee, PSG Institutions, and Dr. R. Rudramoorthy, Principal, PSG College of Technology, Coimbatore, India, for their wholehearted cooperation and great encouragement in this successful endeavor.

Dr. L. Ashok Kumar would like to take this opportunity to acknowledge those people who helped me in completing this book. This book would not have come to its completion without the help of my students, my department staff and my institute, and especially my project staff. I am thankful to all my research scholars and students who are doing their project and research work with me. But the writing of this book is greatly possible mainly because of the support of my family members, parents, and sisters. Most importantly, I am very grateful to my wife, Y. Uma Maheswari, for her constant support during writing. Without her, all these things would not be possible. I would like to express my special gratitude to my daughter A. K. Sangamithra for her smiling face and support; it helped a lot in completing this work.

Dr. M. Senthilkumar would like to thank the management, the principal, PSG Polytechnic College, and all my colleagues who have been with me in all my endeavors with their excellent, unforgettable help and assistance in the successful execution of this work. I am very grateful to my family, especially my son S. Harish, for their support in completing this project.

Authors

Dr. L. Ashok Kumar is a Postdoctoral Research Fellow from San Diego State University, California. He is a recipient of the BHAVAN fellowship from the Indo-US Science and Technology Forum. His current research focuses on integration of renewable energy systems in the smart grid and wearable electronics. He has 3 years of industrial experience and 18 years of academic and research experience. He has published 137 technical papers in international and national journals and presented 107 papers in national and international conferences. He has completed 16 government of India funded projects, and currently 5 projects are in progress. His PhD work on wearable electronics earned him a National Award from ISTE, and he has received 24 awards on the national level. Ashok Kumar has five patents to his credit. He has guided 82 graduate and postgraduate projects. He is a member and in prestigious positions in various national forums. He has visited many countries for institute industry collaboration and as a keynote speaker. He has been an invited speaker in 125 programs. Also he has organized 62 events, including conferences, workshops, and seminars. He completed his graduate program in Electrical and Electronics Engineering from University of Madras and his post-graduate from PSG College of Technology, India, and Masters in Business Administration from IGNOU, New Delhi. After completion of his graduate degree, he joined as project engineer for Serval Paper Boards Ltd., Coimbatore (now ITC Unit, Kovai). Presently he is working as a professor and associate HoD in the Department of EEE, PSG College of Technology and also doing research work in wearable electronics, smart grid, solar PV, and wind energy systems. He is also a Certified Charted Engineer and BSI Certified ISO 500001 2008 Lead Auditor. He has authored the following books in his areas of interest (1) *Computational Intelligence Paradigms for Optimization Problems Using MATLAB®/SIMULINK®*, CRC Press, (2) *Solar PV and Wind Energy Conversion Systems—An Introduction to Theory, Modeling with MATLAB/SIMULINK, and the Role of Soft Computing Techniques*—Green Energy and Technology, Springer, USA (3) *Electronics in Textiles and Clothing: Design, Products and Applications*, CRC Press, (4) *Power Electronics with MATLAB*, Cambridge University Press, London (5) Monograph on *Smart Textiles* (6) Monograph on *Information Technology for Textiles*, and (7) Monograph on *Instrumentation & Textile Control Engineering*.

Dr. M. Senthilkumar obtained his diploma in Textile Technology from PSG Polytechnic College, Coimbatore, India, and pursued his graduate degree, B.Tech (Textile Technology), from Bannari Amman Institute of Technology, Tamilnadu, India. He received best outgoing student award for the academic year 2000–2001. After serving for two years in a knitted garment unit, he qualified in the GATE examination and ranked 75th. He pursued his post graduate degree, M.Tech (Textile Technology), from D.K.T.E Textile and Engineering Institute, Maharashtra, India. He has been involved in teaching and research activities and completed his PhD in the area of Dynamics of elastic fabrics under the guidance of Dr. N. Anbumani, Department of Textile Technology, PSG College of Technology, Coimbatore, India, in 2012. He worked as a lecturer in the Department of Apparel and Fashion Technology, Sona College of Technology, Salem, Tamilnadu, from August 2004 to August 2006. Then he joined as a lecturer in the Department of Textile Technology, PSG Polytechnic College, Coimbatore, India, from September 2006 to June 2017 and was promoted to head of the department, his current position. So far he has published nearly

50 research papers in various international and national journals. He has contributed one chapter as a coauthor for the book entitled *Military Textiles*, from Woodhead Publishing, Cambridge, England. He received the Young Engineer Award from the Institution of Engineers, Kolkata, India, in 2014. He is highly interested in academic and research activities in the area of weft knitting and its product development.

1

Control Systems Engineering

LEARNING OBJECTIVES

- To comprehend the fundamentals of electrical and electronics engineering
- To identify electrical terminologies
- To know the concepts of basic electronics
- To understand the basic concepts of control systems
- To recognize the types of controllers

1.1 Introduction

In this text, we present a comprehensive analysis of the current trends and technologies in control systems for the textile industry. Our approach is to dissect the important components of an integrated control system and then to determine if and how the components are converging to provide manageable and reliable systems throughout the chain from fiber to the ultimate customer. Although the implementation of advanced process control strategies is not foreseen in immediate future, it is apparent that the textile industry is slowly moving toward modular machines and systems. The dedicated systems still prevalent today are gradually being replaced by standard units, distributed automation concepts, and an increasing connectivity of the production floor with planning and scheduling systems.

1.2 Electrical Terminology

Charge is the most fundamental concept in electricity. It derives from the properties of elementary particles, with protons (and hence the nucleus of the atom) being positively charged, and electrons negatively charged. The unit of charge is the coulomb (C). The charge on the electron is 1.60219×10^{-19} C (i.e., one coulomb corresponds to about 6×10^{18} electrons). An important quantity in electrochemistry, known as Faraday's Constant (or often just the Faraday) and given the symbol F, is the charge associated with one mole of a singly charged species such as H^+ or Cl^-. As one mole contains Avogadro's number (6.0228×10^{23}) molecules, the Faraday is $6.0228 \times 10^{23} \times 1.60219 \times 10^{-19}$ C, or 96485 C.

Current is the rate of flow of charge along a conductor (note that this charge may be electrons flowing in a metal or ions flowing in solution). One Amp (A) corresponds to a flow of 1 coulomb per second.

Potential is an indication of the potential energy of a unit charge at a particular point in a circuit. Strictly it is the potential energy involved in moving a charge of one coulomb to that point from infinity; therefore, it is quite difficult to measure.

Potential difference or voltage is the difference in potential between two points. There is one volt between two points if one Joule is required to move one coulomb from one point to the other. As with current, potential may apply to charge in the form of ions or in the form of electrons. However, for a valid potential difference, the charge must be the same at each location.

Resistance is the tendency of a conductor to obstruct the flow of current. Ohm's Law states that the voltage (V) across a resistor is proportional to the current (I) flowing through it:

$$V = IR$$

where R is the resistance, which is measured in units of Ohm. Like many laws, this is an approximation, and many conductors, including the metal-solution interface, have a nonlinear resistance.

Capacitance is the tendency of a device incorporating two conductors that are insulated from each other to absorb charge when the voltage between the conductors is changed. The charge, Q, is given by

$$Q = C\Delta V$$

where:
  C is the capacitance (which is measured in Farads)
  ΔV is the change in voltage

It is noted that the symbol C is used conventionally both for capacitance and coulombs.

We can see the effect of trying to pass a current through a capacitor by remembering that current is charge per unit time. Hence:

$$Q = I\Delta t$$

$$I = C\frac{\Delta V}{\Delta t} = C\frac{dV}{dt}$$

where:
  Δt is the incremental change in time
  dV is the instantaneous change in voltage
  dt is the instantaneous change in time

1.2.1 Inductance

Due to the interrelation between electric currents and magnetic fields, there is a tendency for current to flow at a constant rate through a conductor. In most real conductors, this tendency is counteracted by the resistance of the conductor, although in superconductors, which have no resistance, current will flow essentially forever unless the current is caused to change by the application of a voltage. The inductance (*L*) of a particular conductor is a measure of the voltage needed to cause the current to change at a particular rate.

$$L = \frac{V}{dI/dt}$$

where *dI/dt* is the instantaneous rate of current change. The units of inductance are Henrys. One Henry will give a rate of change of current of one A/s with one voltage (*V*) applied across it.

1.2.2 Impedance

Impedance is a general term used to describe the relationship between the voltage across a component (or essentially any device capable of allowing at least some current to flow) and the current flowing through that device. It is normally used in relation to alternating current with a sine waveform, but it is perfectly valid to refer to the impedance at zero frequency (i.e., direct current).

1.2.3 Amplitude

For fluctuating voltage or current, the amplitude describes how large the fluctuations are. There are several ways of describing the amplitude (Figure 1.1):

- *The root mean square (RMS) value*: As implied in the name, this is obtained by taking the square root of the average value of the square of the voltage or current. When applied specifically to ac signals, the dc level (the average value of the voltage or current) may be subtracted before calculating the RMS value. The RMS value indicates the power present in the signal.
- *The peak-to-peak value*: This is simply the maximum value minus the minimum value. While it is a simple value to measure, it has the disadvantage that signals may have the same peak-to-peak voltage, yet deliver very different powers into a load.
- *The power spectrum*: The two previous measurements only give an overall indication of the power present in the signal, with no indication of how that power is distributed in terms of the frequency. The power spectrum presents the power present in the signal at each frequency (known as the power spectral density, with units of V^2/Hz or I^2/Hz).
- Note that both the RMS and the peak-to-peak amplitudes will depend on the frequency response of the measurement system, with a wider bandwidth giving a larger measured amplitude.

1.2.4 Phase

The phase of a sine wave describes its position in time relative to a reference sine wave of the same frequency. The units used are radians or degrees, and relate to the general equation for a sine wave:

$$y = a \sin e(\omega t + \theta)$$

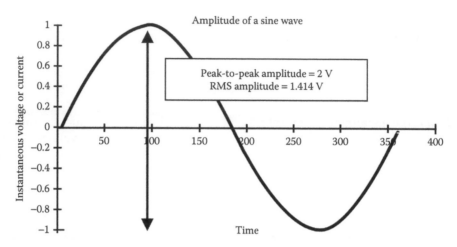

FIGURE 1.1
Amplitude.

where:
  θ is the phase
  ω is the angular frequency
  t is the time
  a is the amplitude

The phase of one signal relative to another may be described as "leading," if the voltage for the signal being considered reaches a specific point in the cycle before the reference sine wave. Conversely, the signal is said to "lag" the reference signal if it reaches a given point in the cycle after the reference signal. Note that whether a lead or lag is observed depends on which signal is taken as the reference (see the Figure 1.2). It is also valid to describe a 90° lag as a 270° lead. When referring to impedance measurements, the current is taken as the reference signal; so the phase of the impedance at a particular frequency will be the phase of the voltage with respect to the current (Figure 1.2).

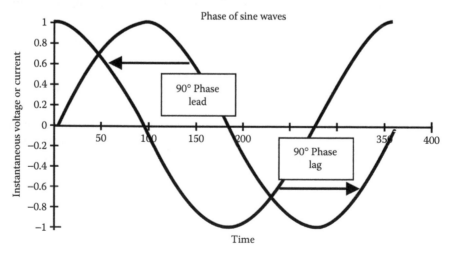

FIGURE 1.2
Phases of sine waves.

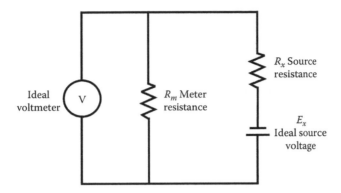

FIGURE 1.3
Measurement of voltage.

1.2.5 Measurement of Voltage

An ideal voltmeter will measure the voltage between its two input terminals, without in any way affecting that voltage. The main problem with voltage measurements is the requirement for current to flow through the voltmeter. Real voltmeters can be treated as an ideal voltmeter, together with a resistor between the two input terminals, which requires a current to flow in order to measure the voltage (Figure 1.3).

R_m in the Figure 1.3 is referred to in voltmeter specifications as the input resistance or input impedance for conventional digital voltmeters or multimeters. It will commonly be 10 MOhm, although it can be up to 1000 MOhm. Electrometers and pH meters are designed to give very high input impedance, typically in the region of 10^{14} Ohm. The effect of these resistances is to allow current to flow, and if there are high resistances associated with the voltages being measured, this may lead to significant errors. For example, if potential of a painted specimen is being measured with a voltmeter with a 10 MOhm input resistance, and the resistance of the paint film is 1 MOhm, this will give an error of 10%. If, as is entirely possible, the resistance of the paint film is 10^8 Ohm, the error will be 90%. For most situations, electrometers are unlikely to give errors of any significance. With modern microelectronic devices, it is very easy to construct an amplifier that will give an input impedance that is comparable to that of an electrometer.

1.2.6 Measurement of Small Voltages

With conventional instrumentation, the smallest voltage that can be resolved accurately is controlled by the input offset voltage of the amplifier used, or the input bias current of the amplifier flowing through the source resistance of the voltage being measured. The most stable amplifiers are automatically zeroed by switching the input between the voltage to be measured and a short circuit. This produces a device with input offset voltage less than 1 μV. The best commercial digital voltmeters give resolutions down to 10 nV. Great care must be taken to minimize noise pickup in such sensitive measurements, and thermal emfs associated with the interconnection of different metals can lead to significant errors (of the order of 1 μV).

1.2.7 Measurement of Current

An ideal ammeter will measure the current flowing between its two terminals while at the same time behaving as a perfect conductor, thus maintaining the two terminals at the same potential. Real ammeters will create a potential difference between the terminals, and this may be represented as a resistor, known as the internal resistance, in series with an ideal ammeter (Figure 1.4).

Typical digital multimeters measure current by measuring the voltage across a resistor, and they will usually develop 10–100 mV across the resistor for a full scale current reading. This doesn't cause any consequences in electrochemical experiments, as the potential drop will have a negligible effect on the current flowing. For example, in a potentiostatic experiment the cell current may be measured in the lead between the potentiostat and the counter electrode, and the potentiostat will provide the extra voltage needed without any difficulty. The main application of current measurements where it is important to minimize the potential drop across the meter is in the study of galvanic corrosion and the measurement of electrochemical current noise. Current can be measured with essentially zero potential drop across the meter (<1 mV) using a current amplifier. This is available within an ammeter, in which case it is known as a zero resistance ammeter, or it can be constructed as an attachment for a conventional multimeter. A potentiostat also can be configured to operate as a zero resistance ammeter.

1.2.8 Measurement of Small Currents

Current amplifiers are the most sensitive devices for measuring very small currents. In this device, the lower limit to the measurable current is controlled by the input bias currents of the amplifier used, or by the input offset voltage of the amplifier acting on the source impedance of the current source being measured. Devices are currently available with maximum input bias currents of 75 fA (75×10^{-15} A). Remembering that the charge on the electron is 1.6×10^{-19} C, this corresponds to only about 5×10^5 electrons per second, or one electron every 2 μs. Similar performance can also be achieved with commercial electrometers, which provide very high quality zero resistance ammeters. In order to maintain the very low leakage currents, it is essential to take great care with connections to the amplifier input, as leakage through dirty or poor quality insulators can far exceed the amplifier input leakage current.

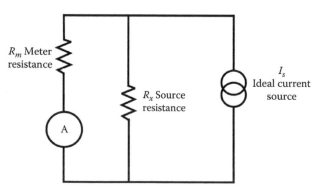

FIGURE 1.4
Measurement of current.

1.2.9 Noise

Noise may be defined as an unwanted signal superimposed on a signal of interest, although the term may also be used to describe a signal consisting of apparently random fluctuations. The latter may be of interest, as in the case of electrochemical noise. However, in this section we are concerned with noise as an unwanted component of a signal, and the ways in which we can reduce noise or cope with it.

We shall typically find two types of noises. They are random noise and interference noise. "Random" noise has a wide frequency content, which derives from the properties of conductors and electronic devices, and interference noise is derived from man-made sources, such as radio-frequency emissions, mains-frequency pick up, spikes due to fridges switching on or off, and so forth.

1.2.10 Interference Noise

As far as noise derived from interference is concerned, there are two basic rules to minimize the amount of noise coupled into the measuring circuits.

1.2.11 Screen Circuits

The principal of screening is to surround the signal circuits with conductors held at ground potential. Then electromagnetic radiation will couple with the screening, where it will be harmlessly adsorbed, rather than the signal circuits. The simplest form of screening is to use screened cables, in which the signal cable is surrounded by braiding that is connected to the ground. This is reasonably effective, although it is difficult to extend the screening to cover all parts of the circuit, in particular the electrochemical cell itself.

For sensitive measurements, an approach is to surround the entire experiment by a "Faraday cage." Essentially this is a grounded conductive box that shields its contents from electromagnetic radiation. Since the Faraday cage only shields against radiation coming from outside the box, it is usually best to use only battery-powered instruments inside the Faraday cage, in order to avoid introducing mains-frequency noise into the cage with mains-powered instruments. Outputs from these devices should be carried through the wall of the cage by screened cables with the screen connected to the cage.

1.2.12 Avoid Signal or Ground Loops

One of the main "man-made" noise problems is the pick-up of noise at mains frequency. This is most severe when a loop of wire is exposed to a mains-frequency electromagnetic field, which, in effect, forms a single turn transformer. If the loop is complete, large currents can flow through it, developing significant potential differences between different parts of the loop. Loops in ground circuits can be particularly insidious because instruments frequently have parts of their circuitry connected to ground.

For example, the working electrode terminal of a potentiostat is commonly connected to ground, as is the negative input of an oscilloscope. However, if both of these connections remain in place while trying to monitor the cell potential, this will establish a ground loop, which will increase the noise in the circuit due to circulating ac currents in the ground loop. The remedy to this problem is to disconnect all but one of the connections to ground. In doing this, one must, of course, pay attention to safety requirements, since the ground connections serve to protect against hazardous voltages in addition to controlling noise.

1.2.13 Electronic Noise

There are various sources of noise in electronic components. The most fundamental of these is due to the random motions of electrons in a conductor. The random motion corresponds to a fluctuating current, and this current will develop a voltage across the resistance of the conductor. For good conductors, the amplitude of this noise is very small, but for large resistances it can become significant. The phenomenon is known as Johnson noise, and the rms noise voltage (e_x) is given by

$$e_x = \sqrt{4KTbR} \tag{1.1}$$

where:
  K is the Boltzmann's constant $(1.38 \times 10^{-23}\,\text{J/K})$
  T is the temperature (K)
  b is the bandwidth (Hz)
  R is the resistance (Ω)

In much electrochemistry we are concerned with measurements from dc to about 10 Hz, hence we have a 10 Hz bandwidth. If our source resistance is 1 MOhm, this will give an rms noise of approximately 0.4 µV. This would usually be tolerable, but we should be aware that noise of this level is unavoidable. Equation 1.1 shows that Johnson noise will create the most severe problems when we wish to measure over a wide frequency range (increasing b) with very high source impedance.

In "real" electronic components, other forms of noise are possible. When a current is flowing through a circuit, there will inevitably be a level of shot noise as a result of the quantized nature of electrical charge. Additionally, semiconductors are subject to *flicker noise*, which gives particularly strong noise at low frequencies.

1.2.14 Frequency Response and Filtering

In many electrochemical measurements, we are concerned with very low frequency measurements, and there is a tendency to ignore the complications that result from the limited frequency response of the instruments being used. However, these may become significant in some circumstances. Frequency determining components tend to take the form of resistor–capacitor combinations, such as those shown in Figure 1.5.

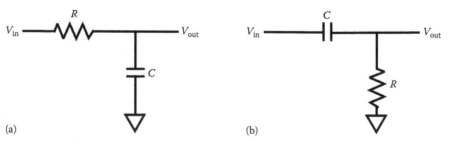

(a)   (b)

FIGURE 1.5
Filters: (a) low-pass filter and (b) high-pass filter.

These two configurations are described as low and high-pass filters on the basis of those frequencies that the filter allows to pass through it. Considering the low-pass filter, the impedance of the resistor will be constant at R, while the impedance of the capacitor will be $1/2\pi fC$, where f is the frequency and C is the capacitance.

1.2.15 Potential Divider

A potential divider is simply two resistors, across which a voltage is applied, with the output being taken from the junction of the resistors. Since the same current must be flowing through both resistors (assuming that the output is connected to a high impedance device), we have $V_1/R_1 = I = V_2/R_2$. Hence $V_1/V_2 = R_1/R_2$, that is, the voltage applied across the resistors is divided into two parts according to the resistor values. If a single variable resistor is used for R_1 and R_2, then the output voltage can be varied, and this is the basis of the volume control.

These two impedances will act as a potential divider, giving

$$V_{\text{out}} = \frac{V_{\text{in}}/2\pi fC}{\left(R + \dfrac{1}{2\pi fC}\right)} = \frac{V_{\text{in}}}{(2\pi fRC + 1)} \tag{1.2}$$

For frequencies very much less than $1/2\pi RC$, V_{out} will be approximately equal to V_{in}, whereas for frequencies very much greater than $1/2\pi RC$, V_{out} will be given by $V_{\text{in}}/2\pi RC$. This may be seen by plotting $V_{\text{out}}/V_{\text{in}}$ on a logarithmic plot of $V_{\text{out}}/V_{\text{in}}$ versus frequency (Figure 1.6). $V_{\text{out}}/V_{\text{in}}$ may be described as the gain of the filter and is frequently described in units of decibels (dB). These are slightly odd logarithmic units that happen to be convenient in signal processing. One decibel corresponds to a factor of ten increase in the output power of a circuit as compared to the input circuit. Assuming a constant load resistance, the power is proportional to the square of the voltage; hence a gain of 20 dB means that the output voltage is ten times the input voltage.

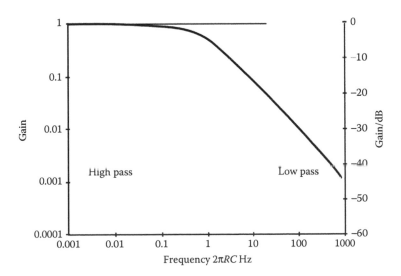

FIGURE 1.6
Frequency Gain.

The gain of a low-pass filter is zero dB at low frequencies. At $f = 1/2\pi RC$, the gain will be 1/2 (since the impedance of the resistor and the capacitor are equal at this frequency), hence the gain will be −6 dB (20 × log 0.5). At higher frequencies, the gain will fall by a factor of 10 for each decade of frequency, that is, −20 dB/decade of frequency (also described as −6 dB/octave).

1.2.16 Operational Amplifiers

Modern microelectronic technology has made a wide range of extraordinarily sophisticated devices available to the engineer. Fortunately, it has also made some of the simpler devices very easy to use, and you do not need a degree in electronic engineering to produce some very useful results. The basis of many devices is a class of general-purpose amplifiers called operational amplifiers. These were originally developed as building blocks for analogue computers, but they have since proved to be extremely versatile devices and will be found in most instruments. A typical operational amplifier has five connections (Figure 1.7).

Two of these connections supply the power to the amplifier (shown here as the common values of +15 and −15 V, although other voltages are also used); two give inputs to the amplifier, while one provides the amplifier output. The operational amplifier is a voltage-controlled device. The output voltage (measured relative to the midpoint of the power supply voltages) is a function of the difference between the two input voltages according to the equation:

$$e_0 = A(e_+ - e_-) \tag{1.3}$$

where A is known as the *gain* of the amplifier and is very large. Ideally the gain is infinite; practical gains are in the region of 10^5–10^7. When drawing circuits using operational amplifiers, it is common to omit the power supply connections for clarity, but these will always be required.

1.2.17 The Non-Inverting Buffer

An amplifier with such a high gain is of no use as it stands, but it proves to be very easy to modify the characteristics of the amplifier by connecting the output back to one or other of the inputs. The simplest example is obtained by connecting the output directly to the non-inverting input (Figure 1.8):

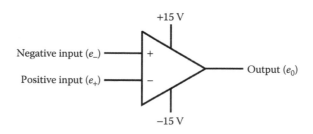

FIGURE 1.7
Operational amplifier.

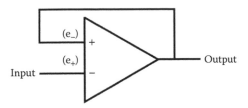

FIGURE 1.8
Non inverting buffer.

We have $e_0 = A(e_+ - e_-)$

$$e_- = e_0$$

$$e_0 = A(e_+ - e_-)$$

$$e_0 = \frac{A}{A+1} e_+ \dots \dots \dots \dots \dots \tag{1.4}$$

Since $A \gg 1$, $e_0 \cong e_+$

Thus with this configuration, the output voltage will be almost exactly the same as the input voltage. Again, this may not seem terribly useful until we examine the properties of the operational amplifier in a little more detail. For the ideal operational amplifier, the input impedance at both the inverting and the non-inverting inputs is infinite (i.e., no current will flow into the input terminals). Conversely the output impedance is zero (i.e., the voltage at the output will remain the same whatever current is drawn from the output). Now we can see that this circuit can be very useful. For example, if we have a voltmeter with an input resistance of 10 MOhms, and we wish to measure the potential of a painted specimen, we have seen that we are liable to get measurement errors due to the high resistance of the paint film. If we place our circuit between the lead from the reference electrode and our voltmeter, we will be able to make accurate measurements because the circuit will accurately reflect the reference electrode potential without allowing any current to flow. So far, we have been simplifying the discussion slightly by talking in terms of the properties of an ideal amplifier. In reality, of course, we can only approach the ideal, but in many cases, developers of microelectronic devices have got remarkably close to the ideal. The properties of typical devices are shown in Table 1.1.

1.2.18 Operational Amplifier Properties

Gain: The ratio of the output voltage (relative to the mid-point of the power supply) to the difference between the voltages at the non-inverting and the inverting inputs. The gain is specified for dc voltages (it is typically constant below about 1 Hz), and the gain for ac signals will decrease as the frequency increases.

Input offset voltage: The difference in voltage between the two input terminals when the output is actually zero (this can often be adjusted to near zero if required).

Input offset voltage drift

The change in input offset voltage as a function of temperature.

Input bias current—The current that flows into either of the input terminals.

TABLE 1.1

Ideal and Typical Operational Amplifier Properties

Properties	Ideal	Typical Device	Best Device
Gain	∞	10^6	5×10^6
Input offset voltage	0	25 μV	0.7 μV
Input Offset Voltage Temperature Drift	0	0.6 μV/°C	0.01 μV/°C
Input bias current	0	40 nA	75 fA
Input bias current temperature drift	0	<1 nA/°C	×2/10°C
Output impedance	0	70 Ohms	0.1 Ohms
Maximum output current	∞	17 mA	15 A
Maximum output voltage	∞	10 V	145 V
Unity gain band-width	∞	8 MHz	5 GHz
Slew rate	∞	1.9 V/μs	500 V/μs

Input bias current drift: The variation of the input bias current with temperature. The nature of this will vary according to the technology used for the device. Bipolar devices will typically provide a drift that is reasonably constant over a moderate temperature range, and this will be specified as A/°C. Devices based on field-effect transistors will typically have an input bias drift that doubles about every 10°C.

Output impedance: The effective impedance at the output (this may be thought of as a resistor between the output of an ideal amplifier and the real amplifier). It is not usually very important as the feedback circuits tend to compensate for any errors, although it will place one limit on the maximum current that the device can deliver.

Maximum output current: Standard low-cost amplifiers will usually deliver at least 10 mA without problems. Larger current can be obtained using power amplifiers. These will often require heat-sinks to prevent the device overheating when delivering a large current.

Maximum output voltage: Standard low-cost amplifiers will usually provide an output voltage of ±10 V without problems when operating from ±15 V power supplies. Higher output voltages require special-purpose amplifiers or discrete transistor output stages.

Unity gain band-width: The gain of a typical operational amplifier is inversely proportional to frequency. The unity-gain band-width is that frequency at which the gain has fallen to one, and indicates the frequency response of the amplifier.

Slew rate: This is the rate at which the output voltage will change when the input is instantaneously changed by significant amount.

1.2.19 Operational Amplifier Circuits—Unity-Gain Non-Inverting Buffer

The output of this circuit will be the input voltage. The input impedance is very high (the effective input impedance of the amplifier is increased by the feedback circuit), while the output impedance is very low (the output impedance is similarly reduced by the feedback circuit). The main errors are associated with the input offset voltage of the amplifier, together with the effects of the input bias current when using high source impedances (Figure 1.9).

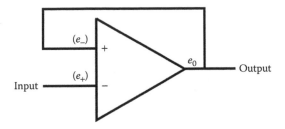

FIGURE 1.9
Unity-gain non-inverting buffer.

1.2.20 Non-Inverting Voltage Amplifier

By reducing the proportion of the output voltage fed back to the inverting input, the gain of the amplifier can be increased. For the circuit shown, the gain will be $R_2/(R_1 + R_2)$. Gains up to about 1000 can be obtained reasonably easily, although the input offset voltage of the amplifier is magnified by the gain, so this may become a significant voltage when large gains are used. Note that the ground connection shown in the figure is the reference point for the measurement of voltages (Figure 1.10).

1.2.21 Differential Voltage Amplifier

Note that the two resistors labeled R_1 and similarly the two labeled R_2 have the same value. The output voltage is given by the difference between the + and − inputs times the gain, where the gain is R_2/R_1. One limitation of this circuit is the low input impedance (approximately $R_1 + R_2$), and if this presents a problem, an instrumentation amplifier should be used. Gains of up to about 1000 may be used without problems (Figure 1.11).

1.2.22 Instrumentation Amplifier

An instrumentation amplifier is essentially a differential voltage amplifier with a high input impedance. It can be constructed using two or three operational amplifiers, but it is simpler to use a ready-made device. An example is the analog devices.

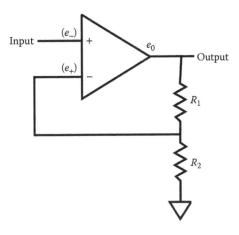

FIGURE 1.10
Non inverting voltage amplifier.

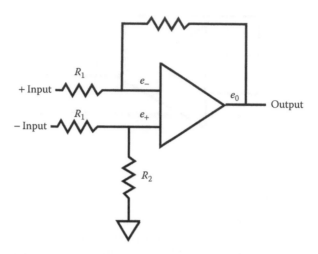

FIGURE 1.11
Differential voltage amplifier.

1.2.23 Current Amplifier

In the current amplifier, the current at the −ve input terminal must be supplied from the amplifier output via R_1 (since the input to the amplifier has a very high impedance). Also, the action of the amplifier will be to deliver sufficient current such that the voltage at the inverting input is the same as that at the non-inverting input. Hence there will be no difference in potential between the two inputs (except for the input offset voltage of the amplifier), and the output voltage will be $I_{in}R_1$ (from Ohm's Law). The maximum sensitivity of the current amplifier is limited by the input bias current of the amplifier or (probably more commonly) the effect of the input offset voltage of the amplifier on the source resistance of the current source (Figure 1.12).

1.2.24 Potentiostat

In the simple potentiostat circuit, the voltage at the counter electrode terminal will be maintained by the amplifier such that the voltage at the reference electrode input is the same as that at the non-inverting input of the amplifier. Hence, the potential of the reference electrode relative to the working electrode will be the same as the control voltage. Note that this circuit connects the working electrode to the ground of the system power supply. Some care may therefore be needed to avoid ground loops or similar problems during the measurement of cell potential and current (Figure 1.13).

1.2.25 Galvanostat

In the Galvanostat, the output current will flow through the external circuit and then through the resistor R. The output voltage will be controlled such that the voltage across R is equal to the control voltage and thereby controlling the current. The main limitations of this circuit are

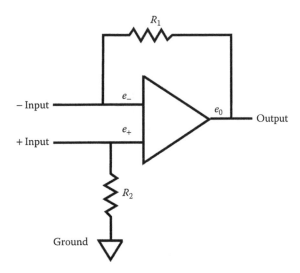

FIGURE 1.12
Circuit diagram of current amplifier.

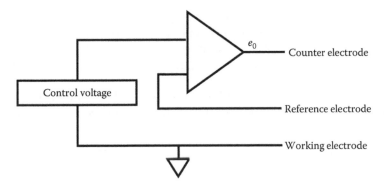

FIGURE 1.13
Circuit diagram of potentiostat.

- The total voltage available will be limited by the maximum output voltage of the amplifier (this is not usually a problem for electrochemical work)
- The current is limited by the maximum output current of the amplifier
- The control of very small currents will become limited by the input bias current of the amplifier and/or the action of the input offset voltage on the resistor R (Figure 1.14)

1.2.26 Active Filter

There are many forms of active filter, but in general the amplifier is used to compensate for the effect of the impedance of passive frequency-determining elements. The simplest active filter consists of a resistor-capacitor high- or low-pass filter followed by a unity-gain non-inverting buffer.

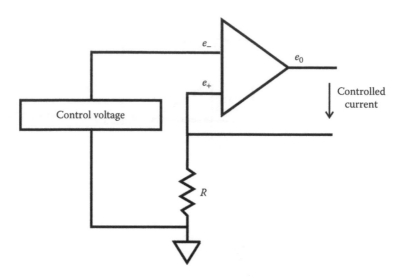

FIGURE 1.14
Circuit of Galvanostat.

This gives what is known as a first order response, with the gain above or below the cut-off frequency being proportional (or inversely proportional) to frequency. Addition of a second resistor and capacitor gives a second order filter, with gain proportional to frequency. Note that changing from a high-pass to a low-pass filter essentially just involves swapping the resistors and capacitors.

The frequency responses for active filters of orders 1–4 are shown in Figure 1.15. The response for high pass filters will be essentially the same but reflected about the y axis.

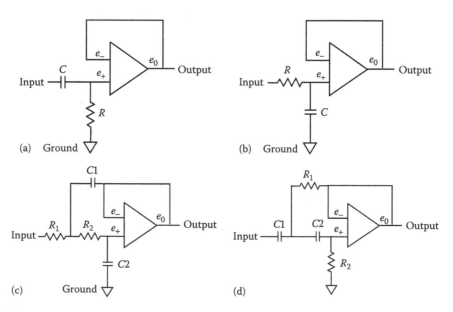

FIGURE 1.15
Circuit diagram of active filters: (a) first order high pass active filter; (b) first order low pass active filter; (c) second order low pass active filter; and (d) second order high pass active filter.

1.3 Cell Design for Electrochemistry

A typical electrochemical cell consists of three electrodes (working, counter, and reference) together with features for the control of solution flow and state of aeration (Figure 1.16).

1.3.1 The Working Electrode

The first stage in the preparation of the working electrode should be to characterize the material in terms of its chemical composition and metallurgical condition. The specimen must then be mounted in some way to expose it to the solution. This may present a number of problems, including the presence of gas-solution interfaces or crevices between the mounting system and the specimen.

Whatever mounting method is used, it is important that there is a sound electrical connection to the specimen, preferably achieved by spot-welding, soldering, or bolting. The point of connection and connecting wire must be isolated from the solution, either by way of the specimen mount or by keeping it out of the solution. Crevices are particularly damaging for passive specimens, where you may obtain a very large error in the observed passive current density. Typical methods that have been used to avoid crevice problems include PTFE compression seals, epoxy resin mounting with careful pretreatment of the metal, and the flushing of a filter paper filled crevice with distilled water.

The surface treatment of the specimen is also important. This should be reproducible and relevant to the intended application, although these two requirements are often incompatible. Typical surface conditions include as received (this is often very relevant, but not very reproducible), ground, and polished. Electrochemical reactions are very sensitive to surface contamination, and it is important to clean the specimen before testing, particularly to remove grease. However, care must also be taken to avoid contamination of the surface with species extracted from mounting materials (e.g., epoxy resins) by powerful solvents such as acetone.

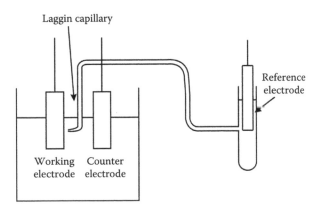

FIGURE 1.16
Diagramic electrochemical cell.

1.3.2 The Counter Electrode (or Secondary or Auxiliary Electrode)

The counter electrode must be inert in the conditions to which it is exposed in order to avoid contamination of the solution. Typically, platinum or graphite are used. It is often specified that the counter electrode should have a large area compared to the working electrode, but this may not be necessary in many cases. The fundamental requirement is that it be able to pass sufficient current into the solution without needing an excessive cell voltage or creating a non-uniform current distribution on the working electrode.

1.3.3 The Reference Electrode

Saturated Calomel Electrode (SCE) is the most commonly used in corrosion studies. The Ag/AgCl electrode is very convenient for small, cheap electrodes (but it is chloride sensitive). Many corrosion systems are very sensitive to chloride, and the Hg/Hg_2SO_4 electrode is often useful to avoid chloride contamination. Recent legislation on the transport by air of objects containing metallic mercury has led to a move from the SCE to Ag/AgCl (Figure 1.17).

A *Luggin* capillary is used to shield the potential measurement from potential drop in solution, such that the potential is effectively measured at the tip of the capillary. This should therefore be close to the working electrode, but not so close as to shield part of the surface from the current flow (the ideal is about 3× the tip diameter of the capillary).

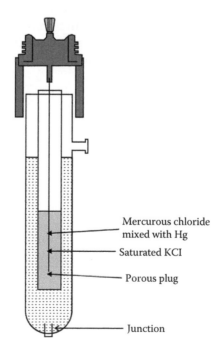

Mercurous chloride mixed with Hg

Saturated KCl

Porous plug

Junction

FIGURE 1.17
Saturated calomel electrode.

1.3.4 Composition

The solution composition, including aeration and pH, must be well-defined and controlled. Similarly, the solution temperature should be controlled, or at least measured, as the rates of electrochemical reactions are typically very temperature dependent.

1.3.5 Solution Flow

For systems that may be subject to mass transport control, it is also important to control the flow of solution past the electrode. Various methods are available to achieve this. For systems with only a moderate tendency to mass transport control, it may be sufficient simply to stir the solution or to rely on agitation from the gas bubbling used to control the oxygen concentration of the solution. These methods will generally give reasonably good mass transport, but the exact conditions will be difficult to define and reproduce.

1.3.6 The Rotating Disk Electrode

The rotating disk electrode (RDE) consists of a disk on the end of an insulated shaft that is rotated at a controlled angular velocity. Providing the flow is laminar over all of the disk, the mathematical description of the flow is surprisingly simple, with the solution velocity toward the disk being a function of the distance from the surface, but it is independent of the radial position. Consequently, the mass transport conditions are uniform over the surface of the disk, and the limiting current is given by

$$i_L = 1.554zFAD^{2/3}V^{-1/6}C_\infty\omega^{1/2} \tag{1.5}$$

The rotating cylinder electrode provides more uniform behavior over the electrode surface than the RDE when turbulent flow conditions are used (the central region of the RDE is always laminar).

Flow channels have the advantage that the electrode remains stationary, and is therefore easily examined during the experiment. The design of a flow channel to achieve well-defined flow conditions requires considerable care. In particular, the electrodes must be accurately aligned with the wall of the channel, with no gaps or steps that could induce turbulence, and the flow pattern must be well-established before reaching the electrodes, which requires a long *entry region* to allow the flow to stabilize.

1.4 Principles of Control Systems

Instrumentation provides the various indications used to operate a nuclear facility. In some cases, operators record these indications for use in day-to-day operation of the facility. The information recorded helps the operator evaluate the current condition of the system and take actions if the conditions are not as expected. Requiring the operator to take all of the required corrective actions is impractical, or sometimes impossible, especially if a large number of indications must be monitored. For this reason, most systems are controlled automatically once they are operating under normal conditions. Automatic controls greatly reduce the burden on the operator and make his or her job manageable. Process variables requiring control in a system include, but are not limited to, flow, level, temperature, and pressure. Some systems do not require all of their process variables to be controlled. A basic heating system operates on temperature and disregards the other atmospheric parameters of the house.

The thermostat monitors the temperature of the house. When the temperature drops to the value selected by the occupants of the house, the system activates to raise the temperature of the house. When the temperature reaches the desired value, the system turns off. Automatic control systems neither replace nor relieve the operator of the responsibility for maintaining the facility. The operation of the control systems is periodically checked to verify proper operation. If a control system fails, the operator must be able to take over and control the process manually. In most cases, understanding how the control system works aids the operator in determining if the system is operating properly and which actions are required to maintain the system in a safe condition.

A *control system* is a system of integrated elements whose function is to maintain a process variable at a desired value or within a desired range of values. The control system monitors a process variable or variables, and then causes some action to occur to maintain the desired system parameter. In the example of the central heating unit, the system monitors the temperature of the house using a thermostat. When the temperature of the house drops to a preset value, the furnace turns on, providing a heat source. The temperature of the house increases until a switch in the thermostat causes the furnace to turn off. Two terms that help define a control system are input and output.

Control system input is the stimulus applied to a control system from an external source to produce a specified response from the control system. In the case of the central heating unit, the control system input is the temperature of the house as monitored by the thermostat.

Control system output is the actual response obtained from a control system. In the example above, the temperature dropping to a preset value on the thermostat causes the furnace to turn on, providing heat to raise the temperature of the house. In the case of nuclear facilities, the input and output are defined by the purpose of the control system. A knowledge of the input and output of the control system enables the components of the system to be identified. A control system may have more than one input or output. Control systems are classified by the control action, which is the quantity responsible for activating the control system to produce the output. The two general classifications are open-loop and closed-loop control systems.

1.4.1 Open-Loop Control System

An *open-loop control system* (Figure 1.18) is the one in which the control action is independent of the output. An example of an open-loop control system is a chemical addition pump with a variable speed control. The feed rate of chemicals that maintain proper chemistry of a system is determined by an operator, who is not part of the control system. If the chemistry of the system changes, the pump cannot respond by adjusting its feed rate (speed) without operator action.

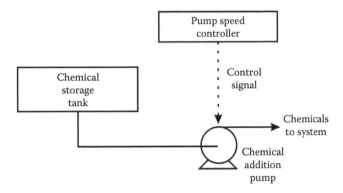

FIGURE 1.18
Open-loop control devices.

1.4.2 Closed-Loop Control System

Systems that utilize feedback are called *closed-loop* control systems. The feedback is used to make decisions about changes to the control signal that drives the plant. An open-loop control system doesn't have or doesn't use feedback.

A basic closed-loop control system is shown in Figure 1.19. This figure can describe a variety of control systems, including those driving elevators, thermostats, and cruise control. Closed-loop control systems typically operate at a fixed frequency. The frequency of changes to the *drive signal* is usually the same as the sampling rate, and certainly not any faster. After reading each new sample from the sensor, the software reacts to the plant's changed state by recalculating and adjusting the drive signal. The plant responds to this change, another sample is taken, and the cycle repeats. Eventually, the plant should reach the desired state and the software will cease making changes.

A closed-loop control system is one in which control action is dependent on the output. Figure 1.19 shows an example of a closed-loop control system. The control system maintains water level in a storage tank. The system performs this task by continuously sensing the level in the tank and adjusting a supply valve to add more or less water to the tank. The desired level is preset by an operator, who is not part of the system.

Feedback is information in a closed-loop control system about the condition of a process variable. This variable is compared with a desired condition to produce the proper control action on the process. Information is continually "fed back" to the control circuit in response to control action. In the previous example, the actual storage tank water level, sensed by the level transmitter, is fed back to the level controller. This feedback is compared with a desired level to produce the required control action that will position the level control as needed to maintain the desired level. Under normal circumstances, cellular control mechanisms lead to the transcription and translation of the gene for the cystic fibrosis transmembrane conductance regulator (CFTR) protein in pulmonary epithelial cells, and this usually results in the insertion of optimal amounts of a camp-regulated chloride channel into the plasma membrane. But there is no mechanism to check whether this has actually taken place, and in individuals homozygous for mutation, a defective protein is assembled and directed to the wrong cellular compartment, with disastrous consequences for the person concerned. The distinguishing feature of closed-loop control systems is that a check is made on the outcome, and corrective measures are initiated if the result differs from the original plan.

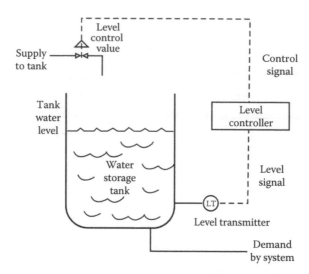

FIGURE 1.19
Closed-loop control system.

Certain fundamental features are common to all closed-loop control systems that keeps a controlled variable C as close as possible to some reference value R despite interference by an external load L which disturbs the result. In order to achieve its objective, the control system *subtracts* C from R so as to generate an *error signal*, E. This error signal regulates the flow of material or energy M into the controlled system so as to minimize E and compensate for the effects of the external load (Figure 1.20).

Every closed-loop system needs a *reference value* that provides a target to aim for. This is true even for biological control systems, although sometimes the targets are obscure. There is no requirement for the target to stay constant, although they often do. Biological reference values may be genetically determined, for example through the amino acid sequences of regulatory proteins, which define their binding constants for allosteric effectors. Behavioral targets for an organism might also reflect the genetically programmed "wiring diagram" for the central nervous system.

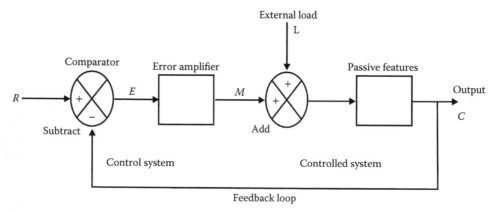

FIGURE 1.20
Block diagram of control systems.

The overall performance of any closed-loop control system can be judged in terms of the following criteria: Accuracy: How closely does it approach the target value? Stability: Is it free from overshoots and oscillations? Resilience: How well does it cope with abnormal loads? Speed: How quickly does it respond to a transient load, or to a change in the target value?

Real systems must compromise between these four conflicting requirements, since the features that make for good performance in one area may well have deleterious effects elsewhere. It is sometimes necessary to distinguish between the control system proper and the *final control elements* that actually deliver energy to the controlled system. The control system itself (e.g., nerves, electronic amplifiers, and office procedures) may exhibit ideal behavior, but the final control elements (e.g., muscles, motorized valves, and bloody-minded employees) may suffer from a variety of imperfections. It may be necessary to build a *secondary control system* around the final control elements to ensure that they obey their instructions with reasonable fidelity. Apparently simple actions, such as reaching for a glass of water, may themselves form part of a larger mechanism for the long-term regulation of salt and water balance. Even the simplest muscular actions depend on a vast network of neural feedback loops that respond to visual, proprioceptive, and tactile error signals until the immediate objective has been achieved. All the individual nerves and muscles involved in the action are themselves dependent on thousands of intracellular control mechanisms that regulate every aspect of their metabolism, and ensure that they will respond quickly and accurately to the instructions received from the higher centers (Figure 1.21).

It is often convenient to divide closed-loop systems into the standard building blocks shown in the figure. H, G1, G2, and G3 are the *transfer functions* that define the relationship between the input and output signals for each of the blocks. H is the feedback path gain and G1, G2, and G3 are individual block gains. Transfer functions in general are mathematical *operators* (analogous to multiplication, differentiation, etc.) but they may sometimes be simple numerical constants. If the blocks function independently (i.e., have zero output impedance and infinite input impedance) then the various transfer functions may be combined in order to calculate composite functions for the complete system. For example:

$$M1 = C1 \cdot E' \quad M3 = G1 \cdot G2 \cdot (R - H \cdot C) + L$$

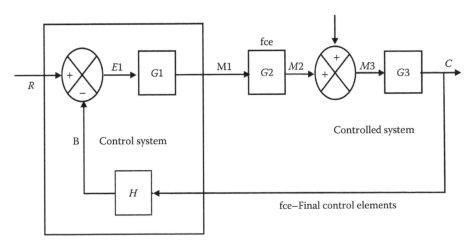

FIGURE 1.21
Block diagram of standard control system.

TABLE 1.2

Names and Symbols for Transfer Function

Mechanical Term		Electrical Term	
Forward loop transfer function	G	Amplifier open-loop gain	A
Feedback loop transfer function	H	Feedback fraction	B
Closed-loop transfer function	C/R	Closed-loop gain	V_{out}/V_{in}

The symbol E' is used above in preference to the error E because the output signal C might be modified within the feedback loop. These considerations lead eventually to the conclusion:

$$C = \frac{G}{(1+GH)}R + \frac{G3}{(1+GH)}L$$

where $G = G1 \cdot G2 \cdot G3...$, and so on. The composite operator G is the resultant transfer function for the whole of the forward loop while H is the resultant transfer function for the feedback loop. Equation given below is the *operational equation* for the system (Table 1.2).

$$\frac{C}{R} = \frac{G}{(1+GH)}[\text{at constant } L] \qquad \frac{C}{L} = \frac{G3}{(1+GH)}[\text{at constant } R]$$

In an ideal world, C/R would be constant and C/L would be zero: the output should exactly follow the reference signal and be independent of the load. This requires that the *loop gain* $G \cdot H$ should be as large as possible when traversing both the forward and the feedback routes. The ratio C/R is variously known as the *closed-loop transfer function* or the *closed-loop gain*. Unfortunately, electrical and mechanical engineers sometimes use different names and symbols for the same basic concepts, although the underlying theory is identical.

1.4.3 Automatic Control System

An automatic control system is a preset closed-loop control system that requires no operator action. This assumes the process remains in the normal range for the control system. An automatic control system has two process variables associated with it: a controlled variable and a manipulated variable. A *controlled variable* is the process variable that is maintained at a specified value or within a specified range. In the previous example, the storage tank level is the controlled variable. A *manipulated variable* is the process variable that is acted on by the control system to maintain the controlled variable at the specified value or within the specified range.

1.4.3.1 Functions of Automatic Control

In any automatic control system, the four basic functions that occur are measurement, comparison, computation, and correction. In the water tank level control system in the example above, the level transmitter measures the level within the tank. The level transmitter sends a signal representing the tank level to the level control device, where it is compared to a desired tank level. The level control device then computes how far to open the supply valve to correct any difference between actual and desired tank levels.

1.4.3.2 Elements of Automatic Control

The three functional elements needed to perform the functions of an automatic control system are a measurement element, an error detection element, and a final control element

Relationships between these elements and the functions they perform in an automatic control system are shown in Figure 1.4. The measuring element performs the measuring function by sensing and evaluating the controlled variable. The error detection element first compares the value of the controlled variable to the desired value, and then signals an error if a deviation exists between the actual and desired values. The final control element responds to the error signal by correcting the manipulated variable of the process (Figure 1.22).

1.4.3.2.1 Feedback Control

An automatic controller is an error-sensitive, self-correcting device. It takes a signal from the process and feeds it back into the process. Therefore, closed-loop control is referred to as feedback control.

1.4.3.2.2 Control Loop Diagrams

A *block diagram* is a pictorial representation of the cause and effect relationship between the input and output of a physical system. A block diagram provides a means to easily identify the functional relationships among the various components of a control system. The simplest form of a block diagram is the *block and arrows diagram*. It consists of a single block with one input and one output. The block normally contains the name of the element or the symbol of a mathematical operation to be performed on the input to obtain the desired output. Arrows identify the direction of information or signal flow.

Although blocks are used to identify many types of mathematical operations, operations of addition and subtraction are represented by a circle, called a *summing point*. As shown

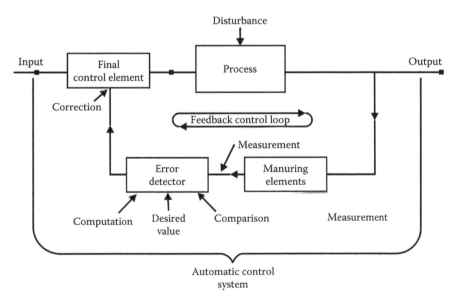

FIGURE 1.22
Relationships of functions and elements in an automatic control system.

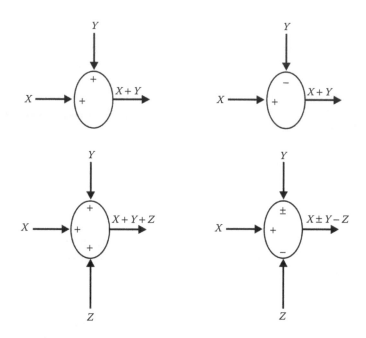

FIGURE 1.23
Summing points.

in Figure 1.23 a summing point may have one or several inputs. Each input has its own appropriate plus or minus sign. A summing point has only one output and is equal to the algebraic sum of the inputs. A *takeoff point* is used to allow a signal to be used by more than one block or summing point.

1.4.3.3 Feedback Control System Block Diagram

The figure shows basic elements of a feedback control system as represented by a block diagram. The functional relationships between these elements are easily seen. An important factor to remember is that the block diagram represents flow paths of control signals, but does not represent flow of energy through the system or process (Figure 1.24).

1.4.3.3.1 Lubricating Oil

Lubricating oil reduces friction between moving mechanical parts and also removes heat from the components. As a result, the oil becomes hot. This heat is removed from the lube oil by a cooler to prevent both breakdown of the oil and damage to the mechanical components it serves. The lube oil cooler consists of a hollow shell with several tubes running through it. Cooling water flows inside the shell of the cooler and around the outside of the tubes. Lube oil flows inside the tubes. The water and lube oil never make physical contact. As the water flows through the shell side of the cooler, it picks up heat from the lube oil through the tubes.

This cools the lube oil and warms the cooling water as it leaves the cooler. The lube oil must be maintained within a specific operating band to ensure optimum equipment performance. This is accomplished by controlling the flow rate of the cooling water with a *temperature control loop*. The temperature control loop consists of a temperature transmitter, a temperature controller, and a temperature control valve. The diagonally crossed lines

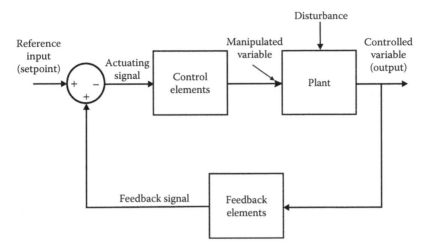

FIGURE 1.24
Feedback control system block diagram.

indicate that the control signals are air (pneumatic). The lube oil temperature is the controlled variable because it is maintained at a desired value (the set point). Cooling water flow rate is the manipulated variable because it is adjusted by the temperature control valve to maintain the lube oil temperature. The temperature transmitter senses the temperature of the lube oil as it leaves the cooler and sends an air signal that is proportional to the temperature controller. Next, the temperature controller compares the actual temperature of the lube oil to the setpoint (the desired value). If a difference exists between the actual and desired temperatures, the controller will vary the control air signal to the temperature control valve. This causes it to move in the direction and by the amount needed to correct the difference.

For example, if the actual temperature is greater than the setpoint value, the controller will vary the control air signal and cause the valve to move in the open direction. This results in more cooling water flowing through the cooler and lowers the temperature of the oil leaving the cooler (Figure 1.25).

The oil cooler is the plant in this example, and its controlled output is the lube oil temperature. The temperature transmitter is the feedback element. It senses the controlled output and oil temperature and produces the feedback signal. The feedback signal is sent to the summing point to be algebraically added to the reference input (the setpoint). Notice the setpoint signal is positive and the feedback signal is negative. This means the resulting actuating signal is the difference between the set point and feedback signals. The actuating signal passes through the two control elements: the temperature controller and the temperature control valve. The temperature control valve responds by adjusting the manipulated variable (the cooling water flow rate). The oil temperature changes in response to the different water flow rate, and the control loop is complete.

1.4.3.3.2 Process Time Lags

In the last example, the control of the lube oil temperature may initially seem easy. Apparently, the operator need only measure the lube oil temperature, compare the actual temperature to the desired (set point), compute the amount of error (if any), and adjust the temperature control valve to correct the error accordingly. However, processes have

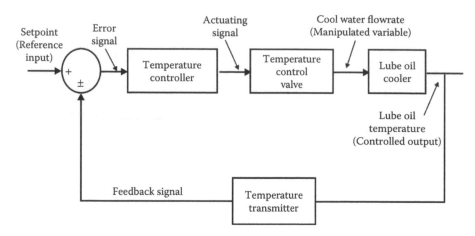

FIGURE 1.25
Block diagram oil temperature control loop.

the characteristic of delaying and retarding changes in the values of the process variables. This characteristic greatly increases the difficulty of control. *Process time lag* is the general term that describes these process delays and retardations.

Process time lags are caused by three properties of the process. They are *capacitance, resistance*, and *transportation time*. *Capacitance is* the ability of a process to store energy. In the figure, for example, the walls of the tubes in the lube oil cooler, the cooling water, and the lube oil can store heat energy. This energy-storing property gives the ability to retard change. If the cooling water flow rate is increased, it will take a period of time for more energy to be removed from the lube oil to reduce its temperature. *Resistance* is that part of the process that opposes the transfer of energy between capacities. In Figure 1.25, the walls of the lube oil cooler oppose the transfer of heat from the lube oil inside the tubes to the cooling water outside the tubes. *Transportation time* is time required to carry a change in a process variable from one point to another in the process. If the temperature of the lube oil is lowered by increasing the cooling water flow rate, some time will elapse before the lube oil travels from the lube oil cooler to the temperature transmitter. If the transmitter is moved farther from the lube oil cooler, the transportation time will increase. This time lag is not just slowing down a change; it is an actual time delay during which no change occurs.

1.4.3.4 Stability of Automatic Control Systems

All control modes previously described can return a process variable to a steady value following a disturbance. This characteristic is called "stability." *Stability* is the ability of a control loop to return a controlled variable to a steady, non-cyclic value, following a disturbance.

Control loops can be either stable or unstable. Instability is caused by a combination of process time lags discussed earlier (i.e., capacitance, resistance, and transport time) and inherent time lags within a control system. This results in slow response to changes in the controlled variable. Consequently, the controlled variable will continuously cycle around the setpoint value. *Oscillations* describes this cyclic characteristic. There are three

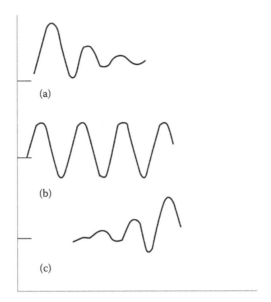

FIGURE 1.26
Types of oscillations: (a) decreasing amplitude; (b) constant amplitude; and (c) increasing amplitude.

types of oscillations that can occur in a control loop. They are *decreasing amplitude, constant amplitude,* and *increasing amplitude.*

Each is shown in Figure 1.26. *Decreasing amplitude.* These oscillations decrease in amplitude and eventually stop with a control system that opposes the change in the controlled variable. This is the condition desired in an automatic control system.

1.4.3.5 Two Position Control Systems

A controller is a device that generates an output signal based on the input signal it receives. The input signal is actually an error signal, which is the difference between the measured variable and the desired value, or setpoint (Figure 1.27).

This input error signal represents the amount of deviation between where the process system is actually operating and where the process system is desired to be operating. The controller provides an output signal to the final control element, which adjusts the process system to reduce this deviation. The characteristic of this output signal is dependent on the type or mode of the controller. This chapter describes the simplest type of controller, which is the two position, or ON–OFF, mode controller.

1.4.3.6 Proportional Control Systems

1.4.3.6.1 Control Mode

In the proportional control mode, the final control element is throttled to various positions that are dependent on the process system conditions. For example, a proportional controller provides a linear stepless output that can position a valve at intermediate positions, as well as "full open" or "full shut." The controller operates within a band that is between the 0% output point and the 100% output point and where the output of the controller is proportional to the input signal (Figure 1.28).

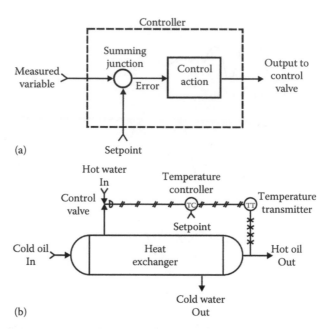

FIGURE 1.27
(a and b) Process control system.

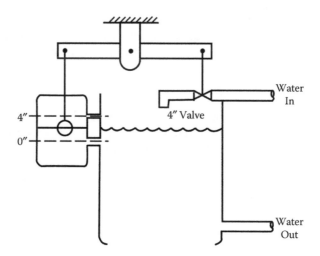

FIGURE 1.28
Proportional system controller.

1.4.3.6.2 *Proportional Band*

With proportional control, the final control element has a definite position for each value of the measured variable. In other words, the output has a linear relationship with the input. Proportional band is the change in input required to produce a full range of change in the output due to the proportional control action. Or simply, it is the percent change of the input signal required to change the output signal from 0% to 100%. The proportional band determines the range of output values from the controller that operate the final control element.

The final control element acts on the manipulated variable to determine the value of the controlled variable. The controlled variable is maintained within a specified band of control points around a setpoint. In this example of a proportional level control system, the flow of supply water into the tank is controlled to maintain the tank water level within prescribed limits. The demand that disturbances placed on the process system are such that the actual flow rates cannot be predicted. Therefore, the system is designed to control tank level within a narrow band in order to minimize the chance of a large demand disturbance causing overflow or run out. A fulcrum and lever assembly is used as the proportional controller. A float chamber is the level measuring element, and a 4-in stroke valve is the final control element. The fulcrum point is set such that a level change of 4-in causes a full 4-in stroke of the valve. Therefore, a 100% change in the controller output equals 4 in. The proportional band is the input band over which the controller provides a proportional output and is defined as follows:

$$\text{Proportional band \% change in input \% change in output} \times 100\% \qquad (1.6)$$

For this example, in the Figure 1.28 the fulcrum point is such that a full 4-in change in float height causes a full 4-in stroke of the valve.

$$\text{P.B.} = 100\% \text{ change in input } 100\% \text{ change in output} \times 100\%$$

Therefore: P.B. = 100%. The controller has a proportional band of 100%, which means the input must change 100% to cause a 100% change in the output of the controller. If the fulcrum setting was changed so that a level change of 2 in, or 50% of the input, causes the full 3-in stroke, or 100% of the output, the proportional band would become 50%. The proportional band of a proportional controller is important because it determines the range of outputs for given inputs.

1.4.3.6.3 Reset Control (Integral)

Integral control describes a controller in which the output rate of change is dependent on the magnitude of the input. Specifically, a smaller amplitude input causes a slower rate of change of the output. This controller is called an integral controller because it approximates the mathematical function of integration. The integral control method is also known as reset control.

1.4.3.6.4 Definition of Integral Control

A device that performs the mathematical function of integration is called an integrator. The mathematical result of integration is called the integral. The integrator provides a linear output with a rate of change that is directly related to the amplitude of the step change input and a constant that specifies the function of integration. The step change has an amplitude of 10%, and the constant of the integrator causes the output to change 0.2% per second for each 1% of the input. The integrator acts to transform the step change into a gradually changing signal. As you can see, the input amplitude is repeated in the output every 5 s. As long as the input remains constant at 10%, the output will continue to ramp up every 5 s until the integrator saturates.

1.4.3.7 Proportional-Integral-Derivative Control Systems

1.4.3.7.1 Proportional-Integral-Derivative

For processes that can operate with continuous cycling, the relatively inexpensive two position controller is adequate. For processes that cannot tolerate continuous cycling, a proportional controller is often employed. For processes that can tolerate neither continuous cycling nor offset error, a proportional plus reset controller can be used. For processes that need improved stability and can tolerate an offset error, a proportional plus rate controller is employed. However, there are some processes that cannot tolerate offset error yet need good stability. The logical solution is to use a control mode that combines the advantages of proportional, reset, and rate action. This chapter describes the mode identified as proportional plus reset plus rate, commonly called Proportional-Integral-Derivative (PID).

1.4.3.7.2 Proportional Plus Reset Plus Rate Controller Actions

When an error is introduced to a PID controller, the controller's response is a combination of the proportional, integral, and derivative actions, as shown in Figure 1.30. Assume the error is due to a slowly increasing measured variable. As the error increases, the proportional action of the PID controller produces an output that is proportional to the error signal. The reset action of the controller produces an output whose rate of change is determined by the magnitude of the error. In this case, as the error continues to increase at a steady rate, the reset output continues to increase its rate of change. The rate action of the controller produces an output whose magnitude is determined by the rate of change.

1.4.3.8 Controllers

Controllers are the controlling element of a control loop. Their function is to maintain process variables such as pressure, temperature, level, and so on, at some desired value. This value may or may not be constant. The function is accomplished by comparing a setpoint signal (desired value) with the actual value (controlled variable). If the two values differ, an error signal is produced. The error signal is amplified (increased in strength) to produce a controller output signal. The output signal is sent to a final control element, which alters a manipulated variable and returns the controlled variable to setpoint. This chapter will describe two controllers commonly found in nuclear facility control rooms. Although plants may have other types of controllers, information presented here will generally apply to those controllers as well.

Control stations perform the function of a controller and provide additional controls and indicators to allow an operator to manually adjust the controller output to the final control element (Figure 1.29).

The *setpoint indicator*, located in the center of the upper half of the controller, indicates the setpoint (desired value) selected for the controller. The scale may be marked 0%–100% or correspond directly to the controlled variable (e.g., 0–1000 psig or −20°F to +180°F). The *setpoint adjustment*, located right of the setpoint indicator, is a thumbwheel type adjustment dial that allows the operator to select the setpoint value. By rotating the thumbwheel, the scale moves under the setpoint index line.

1.4.3.8.1 Self-Balancing Control Station

Self-balancing control system describes a control station in which the non-operating mode output signal follows (tracks) the operating mode output signal. When the control station is in the automatic mode, the manual output signal will follow the

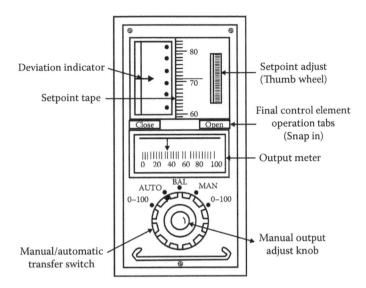

FIGURE 1.29
Typical control station.

automatic output signal. Once the controller is transferred to the manual mode, the output signal will remain at its previous value until one of the manual push buttons is depressed. Then, the output will vary. When the controller is in manual mode, the automatic output signal will track the manual output signal. Once the controller is transferred from the manual to automatic mode, the automatic output signal will initially remain at the manual mode value. If a deviation did exist in the manual mode, the automatic output signal would change slowly and return the controlled variable to setpoint (Figure 1.30).

Manual push buttons: These buttons are located below each end of the output meter and are used in the manual mode of operation. Buttons are labeled to indicate their effect on the final control element. The labels are "open–close" for valves and "slow–fast" for variable speed devices. The left push button decreases the output signal. The right push button increases the output signal. Either button can be depressed at two different positions, half-in and full-in. At the half-in position, the output signal changes slowly. At the full-in position, the output signal changes about ten times faster.

Mode indicating lights: Located directly below the manual push buttons, these lights indicate the operating mode of the controller. When in manual mode, the left light, labeled "M," will be lit; when in the automatic mode, the right light, labeled "A," will be lit

Mode selection buttons: Located directly under each mode indicating light, each button will select its respective mode of control. If the button below the "M" mode light is depressed, the controller will be in the manual mode of operation; if the button below the "A" mode light is depressed, the controller will be in the automatic mode of operation. As previously discussed, a particular plant will probably have controllers different from the two described here. Although most information provided can be generally applied, it is extremely important that the operator know the specific plant's controllers and their applications.

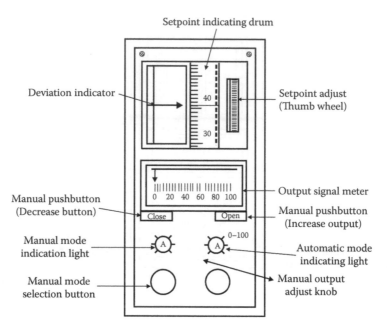

FIGURE 1.30
Self-balancing control station.

Final control elements are devices that complete the control loop. They link the output of the controlling elements with their processes. Some final control elements are designed for specific applications. For example, neutron-absorbing control rods of a reactor are specifically designed to regulate neutron-power level. However, the majority of final control elements are general application devices such as valves, dampers, pumps, and electric heaters. Valves and dampers have similar functions. Valves regulate flow rate of a liquid while dampers regulate flow of air and gases. Pumps, like valves, can be used to control flow of a fluid. Heaters are used to control temperature. These devices can be arranged to provide a type of "on-off" control to maintain a variable between maximum and minimum values. This is accomplished by opening and shutting valves or dampers or energizing and de-energizing pumps or heaters. On the other hand, these devices can be modulated over a given operating band to provide a proportional control. This is accomplished by positioning valves or dampers, varying the speed of a pump, or regulating the current through an electric heater. There are many options to a process control. Of the final control elements discussed, the most widely used in power plants are valves. Valves can be easily adapted to control liquid level in a tank, temperature of a heat exchanger, or flow rate.

1.5 Summary

The fundamental parameters, components, and principles of control systems have been discussed in this chapter. To design control systems, one must bring together various disparate aspects of a system and make them work together, and often that process is highly analytical and mathematical. Monitoring and controlling process variables during the textile manufacturing process minimizes waste, costs, and environmental impact. The terminologies and principles disused in this chapter will be useful in understanding the forthcoming chapters.

References

1. R. C. Dorf and R. H. Bishop. *Modern Control Systems*. Pearson Education, Upper Saddle River, NJ, 11th ed., International Edition, 2008.
2. W. S. Levine. *The Control Handbook: Control System Fundamentals*. The Electrical Engineering Handbook Series. CRC Press, Boca Raton, FL, 2nd ed., 2011.
3. N. S. Nise. *Control Systems Engineering*. John Wiley & Sons, Hoboken, NJ, 6th ed., 2011.
4. K. Ogata. *Modern Control Engineering*. Prentice Hall, Upper Saddle River, NJ, 3rd ed., 1997.
5. G. F. Franklin, J. D. Powell, and A. Emami-Naeini. *Feedback Control of Dynamic Systems*. Prentice-Hall, Upper Saddle River, NJ, 4th ed., International Edition, 2002.
6. J. Dorsey. *Continuous and Discrete Control Systems: Modeling, Identification, Design, and Implementation*. McGraw-Hill Series in Electrical and Computer Engineering. Prentice-Hall, New York, International Edition, 2002.
7. K. Ogata. *Modern Control Engineering*. Prentice Hall PTR, Upper Saddle River, NJ, 2001.
8. R. C. Dorf and R. H. Bishop. *Modern Control Systems*. Prentice-Hall, Upper Saddle River, NJ, 9th ed., 2001.
9. S. Skogestad and I. Postlethwaite. *Multivariable Feedback Control: Analysis and Design*. John Wiley & Sons, Norwegian University of Science and Technology, University of Leicester, 2nd ed., 2005.
10. N. S. Nise. *Control Systems Engineering*. John Wiley & Sons, Hoboken, NJ, 7th ed., 2015.
11. B. C. Kuo. *Automatic Control Systems*. Wiley India, India, 8th ed., 2002.
12. C. S. Phillips. *Digital Control System Analysis and Design*. Prentice Hall, Englewood Cliffs, NJ, 3rd ed., 1994.
13. W. S. Levine. *The Control Handbook*. CRC Press, Taylor & Francis Group, Boca Raton, FL, 1996.
14. J. J. DiStefano, A. R. Stubberud, and I. J. Williams. *Schaum's Outline of Feedback and Control Systems*. Schaums' Engineering, McGraw-Hill Education, New York, 2nd ed., 2013.
15. B. Friedland. *Control System Design: An Introduction to State-space Methods*. Dover Publications, Mineola, NY, 1986.
16. K. J. Åström and R. M. Murray. *Feedback Systems: An Introduction for Scientists and Engineers*. Princeton University Press, Princeton, NJ, 2008.
17. Z. Gajic. *Modern Control Systems Engineering*. Dover Publications, Mineola, NY, 2013.
18. A. Visioli and M. S. Fadali. *Digital Control Engineering: Analysis and Design*. Academic Press, Amsterdam, the Netherlands, 2nd ed., 2012.
19. M. E. El-Hawary. *Control System Engineering*. Reston Publication, Reston, VA, 1984.
20. Francis Harvey Raven. *Automatic Control Engineering*. McGraw-Hill, New York, 1987.
21. R. Burns. *Advanced Control Engineering*. Butterworth-Heinemann, Oxford, UK, 1st ed., 2001.
22. W. S. Levine. *The Control Handbook*. CRC Press, Boca Raton, FL, 2nd ed., 2010.

2

Instrumentation

LEARNING OBJECTIVES

- To list the types of instruments and describe their operation
- To comprehend the concept and working of sensors and transducers
- To explain the method to measure physical and electrical parameters
- To understand the working of meters and measuring devices

2.1 Introduction

Measurement devices perform a complete measuring function, from initial detection to final indication. Two important aspects of a measurement system are the sensor and the transmitter. A third is the transducer.

Here are definitions of these three terms:

- *Sensor*: Primary sensing element
- *Transducer*: Changes one instrument signal value to another instrument signal value
- *Transmitter*: Contains the transducer and produces an amplified, standardized instrument signal

In order to control a dynamic variable in a process, there must be information about the variable itself. This information is obtained from a measurement of the variable. A measurement system is any set of interconnected parts that include one or more measurement devices. Measurement devices such as sensors, or primary elements, measure the variable.

2.1.1 Sensor and Transmitter

Figure 2.1 highlights the role of the primary element and transmitter in the context of the measurement system. The system "starts" with the process shown at the top. The sensor produces a response representing the value of the process variable. A transducer, within the transmitter, converts this response to a standard instrument signal (usually either 3–15 psi or 4–20 mA). The transmitter amplifies this standard signal and sends it to a controller and/or other instruments. Understanding the role of each piece of equipment is important.

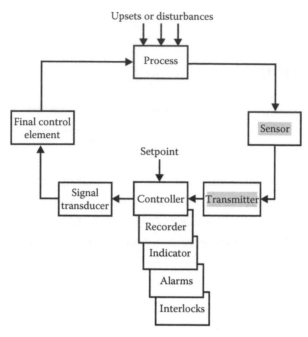

FIGURE 2.1
Block diagram of measurement systems.

2.1.2 Primary Measuring Element Selection and Characteristics

When selecting the sensor, there are a number of factors to keep in mind. Knowing these factors in advance, and the needs of the process, saves a buyer time and money. Before any of these factors can be considered, the intended use of the sensor must be known. Once that is established, several factors should be considered:

Range: What is the normal range over which the controlled variable might vary? Are there extremes to this?

Response time: How much amount of time is required for a sensor to completely respond to a change in its input?

Accuracy: How close does the sensor come to indicating the actual value of the measured variable?

Precision: How consistent is the sensor in measuring the same value under the same operating conditions over a period of time?

Sensitivity: How small a change in the controlled variable can the sensor measure?

Dead band: How much of a change to the process is required before the sensor responds to the change?

Costs: What are the costs involved—not simply the purchase cost, but also the installed/operating costs?

Installation problems: Are there special installation problems, for example, corrosive fluids, explosive mixtures, size and shape constraints, remote transmission questions?

Range is the region, in which the controlled variable might vary, in both normal and extreme situations? For instance, if a process normally has a pressure of between 200 and 300 psi, a sensor that measures from 100 to 400 lb would be desirable; this allows for extreme conditions to be measured as well as normal ones. A broad range also allows the operator time to respond when measurements occur outside the norm. Ideally, a process should be 40%–60%, or 30%–70% of the range most of the time and under normal operating conditions.

2.1.2.1 Response Time

Response time is the amount of time required for a sensor to respond completely to a change in input. The response time of the control loop is the combination of the responses of all the parts, including the sensor. An important objective of control system design is to match correctly the time responses of the control system (and its measurement systems) to that of the process. In general, a system with a quicker response time will be more expensive (Figure 2.2).

This figure shows a key principle of sensor response time. It illustrates that in a fixed amount of time, known as the *time constant*, here shown as 3 s, the sensor registers 63.2% of the total change. In the next time constant, the sensor registers 63.2% of the remaining difference, and so on. In each time constant, the sensor registers the same percent of the remaining difference (Figure 2.3).

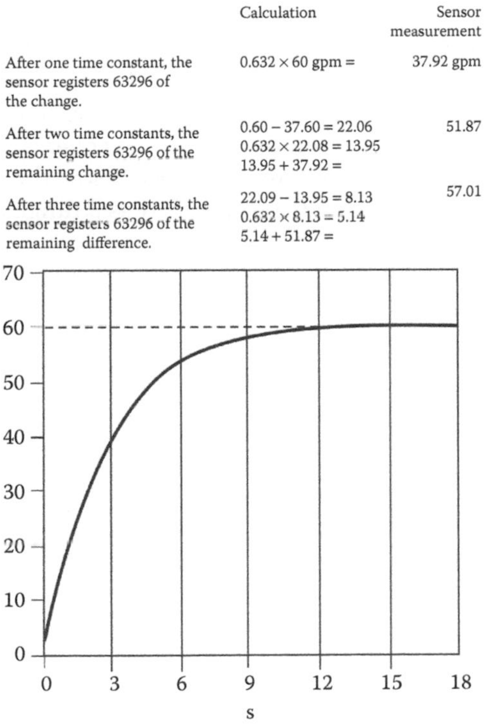

	Calculation	Sensor measurement
After one time constant, the sensor registers 63296 of the change.	0.632×60 gpm =	37.92 gpm
After two time constants, the sensor registers 63296 of the remaining change.	$0.60 - 37.60 = 22.06$ $0.632 \times 22.08 = 13.95$ $13.95 + 37.92 =$	51.87
After three time constants, the sensor registers 63296 of the remaining difference.	$22.09 - 13.95 = 8.13$ $0.632 \times 8.13 = 5.14$ $5.14 + 51.87 =$	57.01

FIGURE 2.2
Sensor response time (time constant).

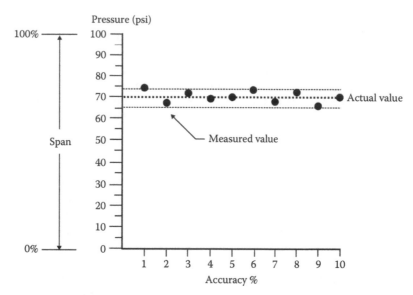

FIGURE 2.3
Precision and accuracy.

These figures contrast the term *precision* as shown above, with the term *accuracy* below. The actual value (70) is shown as a solid line in both figures. The measured values are shown as dots. Accuracy is a measure of how close the sensor comes to indicating the actual value of the measured variable. Here we see that the measured values are scattered above, below, and on the actual value. Accuracy is always given in terms of inaccuracy such as ±2% or +1%, −3%. Precision is a measure of the consistency of a sensor in measuring the same value under the same operating conditions over a period of time. The figure shows precise measured values, all the same distance below the actual value. Precision is synonymous with repeatability and may be specified as a range or value excursions or as a percent.

2.1.2.2 Accuracy

Most of the devices are rated on their accuracy, not their precision. The specifications usually state that the device is accurate to plus or minus some value. Thus, with accuracy, the deviation is known, but not the direction of the deviation. For example, if a watch has an accuracy of 10 min, it means an accuracy of plus or minus 10 min. The owner of this watch has an appointment for 2:30 p.m., but must arrive at 2:20 p.m. in order to be on time because of the rated accuracy of the watch. The owner cannot know if the watch is showing the time ten minutes early or ten minutes late or somewhere in between.

2.1.2.3 Precision

Precision is always within a given value and is always in the same direction. Thus, a precise measurement may be wrong, but it is consistent. For example, another watch has a precision of plus five minutes. The owner of this watch knows that she can arrive at her 2:30 p.m. appointment at 2:35 p.m. (on the watch) and still be on time.

2.1.2.4 Sensitivity

The sensitivity of the sensor is a measurement of how small a change in the *controlled variable* it can actually measure. The greater the sensitivity, the greater the sensor's reaction to an input stimulus.

2.1.2.5 Dead Band

Dead band is the "unresponsiveness" of the sensor. It describes how much change to the process is required before the sensor actually responds to it or even detects it. The term sensitivity has frequently been used to denote dead band, but the terms are not truly interchangeable. Sensitivity refers to the reaction of the sensor. Dead time applies to the time it takes for the sensor to react.

2.1.2.6 Installation Problems

Installation problems can include special problems in the environment such as humidity, vibration, temperature, or dust. Installation problems can also be anything that causes a problem to the devices installed, such as installing the device in a difficult to reach location.

2.1.3 Signal Transmission

As we have mentioned before, the transmitter receives a nonstandard signal and transmits a signal within a standard range. A number of signals can be used to transmit the value of a variable. The most common are pneumatic, electronic, and optical.

2.1.3.1 Signal Types

In most existing plants, *pneumatic* and *electronic* signals are predominant. Pneumatic signals are normally 3–15 psi, and electronic signals are normally 4–20 mA. Optical signals are also used with fiber optic systems or when a direct line of sight exists.

Radio and *hydraulic* signals are also used, though they are not as common because of inherent problems such as radio signal interference and leakage of hydraulic systems. However, radio signals commonly are used when sensors and transmitters are great distances (on pipelines, for example) from control centers.

2.1.3.2 Standard Signal Ranges

Signal ranges vary, but are important to the calculation of process functions. The concept of signal ranges is synonymous with steady-state gain.

2.1.3.3 Electronic Transmitter Adjusted

Figure 2.4 shows how the range of the measured variable, on the left, relates to the range of the transmitted signal, on the right. The measured variable is also the input, and the transmitted signal is the output. The measurement range is 100°C–500°C. Thus, the span of measurement is 400°C. The transmitter range is 4–20 mA, and the transmitter span is 16 mA. At the lowest measurement (100°C), the transmitter output should be 4 mA. At 50% of the measurement (300°C), the transmitter output should be 12 mA.

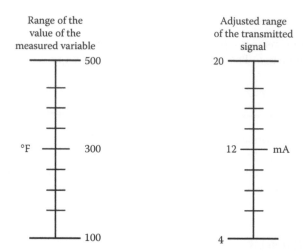

FIGURE 2.4
Electronic transmitter adjusted range.

Using standard signal ranges, current output in a process can be readily calculated using the formula for steady-state gain:

$$\text{Steady-state gain } (K) = \frac{\text{change in output}}{\text{change in Input that caused output change}}$$

The gain for this transmitter is 0.04 (e.g., 16 divided by 400). Once the gain is known, the temperature can be calculated from the output of the transmitter. For example, let's say the transmitter output is 8 mA. Remember, 4 mA is 0% of span, so (subtracting 4 from 8 mA) the transmitter output change is 4 mA. Dividing 4 mA by the gain (0.04) produces an answer of 100. The lowest temperature in the range is 100°C. Adding 100 and 100 produces the actual temperature, 200°C.

2.1.4 Transmission System Dynamics

A major difference between electronic and pneumatic transmission systems is the time required for signal transmission. In an electronic system, there are no moving parts, only the state of the signal changes. This change occurs with virtually no time lost (Figure 2.5).

As we stated previously, mechanical movement takes place whenever any pneumatic process signal changes. When devices move mechanically, time is lost. In addition, pneumatic systems, because they contain moving parts, are higher maintenance and subject to vibration, as well as rotational or gravitational mounting problems. However, pneumatic systems are still in place in many plants because they are safer than electrical systems in certain environments containing potentially explosive atmospheres.

2.1.4.1 Transmission Lag

Figure 2.6 shows the time lost with a pneumatic system. This figure represents a system using 3/16 ID tubing for the transmission line. As shown at the bottom of the graphic, in short distances, the effect of time is small. Under 200 ft, a signal can change 15–3 psi

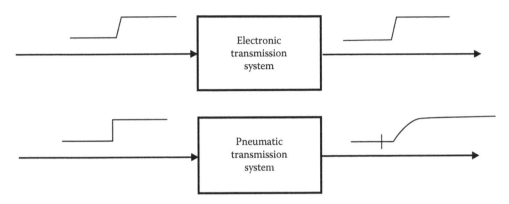

FIGURE 2.5
Signal transmission for electronic and pneumatic signals.

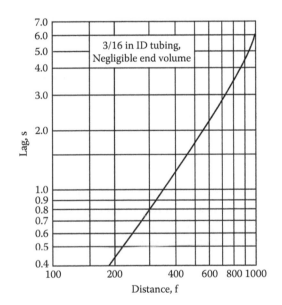

FIGURE 2.6
Pneumatic transmission signal lag.

(the span of a pneumatic device) in roughly 0.4 s or less. This lost time represents the time needed to make up the air volume difference in the line (either replacing or releasing the air volume). A lag of 0.4 s is not critical, but as the distance of the signal line increases, so does the lag. At 400 ft, the lag time rises to about 1.3 s. At 1000 ft, the time is nearly 7 s—in some processes, a critical period of time. Note: This time measurement represents the time required for the signal to travel from the sensing device to the controlling device. If there is a change and the controller responds to it immediately, the amount must be doubled for the signal transmission to reach a final control device. Pneumatic devices are best used for safety applications, simplicity, and for valve *actuators*, always in applications where the line length is kept less than 100 ft; otherwise electronic signals should be used.

$$\text{Steady-state gain} = \frac{\text{Change in output}}{\text{Change in input}}$$

Calculating gain for and electronic transmitter with
an input range of 100–500°F.

$$\text{Gain} = \frac{20-4\text{ mA}}{500-100\text{°F}} = 0.04\text{ (mA/°F)}$$

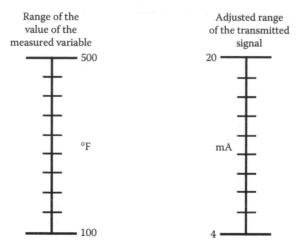

FIGURE 2.7
Transmitter gain for an electronic transmitter.

2.1.4.2 Transmitter Gain

A transmitter's gain, that is the ratio of the output of the *transmitter* to the input signal, is constant regardless of its output. In other words, an electronic transmitter's gain will remain constant whether its output is 0% of span (4 mA) or 100% of span (20 mA) or any other point between those extremes (Figure 2.7).

2.1.4.3 Smart Transmitters

So far, the discussion has centered on electronic and pneumatic transmitters. The input and output of both of these types of transmitters is an *analog* signal—either a mA current or air pressure, both of which are continuously variable. There is another kind of transmitter—the "smart" transmitter.

Figure 2.8 illustrates functions of a *smart transmitter*. They can convert analog signals to digital signals (A/D), making communication swift and easy, and can even send both analog and *digital* signals at the same time as denoted by D/A. A smart transmitter has a number of other capabilities as well. For instance, inputs can be varied, as denoted by A/D. If a temperature transmitter is a smart transmitter, it will accept millivolt signals from *thermocouples* and resistance signals from resistance temperature devices (RTDs) and thermistor.

Components of the smart transmitter are illustrated in the lower figure. The transmitter is built into housing about the size of a softball as seen on the below-mentioned figure. The controller takes the output signal from the transmitter and sends it back to the final control element. The communicator is shown on the right.

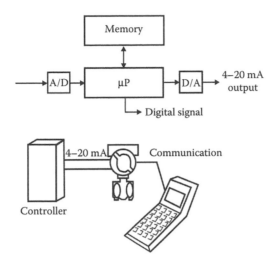

FIGURE 2.8
Smart transmitter components and function.

The communicator is a hand-held interface device that allows digital instructions to be delivered to the smart transmitters. Testing, configuring, and supplying or acquiring data are all accomplished through the communicator. The communicator has a display that lets the technician see the input or output information. The communicator can be connected directly to the smart transmitter, or in parallel anywhere on the loop.

2.1.4.4 Smart Transmitter Microprocessor-Based Features

Smart transmitters also have features such as configuration, re-ranging, characteristics, signal conditioning, and self-diagnosis.

Configuration: Smart transmitters can be configured to meet the demands of the process in which they are used. For example, the same transmitter can be set up to read almost any range or type of thermocouple, RTD, or thermistor. Because of this, they reduce the need for a large number of specific replacement devices.

Re-ranging: The range that the smart transmitter functions under can be easily changed from a remote location, for example by the technician in a control room. The technician or the operator has access to any smart device in the loop and does not even have to be at the transmitter to perform the change. The operator does need to use a communicator, however. A communicator allows the operator to interface with the smart transmitter. The communicator could be a PC, a programmable logic controller (PLC), or a hand held device. The type of communicator depends on the manufacturer. Reranging is simple with the smart transmitter. For instance, using a communicator, the operator can change from a 100 ohm RTD to a type-J thermocouple just by reprogramming the transmitter. The transmitter responds immediately and changes from measuring resistance to measuring millivoltage. There is a wide range of inputs that a smart transmitter will accept. For instance, with pressure units, the operator can determine ahead of time whether to use inches of water, inches of mercury, psi, bars, millibars, Pascals, or Kilopascals.

Characteristics: Another characteristic of a smart transmitter is its ability to act as a stand-alone transmitter. In such a capacity, it sends the output signal to a *distributed control system* (DCS) or a PLC.

Signal conditioning: Smart transmitters can also perform signal conditioning, scanning the average signal and eliminating any "noise" spikes. Signals can also be delayed (dampened) so that the response does not fluctuate. This is especially useful with a rapidly changing process.

Self-diagnosis: Finally, a smart transmitter can diagnose itself and report on any problems in the process. For example, it can report on a circuit board that is not working properly.

Smart transmitter benefits: There are distinct advantages in using a smart transmitter. The most important include ease of installation and communication, self-diagnosis, and improved digital reliability. Smart transmitters are also less subject to effects of temperature and humidity than analog devices. And although vibration can still affect them, the effects are far less than with analog devices. Smart transmitters also provide increased accuracy. And because they can replace several different types of devices, using them allows for inventory reduction.

2.1.5 Characteristics of Instruments

Accuracy and performance of measurement systems are strongly governed by the characteristics of the instruments and transducers used within them. Knowledge of characteristics are essential when designing measurement systems to ensure that the measurement requirements are met and that most appropriate instruments are used in relation to appropriate instruments.

Static describes instrument parameters in steady state, when output has settled to a steady reading. This characteristic has a fundamental effect on the quality of measurement obtained from it.

Dynamic response between time that a measured quality changes and time when the instrument output attains constant value, main consequence finite time must elapse between a measured quantity changing value and the instrument output being read.

2.1.5.1 Static Characteristics

Precision: An instrument's degree of freedom from random variations in its output when measuring a constant quantity; high precision shouldn't be confused with high accuracy

Repeatability: Closeness of output readings when the same input is applied repetitively over a short time period; the same input is applied repetitively over a short period of time with the same conditions

Reproducibility: Closeness of output readings for the same input when there are changes in conditions; both above describe the spread of output readings for the same input

2.2 Order of Control and Measurements Systems

It is instructive to examine the properties of the simple mechanical system shown in Figure 2.9. Very similar considerations apply to complex electrical, biological, economic, and political systems, although these may be more difficult to visualize:

   In this example, a force M is applied to a mass of inertia I attached to a spring with elastic modulus k. The motion of the system is restrained by a damper, which exerts a drag proportional to the velocity of the mass. This arrangement is very similar to the springs and shock absorbers on a motor vehicle suspension. If the damper were omitted, the mass would execute simple harmonic motion on the end of the spring. It is desired to stabilize the mass and to control its displacement C from the equilibrium length of the spring.

   Since *force = mass * acceleration*, we can write down a differential equation:

$$M - a \cdot \frac{dC}{dt} - k \cdot C = I \cdot \frac{d^2C}{dt^2} \tag{2.1}$$

where "a" is the viscous damping coefficient. (The left-hand side of this equation is simply the applied force M minus the contributions from the damper and the spring.) Some of the terms may be negligible in real systems. If we consider M as the restoring force generated by the control system in response to an error signal E, we can distinguish various "orders" of control system, viz.

2.2.1 Zero Order Control Systems

When a and I are very small, the elastic component k may dominate. Such systems have a natural equilibrium position that they adopt in the absence of a control input, M. Some intracellular feedback systems fall into this category.

2.2.2 First Order Control Systems

The inertia I is negligible but some damping is present, and possibly the spring as well. In the simplest case of pure viscous damping with no spring, the system lacks a unique resting position when the control input is removed. Resistive losses provide the damping term

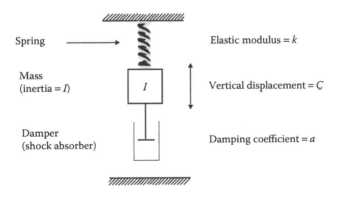

FIGURE 2.9
Dampering system.

in electrical circuits, while in thermal or chemical control systems the transfer of heat (or the material needed to change a chemical concentration) gives rise to analogous behavior. In all such cases, the effect on C depends on the *duration* of the restoring input M, and a larger M is necessary if C is required to change quickly.

2.2.3 Second Order Control Systems

Inertia is present, and possibly the damper and the spring as well. Such systems are the most difficult to control and once pushed may remain in motion for considerable periods of time. Inductance and capacitance are responsible for similar effects in electrical networks. It is hard to visualize momentum and inertia in purely chemical systems. Where a cell is committed to a process that takes time to complete (e.g., mRNA translation), there will be unfinished product in the pipeline with some inertial properties. The real inertias of arms and legs are beautifully compensated by the superb control mechanisms responsible for neuromuscular coordination within the central nervous system.

The complete solution to Equation 2.1 is the sum of the *particular integral* (which depends mainly on the forcing function "M") plus the *complementary function*, which is the solution of the corresponding homogeneous equation. The homogeneous equation is obtained by setting M = 0, after which it is possible to write down an auxiliary equation:

$$I \cdot s^2 + a \cdot s + k = 0 \qquad\qquad (2.2)$$

This is sometimes referred to as the characteristic equation, since the two roots S_1 and S_2 define the principal characteristics of the controlled system. Using the usual formula to find the roots of a quadratic equation,

$$S_1 = -\frac{a}{2I} + \sqrt{\left[\frac{a}{2I}\right]^2 - \frac{k}{I}} \qquad\qquad S_2 = -\frac{a}{2I} - \sqrt{\left[\frac{a}{2I}\right]^2 - \frac{k}{I}}$$

The complementary function is then:

$$c = B_1 \cdot e^{S_1 \cdot t} + B_2 \cdot e^{S_2 \cdot t}$$

where:
  B_1 and B_2 are both arbitrary constants
  t represents time

In the absence of a control input M, both roots must always be negative, and c will eventually settle to some finite value as t increases.

The response of the controlled system to a sudden load or a change in target value depends on the square root term in the roots of the auxiliary equation. If $a^2 < 4kI$, then there will be a pair of complex roots, and we have an *under-damped* system that will repeatedly overshoot the final value whenever it is disturbed. Some time may elapse before things settle down. If $a^2 > 4kI$, then we have an *over-damped* system. There will be no overshoot, but the system will still approach the final position very slowly if the damping term a is large. If $a^2 = 4kI$, then the square root term will vanish from the solution, and we have a *critically damped* system that will stabilize at the final position faster than either of the other two alternatives.

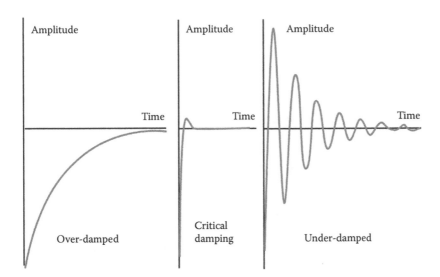

FIGURE 2.10
Different level of damping.

The situation is more complicated when there is an energy input "*M*." There is now a possibility of sustained oscillations. The characteristic equation has a similar form to Equation 2.2, but the coefficients will depend on the feedback network, and there might be additional terms. The system will be stable provided that all the roots are negative. When there are pairs of complex roots, the real parts must be negative or the system will be unstable (Figure 2.10).

Continuous sensors convert physical phenomena to measurable signals, typically voltages or currents. Consider a simple temperature measuring device; there will be an increase in output voltage proportional to a temperature rise. A computer could measure the voltage and convert it to a temperature. The basic physical phenomena typically measured with sensors includes angular or linear position, acceleration, temperature, pressure or flow rates, stress, strain or force, light intensity, and sound.

Most of these sensors are based on subtle electrical properties of materials and devices. As a result, the signals often require *signal conditioners*. These are often amplifiers that boost currents and voltages to larger voltages. Sensors are also called transducers. This is because they convert input phenomena to an output in a different form. This transformation relies upon a manufactured device with limitations and imperfection.

2.3 Temperature Measurement Systems

The hotness or coldness of a piece of material depends upon the molecular activity of the material. Kinetic energy is a measure of the activity of the atoms that make up the molecules of any material. Therefore, temperature is a measure of the kinetic energy of the material in question. Whether you want to know the temperature of the surrounding air, the water cooling a car's engine, or the components of a nuclear facility, you must have some means to measure the kinetic energy of the material. Most temperature measuring devices use the energy of the material or system they are monitoring to raise (or lower) the kinetic energy of the device. A normal household thermometer is one example. The mercury, or other liquid, in the bulb of the thermometer expands as its kinetic energy is raised. By observing how far the liquid rises in the tube, you can tell the temperature of the measured object. Because temperature is one of the most important parameters of a material, many instruments have been developed to measure it. One type of detector used is the resistance temperature detector (RTD). The RTD is used at many DOE nuclear facilities to measure temperatures of the process or materials being monitored.

The RTD incorporates pure metals or certain alloys that increase in resistance as temperature increases and, conversely, decrease in resistance as temperature decreases. RTDs act somewhat like an electrical transducer, converting changes in temperature to voltage signals by the measurement of resistance. The metals that are best suited for use as RTD sensors are pure, of uniform quality, stable within a given range of temperature, and able to give reproducible resistance-temperature readings.

Only a few metals have the properties necessary for use in RTD elements. RTD elements are normally constructed of platinum, copper, or nickel. These metals are best suited for RTD applications because of their linear resistance-temperature characteristics, their high coefficient of resistance, and their ability to withstand repeated temperature cycles. The coefficient of resistance is the change in resistance per degree change in temperature, usually expressed as a percentage per degree of temperature. The material used must be capable of being drawn into fine wire so that the element can be easily constructed (Figure 2.11).

2.3.1 Thermocouple Temperature Detectors

The internal construction: The leads of the thermocouple are encased in a rigid metal sheath. The measuring junction is normally formed at the bottom of the thermocouple housing. Magnesium oxide surrounds the thermocouple wires to prevent vibration that could damage the fine wires and to enhance heat transfer between the measuring junction and the medium surrounding the thermocouple.

Thermocouple Operation: Thermocouples will cause an electric current to flow in the attached circuit when subjected to changes in temperature. The amount of current

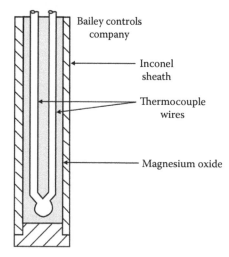

FIGURE 2.11
Internal construction of a typical thermocouple.

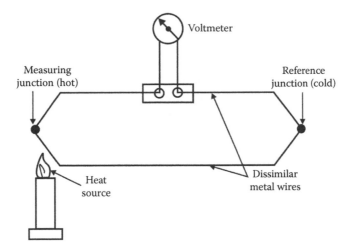

FIGURE 2.12
Simple thermocouple circuit junction.

that will be produced is dependent on the temperature difference between the measurement and reference junction, the characteristics of the two metals used, and the characteristics of the attached circuit. Figure 2.12 illustrates a simple thermocouple circuit.

Heating the measuring of the thermocouple produces a voltage that is greater than the voltage across the reference junction. The difference between the two voltages is proportional to the difference in temperature and can be measured on the voltmeter (in millivolt). For ease of operator use, some voltmeters are set up to read out directly in temperature through use of electronic circuitry.

Temperature Detection Circuit: Figure 2.13 is a block diagram of a typical temperature detection circuit. This represents a balanced bridge temperature detection circuit that has been modified to eliminate the galvanometer.

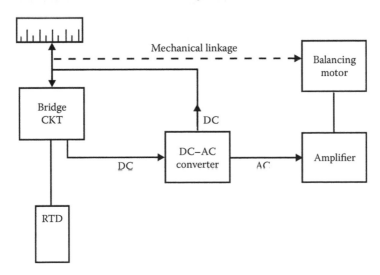

FIGURE 2.13
Block diagram of a typical temperature detection circuit.

The block consists of a temperature detector (RTD) that measures the temperature. The detector is felt as resistance to the bridge network. The bridge network converts this resistance to a DC voltage signal. An electronic instrument has been developed in which the DC voltage of the potentiometer, or the bridge, is converted to an AC voltage. The AC voltage is then amplified to a higher (usable) voltage that is used to drive a bi-directional motor. The bi-directional motor positions the slider on the slide wire to balance the circuit resistance. If the RTD becomes open in either the unbalanced and balanced bridge circuits, the resistance will be infinite, and the meter will indicate a very high temperature. If it becomes shorted, resistance will be zero, and the meter will indicate a very low temperature.

2.4 Instrumentation and Control: Pressure Detectors

2.4.1 Bellows-Type Detectors

The need for a pressure sensing element that was extremely sensitive to low pressures (LPs) and provided power for activating recording and indicating mechanisms resulted in the development of the metallic bellows pressure sensing element. The metallic bellows is most accurate when measuring pressures from 0.5 to 75 psig. However, when used in conjunction with a heavy range spring, some bellows can be used to measure pressures of over 1000 psig. Figure 2.14 shows a basic metallic bellows pressure sensing element.

The bellows is a one-piece, collapsible, seamless metallic unit that has deep folds formed from very thin-walled tubing. The diameter of the bellows ranges from 0.5 to 12 in. and may have as many as 24 folds. System pressure is applied to the internal volume of the bellows. As the inlet pressure to the instrument varies, the bellows will expand or contract. The moving end of the bellows is connected to a mechanical linkage assembly. As the bellows and linkage assembly moves, either an electrical signal is generated or a direct pressure indication is provided. The flexibility of a metallic bellows is similar in character to that of a helical, coiled compression spring. Up to the elastic limit of the bellows, the

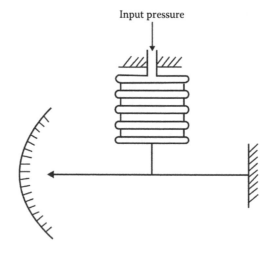

FIGURE 2.14
Basic metallic bellows.

relation between increments of load and deflection is linear. However, this relationship exists only when the bellows is under compression. It is necessary to construct the bellows such that all of the travel occurs on the compression side of the point of equilibrium. Therefore, in practice, the bellows must always be opposed by a spring, and the deflection characteristics will be the resulting force of the spring and bellows.

2.4.2 Bourdon Tube-Type Detectors

The bourdon tube pressure instrument is one of the oldest pressure sensing instruments in use today. The bourdon tube consists of a thin-walled tube that is flattened diametrically on opposite sides to produce a cross-sectional area, elliptical in shape, having two long flat sides and two short round sides. The tube is bent lengthwise into an arc of a circle of 270°–300°. Pressure applied to the inside of the tube causes distention of the flat sections and tends to restore its original round cross-section. This change in cross-section causes the tube to straighten slightly. Since the tube is permanently fastened at one end, the tip of the tube traces a curve that is the result of the change in angular position with respect to the center. Within limits, the movement of the tip of the tube can then be used to position a pointer or to develop an equivalent electrical signal to indicate the value of the applied internal pressure (Figure 2.15).

2.4.3 Resistance-Type Transducers

Included in this category of transducers are strain gauges and moving contacts (slide wire variable resistors). Figure 2.16 illustrates a simple strain gauge. A strain gauge measures the external force (pressure) applied to a fine wire. The fine wire is usually arranged in the

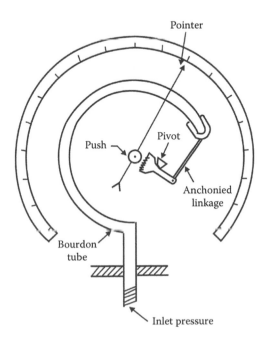

FIGURE 2.15
Bourdon tube.

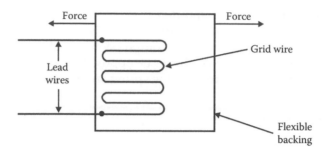

FIGURE 2.16
Strain gauge.

form of a grid. The pressure change causes a resistance change due to the distortion of the wire. The value of the pressure can be found by measuring the change in resistance of the wire grid. Equation below shows the pressure to resistance relationship.

$$R = KL\ A$$

where:
  R is the resistance of the wire grid in ohms
  K is the resistivity constant for the particular type of wire grid
  L is the length of wire grid
  A is the cross-sectional area of wire grid

2.4.4 Strain Gauge Pressure Transducer

As the wire grid is distorted by elastic deformation, its length is increased, and its cross-sectional area decreases. These changes cause an increase in the resistance of the wire of the strain gauge. This change in resistance is used as the variable resistance in a bridge circuit that provides an electrical signal for indication of pressure. Figure 2.17 illustrates a strain gauge pressure transducer.

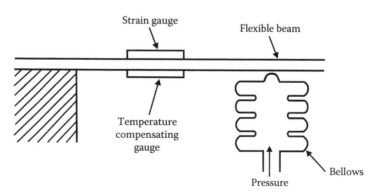

FIGURE 2.17
Strain gauge pressure transducer.

2.4.5 Strain Gauge Used in a Bridge Circuit

An increase in pressure at the inlet of the bellows causes the bellows to expand. The expansion of the bellows moves a flexible beam to which a strain gauge has been attached. The movement of the beam causes the resistance of the strain gauge to change. The temperature compensating gauge compensates for the heat produced by current flowing through the fine wire of the strain gauge (Figure 2.18).

Strain gauges, which are nothing more than resistors, are used with bridge circuits as shown in Figure 2.18. Alternating current is provided by an exciter that is used in place of a battery to eliminate the need for a galvanometer. When a change in resistance in the strain gauge causes an unbalanced condition, an error signal enters the amplifier and actuates the balancing motor. The balancing motor moves the slider along the slide wire, restoring the bridge to a balanced condition. The slider's position is noted on a scale marked in units of pressure.

2.4.6 Resistance-Type Transducers

Other resistance-type transducers combine a bellows or a bourdon tube with a variable resistor, as shown in Figure 2.19. As pressure changes, the bellows will either expand or contract. This expansion and contraction causes the attached slider to move along the slide wire, increasing or decreasing the resistance, and thereby indicating an increase or decrease in pressure.

2.4.7 Inductance-Type Transducers

The inductance-type transducer consists of three parts: a coil, a movable magnetic core, and a pressure sensing element. The element is attached to the core, and, as pressure varies, the element causes the core to move inside the coil. An AC voltage is applied to the coil, and, as the core moves, the inductance of the coil changes. The current through the coil

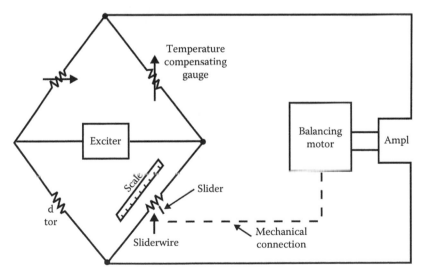

FIGURE 2.18
Strain gauge used in a bridge circuit.

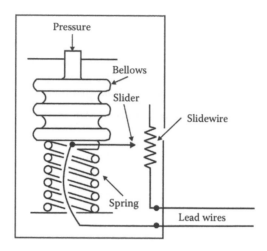

FIGURE 2.19
Bellows resistance transducer.

will increase as the inductance decreases. For increased sensitivity, the coil can be separated into two coils by utilizing a center tap, as shown in Figure 2.20. As the core moves within the coils, the inductance of one coil will increase, while the other will decrease.

2.4.8 Differential Transformer

Another type of inductance transducer, illustrated in Figure 2.21, utilizes two coils wound on a single tube and is commonly referred to as a differential transformer. The primary coil is wound around the center of the tube. The secondary coil is divided with one half wound around each end of the tube. Each end is wound in the opposite direction, which

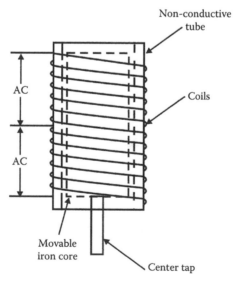

FIGURE 2.20
Inductance-type pressure transducer coil.

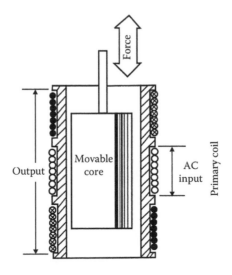

FIGURE 2.21
Differential transformer.

causes the voltages induced to oppose one another. A core, positioned by a pressure element, is movable within the tube. When the core is in the lower position, the lower half of the secondary coil provides the output. When the core is in the upper position, the upper half of the secondary coil provides the output. The magnitude and direction of the output depends on the amount the core is displaced from its center position. When the core is in the mid-position, there is no secondary output.

2.4.9 Capacitive-Type Transducers

Capacitive-type transducers, illustrated in Figure 2.22, consist of two flexible conductive plates and a dielectric. In this case, the dielectric is the fluid.

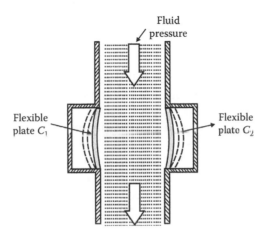

FIGURE 2.22
Capacitive pressure transducer.

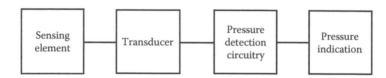

FIGURE 2.23
Typical pressure detection lock diagram.

As pressure increases, the flexible conductive plates will move farther apart, changing the capacitance of the transducer. This change in capacitance is measurable and is proportional to the change in pressure.

2.4.9.1 Detection Circuitry

Figure 2.23 shows a block diagram of a typical pressure detection circuit. The sensing element senses the pressure of the monitored system and converts the pressure to a mechanical signal. The sensing element supplies the mechanical signal to a transducer, as discussed above. The transducer converts the mechanical signal to an electrical signal that is proportional to system pressure. If the mechanical signal from the sensing element is used directly, a transducer is not required and therefore not used. The detector circuitry will amplify and/or transmit this signal to the pressure indicator. The electrical signal generated by the detection circuitry is proportional to system pressure. The exact operation of detector circuitry depends upon the type of transducer used. The pressure indicator provides remote indication of the system pressure being measured.

2.4.10 Pressure Detector Functions

Although the pressures that are monitored vary slightly depending on the details of facility design, all pressure detectors are used to provide up to three basic functions: indication, alarm, and control. Since the fluid system may operate at both saturation and sub cooled conditions, accurate pressure indication must be available to maintain proper cooling. Some pressure detectors have audible and visual alarms associated with them when specified preset limits are exceeded. Some pressure detector applications are used as inputs to protective features and control functions.

Detector failure: If a pressure instrument fails, spare detector elements may be utilized if installed. If spare detectors are not installed, the pressure may be read at an independent local mechanical gauge, if available, or a precision pressure gauge may be installed in the system at a convenient point. If the detector is functional, it may be possible to obtain pressure readings by measuring voltage or current values across the detector leads and comparing this reading with calibration curves.

Environmental concerns: Pressure instruments are sensitive to variations in the atmospheric pressure surrounding the detector. This is especially apparent when the detector is located within an enclosed space. Variations in the pressure surrounding the detector will cause the indicated pressure from the detector to change. This will greatly reduce the accuracy of the pressure instrument and should be considered when installing and maintaining these instruments.

2.5 Angular Displacement

2.5.1 Potentiometers

Potentiometers measure the angular position of a shaft using a variable resistor. A potentiometer is shown in Figure 2.24. The potentiometer is a resistor, normally made with a thin film of resistive material. A wiper can be moved along the surface of the resistive film. As the wiper moves toward one end, there will be a change in resistance proportional to the distance moved. If a voltage is applied across the resistor, the voltage at the wiper interpolates the voltages at the ends of the resistor.

The potentiometer in Figure 2.25 is being used as a voltage divider. As the wiper rotates, the output voltage will be proportional to the angle of rotation.

Potentiometers are popular because they are inexpensive and don't require special signal conditioners. But, they have limited accuracy, normally in the range of 1%, and they are subject to mechanical wear. Potentiometers measure absolute position, and they are

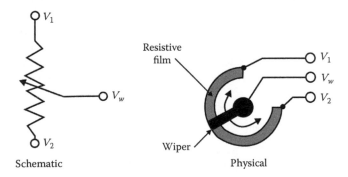

Schematic   Physical

FIGURE 2.24
Potentiometer.

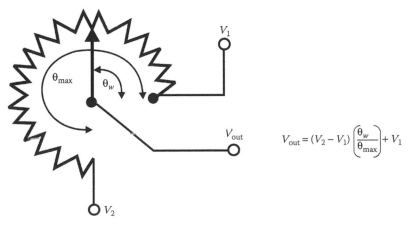

$$V_{out} = (V_2 - V_1)\left(\frac{\theta_w}{\theta_{max}}\right) + V_1$$

FIGURE 2.25
Potentiometer as a voltage divider.

calibrated by rotating them in their mounting brackets and then tightening them in place. The range of rotation is normally limited to less than 360° or multiples of 360°. Some potentiometers can rotate without limits, and the wiper will jump from one end of the resistor to the other. Faults in potentiometers can be detected by designing the potentiometer to never reach the ends of the range of motion. If an output voltage from the potentiometer ever reaches either end of the range, then a problem has occurred, and the machine can be shut down. Two examples of problems that might cause this are wires that fall off or the potentiometer rotates in its mounting.

2.6 Encoders

Encoders use rotating disks with optical windows, as shown in Figure 2.26. The encoder contains an optical disk with fine windows etched into it. Light from emitters passes through the openings in the disk to detectors. As the encoder shaft is rotated, the light beams are broken.

There are two fundamental types of encoders; absolute and incremental. An absolute encoder will measure the position of the shaft for a single rotation. The same shaft angle will always produce the same reading. The output is normally a binary or grey code number. An incremental (or relative) encoder will output two pulses that can be used to

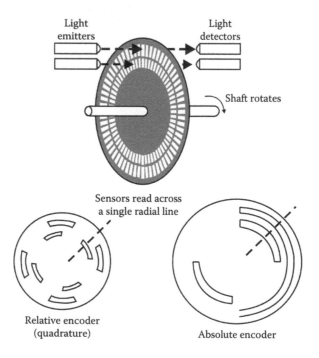

FIGURE 2.26
Encoder disks.

determine displacement. Logic circuits or software is used to determine the direction of rotation and count pulses to determine the displacement. The velocity can be determined by measuring the time between pulses. Encoder disks are shown in the figure. The absolute encoder has two rings; the outer ring is the most significant digit of the encoder, and the inner ring is the least significant digit. The relative encoder has two rings, with one ring rotated a few degrees ahead of the other, but otherwise the same. Both rings detect position to a quarter of the disk. To add accuracy to the absolute encoder, more rings must be added to the disk, and more emitters and detectors. To add accuracy to the relative encoder, we only need to add more windows to the existing two rings. Typical encoders will have from two to thousands of windows per ring.

When using absolute encoders, the position during a single rotation is measured directly. If the encoder rotates multiple times, then the total number of rotations must be counted separately. When using a relative encoder, the distance of rotation is determined by counting the pulses from one of the rings. If the encoder only rotates in one direction, then a simple count of pulses from one ring will determine the total distance. If the encoder can rotate both directions, a second ring must be used to determine when to subtract pulses. The quadrature scheme, using two rings, is shown in Figure 2.26. The signals are set up so that one is out of phase with the other. Notice that for different directions of rotation, input B either leads or lags A.

2.6.1 Tachometers

Tachometers measure the velocity of a rotating shaft. A common technique is to mount a magnet to a rotating shaft. When the magnet moves past a stationary pick-up coil, current is induced. For each rotation of the shaft there is a pulse in the coil, as shown in the Figure 2.27. When the time between the pulses is measured, the period for one rotation can be found, and the frequency calculated. This technique often requires some signal conditioning circuitry (Figure 2.27).

Another common technique uses a simple permanent magnet DC generator. (Note: you can also use a small DC motor.) The generator is hooked to the rotating shaft. The rotation of a shaft will induce a voltage proportional to the angular velocity. This technique will introduce some drag into the system and is used where efficiency is not an issue. Both of these techniques are common and inexpensive.

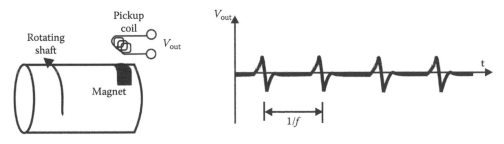

FIGURE 2.27
Magnetic tachometer.

2.7 Linear Position

2.7.1 Potentiometers

Rotational potentiometers were discussed previously, but potentiometers are also available in linear/sliding form. These are capable of measuring linear displacement over long distances. Figure 2.28 shows the output voltage when using the potentiometer as a voltage divider.

Linear/sliding potentiometers have the same general advantages and disadvantages as rotating potentiometers.

2.8 Level Detectors

Liquid level measuring devices are classified into two groups:

1. Direct method
2. Inferred method

Gauge glass: A very simple means by which liquid level is measured in a vessel is by the gauge glass method.

In the gauge glass method, a transparent tube is attached to the bottom and top (top connection not needed in a tank open to atmosphere) of the tank that is monitored. The height of the liquid in the tube will be equal to the height of water in the tank (Figure 2.29).

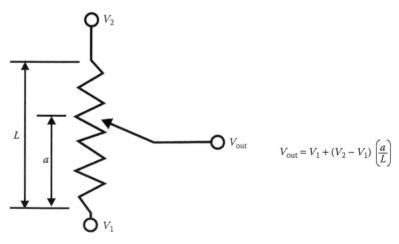

$$V_{out} = V_1 + (V_2 - V_1)\left(\frac{a}{L}\right)$$

FIGURE 2.28
Linear potentiometer.

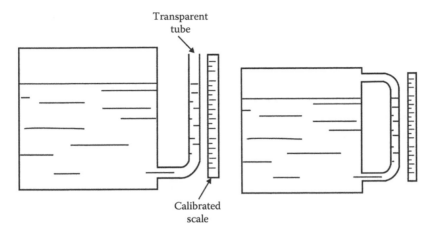

FIGURE 2.29
Transparent tube.

2.8.1 Gauge Glass

A gauge glass is used for vessels where the liquid is at ambient temperature and pressure conditions. A gauge glass is used for vessels where the liquid is at an elevated pressure or a partial vacuum. Notice that the gauge glasses in Figure 2.30 effectively form a "U" tube manometer where the liquid seeks its own level due to the pressure of the liquid in the vessel. Gauge glasses made from tubular glass or plastic are used for service up to 450 psig and 400°F. If it is desired to measure the level of a vessel at higher temperatures and pressures, a different type of gauge glass is used. The type of gauge glass utilized in this instance has a body made of metal with a heavy glass or quartz section for visual observation of the liquid level. The glass section is usually flat to provide strength and safety.

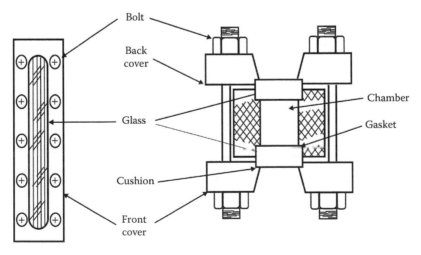

FIGURE 2.30
Typical transparent gauge glass.

2.8.2 Reflex Gauge Glass

Another type of gauge glass is the reflex gauge glass (Figure 2.31). In this type, one side of the glass section is prism-shaped. The glass is molded such that one side has 90°-angles that run lengthwise. Light rays strike the outer surface of the glass at a 90°-angle. The light rays travel through the glass, striking the inner side of the glass at a 45°-angle. The presence or absence of liquid in the chamber determines if the light rays are refracted into the chamber or reflected back to the outer surface of the glass. When the liquid is at an intermediate level in the gauge glass, the light rays encounter an air-glass interface in one portion of the chamber and a water-glass interface in the other portion of the chamber. Where an air-glass interface exists, the light rays are reflected back to the outer surface of the glass because the critical angle for light to pass from air to glass is 42°. This causes the gauge glass to appear silvery-white. In the portion of the chamber with the water-glass interface, the light is refracted into the chamber by the prisms. Reflection of the light back to the outer surface of the gauge glass does not occur because the critical angle for light to pass from glass to water is 62°. This results in the glass appearing black, since it is possible to see through the water to the walls of the chamber, which are painted black. A third type of gauge glass is the refraction type. This type is especially useful in areas of reduced lighting; lights are usually attached to the gauge glass. Operation is based on the principle that the bending of light, or refraction, will be different as light passes through (Figure 2.32).

Light is bent, or refracted, to a greater extent in water than in steam. For the portion of the chamber that contains steam, the light rays travel relatively straight, and the red lens is illuminated. For the portion of the chamber that contains water, the light rays are bent, causing the green lens to be illuminated. The portion of the gauge containing water appears green; the portion of the gauge from that level upward appears red.

2.8.3 Ball Float

The ball float method is a direct reading liquid level mechanism. The most practical design for the float is a hollow metal ball or sphere. However, there are no restrictions to the size, shape, or material used. The design consists of a ball float attached to a rod, which in turn is connected to a rotating shaft that indicates level on a calibrated scale. The operation of

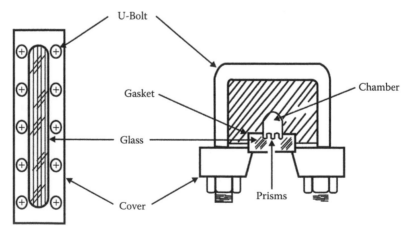

FIGURE 2.31
Reflex gauge glass.

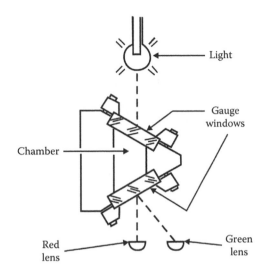

FIGURE 2.32
Refraction gauge glass.

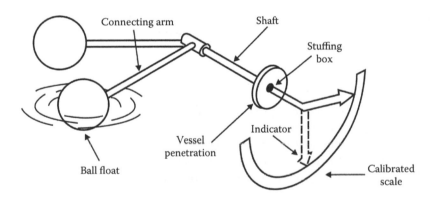

FIGURE 2.33
Ball float.

the ball float is simple. The ball floats on top of the liquid in the tank. If the liquid level changes, the float will follow and change the position of the pointer attached to the rotating shaft (Figure 2.33).

The travel of the ball float is limited by its design to be within ±30° from the horizontal plane, which results in optimum response and performance. The actual level range is determined by the length of the connecting arm. The stuffing box is incorporated to form a watertight seal around the shaft to prevent leakage from the vessel.

2.8.4 Chain Float

This type of float gauge has a float ranging in size up to 12 inches in diameter and is used where small level limitations imposed by ball floats must be exceeded. The range of level measured will be limited only by the size of the vessel. The operation of the chain float is

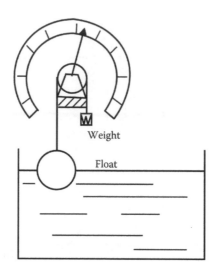

FIGURE 2.34
Chain float gauge.

similar to the ball float except in the method of positioning the pointer and in its connection to the position indication. The float is connected to a rotating element by a chain with a weight attached to the other end to provide a means of keeping the chain taut during changes in level (Figure 2.34).

2.8.5 Magnetic Bond Method

The magnetic bond method was developed to overcome the problems of cages and stuffing boxes. The magnetic bond mechanism consists of a magnetic float that rises and falls with changes in level. The float travels outside of a non-magnetic tube, which houses an inner magnet connected to a level indicator. When the float rises and falls, the outer magnet will attract the inner magnet, causing the inner magnet to follow the level within the vessel (Figure 2.35).

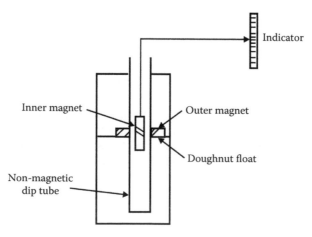

FIGURE 2.35
Magnetic bond detector.

2.8.6 Conductivity Probe Method

Figure 2.36 illustrates a conductivity probe level detection system. It consists of one or more level detectors, an operating relay, and a controller. When the liquid makes contact with any of the electrodes, an electric current will flow between the electrode and ground. The current energizes a relay, which causes the relay contacts to open or close depending on the state of the process involved.

The relay in turn will actuate an alarm, a pump, a control valve, or all three. A typical system has three probes: a low-level probe, a high-level probe, and a high-level alarm probe.

2.8.7 Differential Pressure Level Detectors

The differential pressure (dP) detector method of liquid level measurement uses a dP detector connected to the bottom of the tank being monitored. The higher pressure, caused by the fluid in the tank, is compared to a lower reference pressure (usually atmospheric). This comparison takes place in the dP detector. The tank is open to the atmosphere; therefore, it is necessary to use only the high pressure (HP) connection on the dP transmitter. The LP side is vented to the atmosphere; therefore, the pressure differential is the hydrostatic head, or weight, of the liquid in the tank. The maximum level that can be measured by the dP transmitter is determined by the maximum height of liquid above the transmitter. The minimum level that can be measured is determined by the point where the transmitter is connected to the tank. Not all tanks or vessels are open to the atmosphere. Many are totally enclosed to prevent vapors or steam from escaping, or to allow pressurizing the contents of the tank. When measuring the level in a tank that is pressurized, or the level that can become pressurized by vapor pressure from the liquid, both the HP and LP sides of the dP transmitter must be connected (Figure 2.37).

2.8.8 Closed Tank, Dry Reference Leg

The HP connection is connected to the tank at or below the lower range value to be measured. The LP side is connected to a "reference leg" that is connected at or above the upper range value to be measured. The reference leg is pressurized by the gas or vapor pressure,

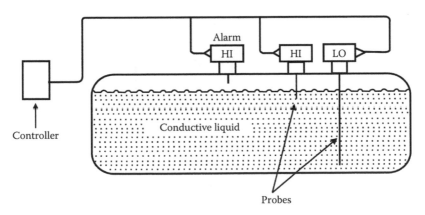

FIGURE 2.36
Probe level detection system.

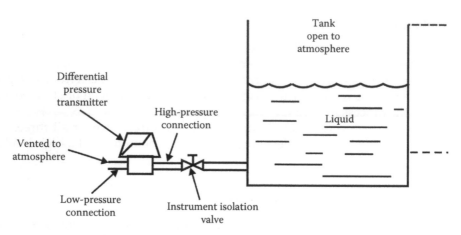

FIGURE 2.37
Open tank differential pressure detector.

but no liquid is permitted to remain in the reference leg. The reference leg must be maintained dry so that there is no liquid head pressure on the LP side of the transmitter. The
HP side is exposed to the hydrostatic head of the liquid plus the gas or vapor pressure
exerted on the liquid's surface. The gas or vapor pressure is equally applied to the LP and
HP sides. Therefore, the output of the dP transmitter is directly proportional to the hydrostatic head pressure, that is, the level in the tank. Where the tank contains a condensable
fluid, such as steam, a slightly different arrangement is used. In applications with condensable fluids, condensation is greatly increased in the reference leg. To compensate for this
effect, the reference leg is filled with the same fluid as the tank. The liquid in the reference
leg applies a hydrostatic head to the HP side of the transmitter, and the value of this level
is constant as long as the reference leg is maintained full. If this pressure remains constant,
any change in ΔP is due to a change on the LP side of the transmitter (Figure 2.38).

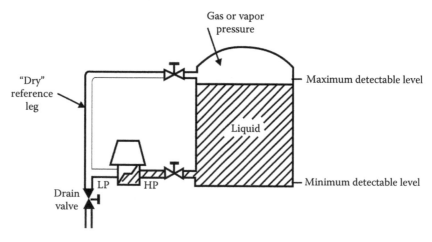

FIGURE 2.38
Closed tank dry reference leg.

2.8.9 Closed Tank, Wet Reference Leg

The filled reference leg applies a hydrostatic pressure to the HP side of the transmitter, which is equal to the maximum level to be measured. The dP transmitter is exposed to equal pressure on the HP and LP sides when the liquid level is at its maximum; therefore, the differential pressure is zero. As the tank level goes down, the pressure applied to the LP side goes down also, and the differential pressure increases. As a result, the differential pressure and the transmitter output are inversely proportional to the tank level (Figure 2.39).

2.8.10 Density Compensation

2.8.10.1 Specific Volume

Before examining an example that shows the effects of density, the unit "specific volume" must be defined. Specific volume is defined as volume per unit mass as shown in the following equation:

$$\text{Specific volume} = \frac{\text{volume}}{\text{mass}}$$

Specific volume is the reciprocal of density as shown in the following equation:

$$\text{Specific volume} = \frac{1}{\text{density}}$$

Specific volume is the standard unit used when working with vapors and steam that have low values of density. For the applications that involve water and steam, specific volume can be found using "Saturated Steam Tables," which list the specific volumes for water and saturated steam at different pressures and temperatures. The density of steam (or vapor) above the liquid level will have an effect on the weight of the steam or vapor bubble and the hydrostatic head pressure. As the density of the steam or vapor increases, the weight increases and causes an increase in hydrostatic head even though the actual level of the

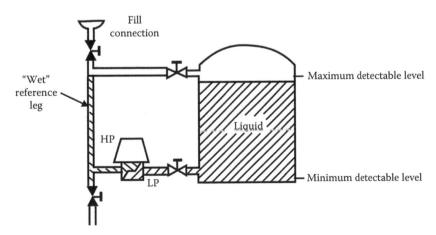

FIGURE 2.39
Closed tank wet reference leg.

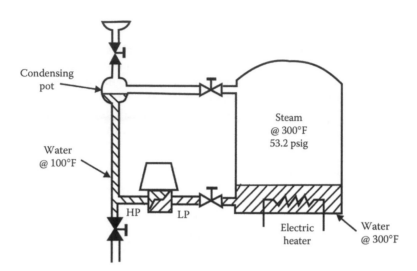

FIGURE 2.40
Effects of fluid density.

tank has not changed. The larger the steam bubble, the greater the change in hydrostatic head pressure (Figure 2.40).

A condensing pot at the top of the reference leg is incorporated to condense the steam and maintain the reference leg filled. As previously stated, the effect of the steam vapor pressure is canceled at the dP transmitter due to the fact that this pressure is equally applied to both the LP and HP sides of the transmitter. The differential pressure to the transmitter is due only to hydrostatic head pressure, as stated in the following equation:

$$\text{Hydrostatic head pressure} = \text{density} \times \text{height}$$

2.8.10.2 Reference Leg Temperature Considerations

When the level to be measured is in a pressurized tank at elevated temperatures, a number of additional consequences must be considered. As the temperature of the fluid in the tank is increased, the density of the fluid decreases. As the fluid's density decreases, the fluid expands, occupying more volume. Even though the density is less, the mass of the fluid in the tank is the same. The problem is that, as the fluid in the tank is heated and cooled, the density of the fluid changes, but the reference leg density remains relatively constant, which causes the indicated level to remain constant. The density of the fluid in the reference leg is dependent upon the ambient temperature of the room in which the tank is located; therefore, it is relatively constant and independent of tank temperature. If the fluid in the tank changes temperature, and therefore density, some means of density compensation must be incorporated in order to have an accurate indication of tank level. This is the problem encountered when measuring pressurizer water level or steam generator water level in pressurized water reactors, and when measuring reactor vessel water level in boiling water reactors.

2.8.10.2.1 Pressurizer Level Instruments

Pressurizer temperature is held fairly constant during normal operation. The dP detector for level is calibrated with the pressurizer hot, and the effects of density changes do not occur. The pressurizer will not always be hot. It may be cooled down for non-operating

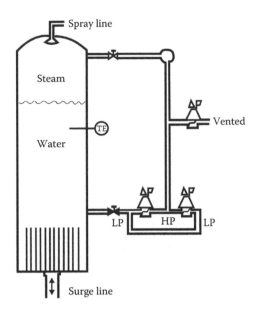

FIGURE 2.41
Pressurizer level system.

maintenance conditions, in which case a second dP detector, calibrated for level measurement at low temperatures, replaces the normal dP detector. The density has not really been compensated for; it has actually been aligned out of the instrument by calibration. Density compensation may also be accomplished through electronic circuitry. Some systems compensate for density changes automatically through the design of the level detection circuitry. Other applications compensate for density by manually adjusting inputs to the circuit as the pressurizer cools down and depressurizes, or during heat-up and pressurization. Calibration charts are also available to correct indications for changes in reference leg temperature (Figure 2.41).

2.8.10.2.2 Steam Generator Level Instrument

The dP detector measures actual differential pressure. A separate pressure detector measures the pressure of the saturated steam. Since saturation pressure is proportional to saturation temperature, a pressure signal can be used to correct the differential pressure for density. An electronic circuit uses the pressure signal to compensate for the difference in density between the reference leg water and the steam generator fluid. As the saturation temperature and pressure increase, the density of the steam generator water will decrease. The dP detector should now indicate a higher level, even though the actual dP has not changed. The increase in pressure is used to increase the output of the dP level detector in proportion to saturation pressure to reflect the change in actual level (Figure 2.42).

2.8.11 Level Detection Circuitry

2.8.11.1 Remote Indication

Remote indication is necessary to provide transmittal of vital level information to a central location, such as the control room, where all level information can be coordinated and evaluated. There are three major reasons for utilizing remote level indication: (1) Level

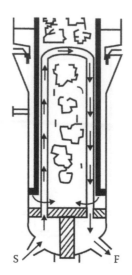

FIGURE 2.42
Steam generator level system detection arrangement.

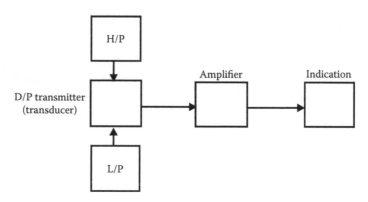

FIGURE 2.43
Block diagram of a differential pressure level detection.

measurements may be taken at locations far from the main facility. (2) The level to be controlled may be a long distance from the point of control. (3) The level being measured may be in an unsafe/radioactive area (Figure 2.43).

It consists of a differential pressure (D/P) transmitter (transducer), an amplifier, and level indication. The D/P transmitter consists of a diaphragm with the high pressure (H/P) and low pressure (L/P) inputs on opposite sides. As the differential pressure changes, the diaphragm will move. The transducer changes this mechanical motion into an electrical signal. The electrical signal generated by the transducer is then amplified and passed on to the level indicator for level indication at a remote location. Using relays, this system provides alarms on high and low level. It may also provide control functions such as repositioning a valve and protective features such as tripping a pump.

2.8.11.2 *Environmental Concerns*

Density of the fluid to be measured can have a large effect on level detection instrumentation. It primarily affects level sensing instruments that utilize a wet reference leg. In these instruments, it is possible for the reference leg temperature to be different from the temperature of the fluid whose level is to be measured. An example of this is the level detection instrumentation for a boiler steam drum. The water in the reference leg is at a lower temperature than the water in the steam drum. Therefore, it is denser, and must be compensated for to ensure the indicated steam drum level is accurately indicated. Ambient temperature variations will affect the accuracy and reliability of level detection instrumentation. Variations in ambient temperature can directly affect the resistance of components in the instrumentation circuitry, and, therefore, affect the calibration of electric/electronic equipment. The effects of temperature variations are reduced by the design of the circuitry and by maintaining the level detection instrumentation in the proper environment. The presence of humidity will also affect most electrical equipment, especially electronic equipment. High humidity causes moisture to collect on the equipment. This moisture can cause short circuits, grounds, and corrosion, which, in turn, may damage components. The effects due to humidity are controlled by maintaining the equipment in the proper environment.

2.9 Instrumentation and Control Module on Flow Detectors

2.9.1 Head Flow Meters

The flow path restriction, such as an orifice, causes a differential pressure across the orifice. This pressure differential is measured by a mercury manometer or a differential pressure detector. From this measurement, flow rate is determined from known physical laws. The head flow meter actually measures volume flow rate rather than mass flow rate. Mass flow rate is easily calculated or computed from volumetric flow rate by knowing or sensing temperature and/or pressure. Temperature and pressure affect the density of the fluid and, therefore, the mass of fluid flowing past a certain point. If the volumetric flow rate signal is compensated for changes in temperature and/or pressure, a true mass flow rate signal can be obtained. Thermodynamics states that temperature and density are inversely proportional, while pressure and density are directly proportional. To show the relationship between temperature and pressure, the mass flow rate equation is often written as either of the following equations:

$$m = \frac{KA}{dP(P)}$$

$$m = \frac{KA}{dP(1/T)}$$

where:
   m is the mass flow rate (lbm/sec)
   A is the area (ft^2)
   dP is the differential pressure (lbf/ft^2)
   P is the pressure (lbf/ft^2)
   T is the temperature (°F)
   K is the flow coefficient

The flow coefficient is constant for the system based mainly on the construction character-istics of the pipe and type of fluid flowing through the pipe. The flow coefficient in each equation contains the appropriate units to balance the equation and provide the proper units for the resulting mass flow rate. The area of the pipe and differential pressure are used to calculate volumetric flow rate. As stated above, this volumetric flow rate is con-verted to mass flow rate by compensating for system temperature or pressure.

2.9.1.1 Orifice Plate

The orifice plate is the simplest of the flow path restriction used in flow detection, as well as the most economical. Orifice plates are flat plates 1/16–1/4 in. thick. They are normally mounted between a pair of flanges and are installed in a straight run of smooth pipe to avoid disturbance of flow patterns from fittings and valves. Three kinds of orifice plates are used: concentric, eccentric, and segmental (as shown in Figure 2.44).

The concentric orifice plate is the most common of the three types. As shown, the orifice is equidistant (concentric) to the inside diameter of the pipe. Flow through a sharp-edged orifice plate is characterized by a change in velocity. As the fluid passes through the orifice, the fluid converges, and the velocity of the fluid increases to a maximum value. At this point, the pressure is at a minimum value. As the fluid diverges to fill the entire pipe area, the velocity decreases back to the original value. The pressure increases to about 60%–80% of the original input value. The pressure loss is irrecoverable; therefore, the output pres-sure will always be less than the input pressure. The pressures on both sides of the orifice are measured, resulting in a differential pressure that is proportional to the flow rate.

Segmental and eccentric orifice plates are functionally identical to the concentric orifice. The circular section of the segmental orifice is concentric with the pipe. The segmental portion of the orifice eliminates damming of foreign materials on the upstream side of the orifice when mounted in a horizontal pipe. Depending on the type of fluid, the segmental section is placed on either the top or bottom of the horizontal pipe to increase the accuracy of the measurement. Eccentric orifice plates shift the edge of the orifice to the inside of the pipe wall. This design also prevents upstream damming and is used in the same way as the segmental orifice plate. Orifice plates have two distinct disadvantages: they cause a high permanent pressure drop (outlet pressure will be 60%–80% of inlet pressure), and they are subject to erosion, which will eventually cause inaccuracies in the measured dif-ferential pressure.

Concentric

Eccentric

FIGURE 2.44
Orifice plates.

2.9.1.2 Venturi Tube

The Venturi tube, illustrated in Figure 2.45, is the most accurate flow-sensing element when properly calibrated. The Venturi tube has a converging conical inlet, a cylindrical throat, and a diverging recovery cone. It has no projections into the fluid, no sharp corners, and no sudden changes in contour.

The inlet section decreases the area of the fluid stream, causing the velocity to increase and the pressure to decrease. The LP is measured in the center of the cylindrical throat since the pressure will be at its lowest value, and neither the pressure nor the velocity is changing. The recovery cone allows for the recovery of pressure such that total pressure loss is only 10%–25%. The HP is measured upstream of the entrance cone. The major disadvantages of this type of flow detection are the high initial costs for installation and difficulty in installation and inspection.

2.9.1.3 Pitot Tube

The pitot tube, illustrated in Figure 2.46, is another primary flow element used to produce a differential pressure for flow detection. In its simplest form, it consists of a tube with an opening at the end. The small hole in the end is positioned such that it faces the flowing fluid. The velocity of the fluid at the opening of the tube decreases to zero. This provides for the HP input to a differential pressure detector. A pressure tap provides the LP input.

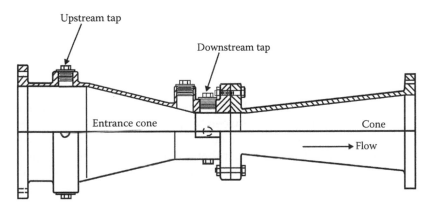

FIGURE 2.45
Venturi tube.

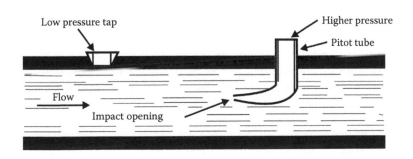

FIGURE 2.46
Pitot tube.

The pitot tube actually measures fluid velocity instead of fluid flow rate. However, volumetric flow rate can be obtained using the following equation:

$$V = KAV$$

where:
  V is the volumetric flow rate (ft^3/s)
  A is the area of flow cross-section (ft^2)
  V is the velocity of flowing fluid (ft/s)
  K is the flow coefficient (normally about 0.8)

Pitot tubes must be calibrated for each specific application, as there is no standardization. This type of instrument can be used even when the fluid is not enclosed in a pipe or duct.

2.9.2 Hot-Wire Anemometer

The hot-wire anemometer, principally used in gas flow measurement, consists of an electrically heated, fine platinum wire, which is immersed into the flow. As the fluid velocity increases, the rate of heat flow from the heated wire to the flow stream increases. Thus, a cooling effect on the wire electrode occurs, causing its electrical resistance to change. In a constant-current anemometer, the fluid velocity is determined from a measurement of the resulting change in wire resistance. In a constant-resistance anemometer, fluid velocity is determined from the current needed to maintain a constant wire temperature and, thus, the resistance constant.

2.9.3 Electromagnetic Flowmeter

The electromagnetic flowmeter is similar in principle to the generator. The rotor of the generator is replaced by a pipe placed between the poles of a magnet so that the flow of the fluid in the pipe is normal to the magnetic field. As the fluid flows through this magnetic field, an electromotive force is induced in it that will be mutually normal (perpendicular) to both the magnetic field and the motion of the fluid. This electromotive force may be measured with the aid of electrodes attached to the pipe and connected to a galvanometer or an equivalent. For a given magnetic field, the induced voltage will be proportional to the average velocity of the fluid. However, the fluid should have some degree of electrical conductivity.

2.9.4 Ultrasonic Flow Equipment

Devices such as ultrasonic flow equipment use the Doppler frequency shift of ultrasonic signals reflected from discontinuities in the fluid stream to obtain flow measurements. These discontinuities can be suspended solids, bubbles, or interfaces generated by turbulent eddies in the flow stream. The sensor is mounted on the outside of the pipe, and an ultrasonic beam from a piezoelectric crystal is transmitted through the pipe wall into the fluid at an angle to the flow stream. Signals reflected off flow disturbances are detected by a second piezoelectric crystal located in the same sensor. Transmitted and reflected signals are compared in an electrical circuit, and the corresponding frequency shift is proportional to the flow velocity.

2.9.5 Steam Flow Detection

The flow nozzle is commonly used for the measurement of steam flow and other high velocity fluid flow measurements where erosion may occur. It is capable of measuring approximately 60% higher flow rates than an orifice plate with the same diameter. This is due to the streamlined contour of the throat, which is a distinct advantage for the measurement of high velocity fluids (Figure 2.47).

The flow nozzle requires less straight run piping than an orifice plate. However, the pressure drop is about the same for both.

2.9.6 Simple Mass Flow Detection System

As the previous equations demonstrate, temperature and pressure values can be used to electronically compensate flow for changes in density. A simple mass flow detection system is illustrated by Figure 2.48, where measurements of temperature and pressure are made with commonly used instruments.

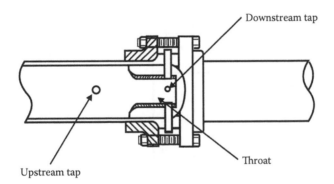

FIGURE 2.47
Flow nozzle.

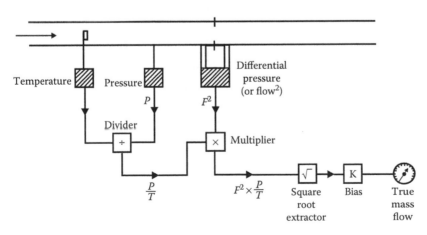

FIGURE 2.48
Simple mass flow detection system.

For the precise measurement of gas flow (steam) at varying pressures and temperatures, it is necessary to determine the density, which is pressure and temperature dependent, and from this value to calculate the actual flow. The use of a computer is essential to measure flow with changing pressure or temperature.

Figure 2.49 illustrates an example of a computer specifically designed for the measurement of gas flow. The computer is designed to accept input signals from commonly used differential pressure detectors, or from density or pressure plus temperature sensors, and to provide an output that is proportional to the actual rate of flow. The computer has accuracy better than +0.1% at flow rates of 10%–100%.

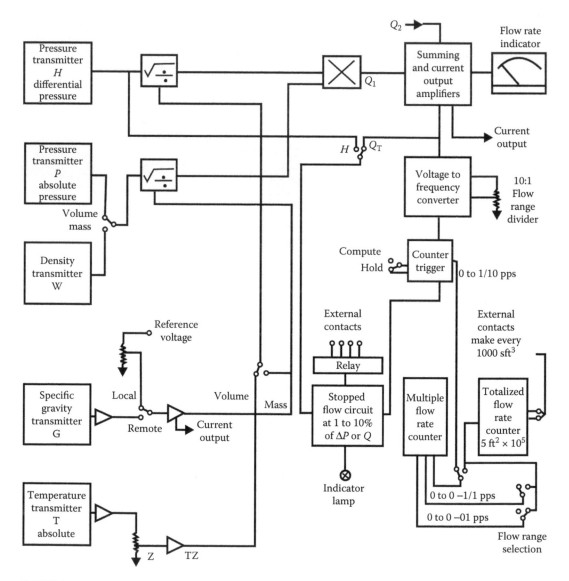

FIGURE 2.49
Gas flow computer.

2.9.6.1 Flow Circuitry

Figure 2.50 shows a block diagram of a typical differential pressure flow detection circuit. The dP transmitter operation is dependent on the pressure difference across an orifice, Venturi, or flow tube. This differential pressure is used to position a mechanical device such as a bellows. The bellows acts against spring pressure to reposition the core of a differential transformer. The transformer's output voltage on each of two secondary windings varies with a change in flow.

A loss of differential pressure integrity of the secondary element, the dP transmitter, will introduce an error into the indicated flow. This loss of integrity implies an impaired or degraded pressure boundary between the HP and LP sides of the transmitter. A loss of differential pressure boundary is caused by anything that results in the HP and LP sides of the dP transmitter being allowed to equalize pressure. As previously discussed, flow rate is proportional to the square root of the differential pressure. The extractor is used to electronically calculate the square root of the differential pressure and provide an output proportional to system flow. The constants are determined by selection of the appropriate electronic components. The extractor output is amplified and sent to an indicator. The indicator provides either a local or a remote indication of system flow.

2.9.6.2 Use of Flow Indication

The flow of liquids and gases carries energy through the piping system. In many situations, it is very important to know whether there is flow and the rate at which the flow is occurring. An example of flow that is important to a facility operator is equipment cooling flow. The flow of coolant is essential in removing the heat generated by the system, thereby preventing damage to the equipment. Typically, flow indication is used in protection systems and control systems that help maintain system temperature. Another method of determining system coolant flow is by using pump differential pressure. If all means of flow indication are lost, flow can be approximated using pump differential pressure. Pump differential pressure is proportional to the square of pump flow.

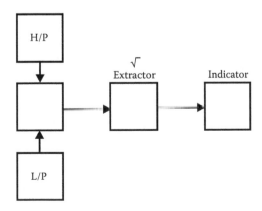

FIGURE 2.50
Differential pressure flow detection block diagram.

2.9.6.3 *Environmental Concerns*

As previously discussed, the density of the fluid whose flow is to be measured can have a large effect on flow sensing instrumentation. The effect of density is most important when the flow sensing instrumentation is measuring gas flows, such as steam. Since the density of a gas is directly affected by temperature and pressure, any changes in either of these parameters will have a direct effect on the measured flow. Therefore, any changes in fluid temperature or pressure must be compensated for to achieve an accurate measurement of flow. Ambient temperature variations will affect the accuracy and reliability of flow sensing instrumentation. Variations in ambient temperature can directly affect the resistance of components in the instrumentation circuitry, and, therefore, affect the calibration of electric/electronic equipment. The effects of temperature variations are reduced by the design of the circuitry and by maintaining the flow sensing instrumentation in the proper environment. The presence of humidity will also affect most electrical equipment, especially electronic equipment. High humidity causes moisture to collect on the equipment. This moisture can cause short circuits, grounds, and corrosion, which, in turn, may damage components. The effects due to humidity are controlled by maintaining the equipment in the proper environment.

2.10 Instrumentation and Control Module on Position Indicators

Synchro equipment remote indication or control may be obtained by the use of self-synchronizing motors, called synchro equipment. Synchro equipment consists of synchro units, which electrically govern or follow the position of a mechanical indicator or device. An electrical synchro has two distinct advantages over mechanical indicators: (1) greater accuracy, and (2) simpler routing of remote indication. There are five basic types of synchros, which are designated according to their function. The basic types are transmitters, differential transmitters, receivers, differential receivers, and control transformers.

Figure 2.51 illustrates schematic diagrams used to show external connections and the relative positions of synchro windings. If the power required to operate a device is higher than the power available from a synchro, power amplification is required. Servomechanism is a term that refers to a variety of power-amplifiers. These devices are incorporated into synchro systems for automatic control rod positioning in some reactor facilities. The transmitter, or synchro generator, consists of a rotor with a single winding and a stator with three windings placed 120° apart. When the mechanical device moves, the mechanically attached rotor moves. The rotor induces a voltage in each of the stator windings based on the rotor's angular position. Since the rotor is attached to the mechanical device, the induced voltage represents the position of the attached mechanical device. The voltage produced by each of the windings is utilized to control the receiving synchro position.

The receiver, or synchro motor, is electrically similar to the synchro generator. The synchro receiver uses the voltage generated by each of the synchro generator windings to position the receiver rotor. Since the transmitter and receiver are electrically similar, the angular position of the receiver rotor corresponds to that of the synchro transmitter rotor. The receiver differs mechanically from the transmitter in that it incorporates a damping device to prevent hunting. Hunting refers to the overshoot and undershoot that occur as the receiving device tries to match the sending device. Without the damping device, the receiver would go past the desired point slightly, and then return past the desired point slightly in the other direction. This would continue, by smaller amounts each time, until the receiver

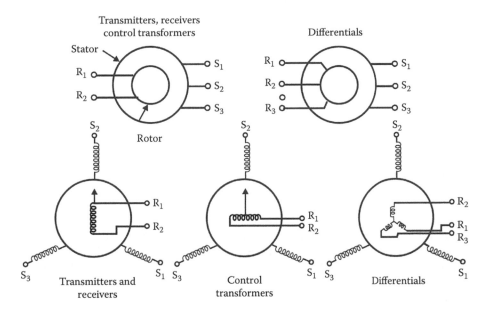

FIGURE 2.51
Synchro schematics.

came to rest at the desired position. The damper prevents hunting by feeding some of the signal back, thus slowing down the approach to the desired point. Differential synchros are used with transmitter and receiver synchros to insert a second signal. The angular positions of the transmitter and the differential synchros are compared, and the difference or sum is transmitted to the receiver. This setup can be used to provide a feedback signal to slow the response time of the receiver, thus providing a smooth receiver motion.

Control transformer synchros are used when only a voltage indication of angular position is desired. It is similar in construction to an ordinary synchro except that the rotor windings are used only to generate a voltage, which is known as an error voltage. The rotor windings of a control transformer synchro are wound with many turns of fine wire to produce a high impedance. Since the rotor is not fed excitation voltage, the current drawn by the stator windings would be high if they were the same as an ordinary synchro; therefore, they are also wound with many turns of fine wire to prevent excessive current. During normal operation, the output of a control transformer synchro is nearly zero (nulled) when its angular position is the same as that of the transmitter. A simple synchro system, consisting of one synchro transmitter (or generator) connected to one synchro receiver (or motor), is shown in Figure 2.52.

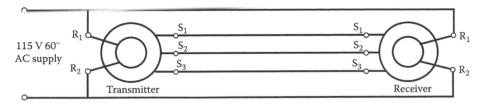

FIGURE 2.52
Simple synchro system.

When the transmitter's shaft is turned, the synchro receiver's shaft turns such that its "electrical position" is the same as the transmitters. What this means is that when the transmitter is turned to electrical zero, the synchro receiver also turns to zero. If the transmitter is disconnected from the synchro receiver and then reconnected, its shaft will turn to correspond to the position of the transmitter shaft.

2.11 Switches

2.11.1 Limit Switches

A limit switch is a mechanical device that can be used to determine the physical position of equipment. For example, an extension on a valve shaft mechanically trips a limit switch as it moves from open to shut or shut to open. The limit switch gives ON/OFF output that corresponds to valve position. Normally, limit switches are used to provide full open or full shut indications as illustrated in Figure 2.53.

Many limit switches are the push-button variety. When the valve extension comes in contact with the limit switch, the switch depresses to complete, or turn on, the electrical circuit. As the valve extension moves away from the limit switches, spring pressure opens the switch, turning off the circuit. Limit switch failures are normally mechanical in nature. If the proper indication or control function is not achieved, the limit switch is probably faulty. In this case, local position indication should be used to verify equipment position.

2.11.2 Reed Switches

Reed switches, illustrated in Figure 2.54, are more reliable than limit switches, due to their simplified construction. The switches are constructed of flexible ferrous strips (reeds) and are placed near the intended travel of the valve stem or control rod extension.

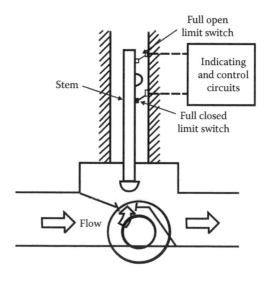

FIGURE 2.53
Limit switches.

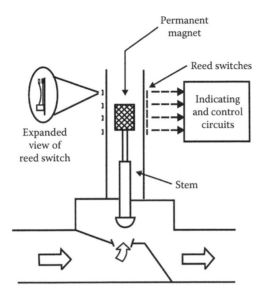

FIGURE 2.54
Reed switches.

When using reed switches, the extension used is a permanent magnet. As the magnet approaches the reed switch, the switch shuts. When the magnet moves away, the reed switch opens. This ON/OFF indicator is similar to mechanical limit switches. By using a large number of magnetic reed switches, incremental position can be measured. This technique is sometimes used in monitoring a reactor's control rod position.

2.12 Variable Output Devices

2.12.1 Potentiometer

Potentiometer valve position indicators provide an accurate indication of position throughout the travel of a valve or control rod. The extension is physically attached to a variable resistor. As the extension moves up or down, the resistance of the attached circuit changes, changing the amount of current flow in the circuit. The amount of current is proportional to the valve position (Figure 2.55).

Potentiometer valve position indicator failures are normally electrical in nature. An electrical short or open will cause the indication to fail at one extreme or the other. If an increase or decrease in the potentiometer resistance occurs, erratic indicated valve position occurs.

2.12.2 Linear Variable Differential Transformers

A linear variable differential transformer (LVDT), illustrated in Figure 2.56, is a device that provides accurate position indication throughout the range of valve or control rod travel. Unlike the potentiometer position indicator, no physical connection to the extension is required.

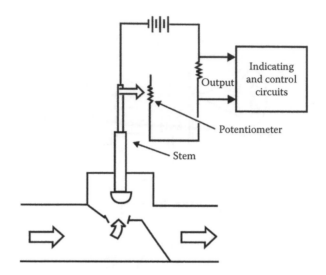

FIGURE 2.55
Potentiometer valve position indicator.

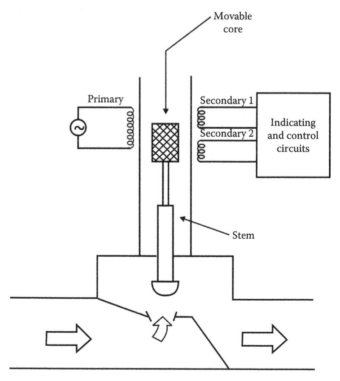

FIGURE 2.56
Linear variable differential transformer.

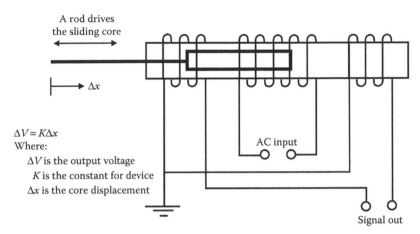

FIGURE 2.57
Circuit diagram of LVDT.

The extension valve shaft, or control rod, is made of a metal suitable for acting as the movable core of a transformer. Moving the extension between the primary and secondary windings of a transformer causes the inductance between the two windings to vary, thereby varying the output voltage proportional to the position of the valve or control rod extension (Figure 2.57).

Figure 2.57 illustrates a valve whose position is indicated by an LVDT. If the open and shut position is all that is desired, two small secondary coils could be utilized at each end of the extension's travel. LVDTs are extremely reliable. As a rule, failures are limited to rare electrical faults, which cause erratic or erroneous indications. An open primary winding will cause the indication to fail to some predetermined value equal to zero differential voltage. This normally corresponds to mid-stroke of the valve. A failure of either secondary winding will cause the output to indicate either full open or full closed.

2.12.3 Position Indication Circuitry

As described above, position detection devices provide a method to determine the position of a valve or control rod. The four types of position indicators discussed were limit switches, reed switches, potentiometer valve position indicators, and LVDTs. Reed and limit switches act as ON/OFF indicators to provide open and closed indications and control functions. Reed switches can also be used to provide coarse, incremental position indication. Potentiometer and LVDT position indicators provide accurate indication of valve and rod position throughout their travel. In some applications, LVDTs can be used to indicate open and closed positions when small secondary windings are used at either end of the valve stem stroke. The indicating and control circuitry provides for remote indication of valve or rod position and/or various control functions. Position indications vary from simple indications such as a light to meter indications showing exact position. Control functions are usually in the form of interlocks. Pump isolation valves are sometimes interlocked with the pump. In some applications, these interlocks act to prevent the pump from being started with the valves shut. The pump/valve interlocks can also be used to automatically turn off the pump if one of its isolation valves goes shut or to open a discharge valve at some time interval after the pump starts. Valves are sometimes interlocked with each other. In some systems, two valves may be interlocked to prevent both of the valves from being opened at the same time. This feature is used to prevent undesirable system flow paths. Control rod interlocks are normally used to prevent outward motion of certain rods unless certain conditions are met. One such interlock does not allow outward motion of control rods until the rods used to scram the reactor have been withdrawn to a predetermined height.

2.13 Summary

On completion of this chapter, the user will be able to understand the classification of different types of instruments. The working and methods of measurement of electrical and non-electrical parameters is explained. The concept of working of sensors and transducers and their classifications have been elaborated in detail.

References

1. W. Bolton. *Instrumentation and Control Systems.* 1st ed., 2004.
2. B. G. Liptak. *Instrument Engineers' Handbook, Volume 1, Fourth Edition: Process Measurement and Analysis.* CRC Press, Boca Raton, FL, 4th ed., 2003.
3. P. R. N. Childs. *Practical Temperature Measurement.* Butterworth-Heinemann, Boston, MA, 1st ed., 2001.
4. D. W. Spitzer. *Flow Measurement: Practical Guides for Measurement and Control.* Instrument Society of America, Research Triangle Park, NC, 2nd ed., 2001.
5. E. O. Doebelin. *Measurement Systems: Application and Design.* McGraw Hill College Division, Boston, MA, 1989.
6. T. A. Hughes. *Measurement and Control Basics.* ISA—The Instrumentation, Systems, and Automation Society, Research Triangle Park, NC, 3rd ed., 2002.
7. S. A. Dyer. *Wiley Survey of Instrumentation and Measurement.* John Wiley & Sons, New York, 1st ed., 2004.
8. N. Waldemar. *Measurement Systems and Sensors.* Artech House Publishers, Boston, MA, 2015.
9. N. A. Anderson. *Instrumentation for Process Measurement and Control.* CRC Press, Boca Raton, FL, 3rd ed., 2013.
10. E. O. Doebelin. *Doebelin's Measurement Systems.* McGraw Hill Education, Boston, MA, 6th ed., 2017.
11. N. Mohan, T. M. Undeland. *Power Electronics: Converters Applications and Design.* Media Enhanced, Wiley, New Delhi, India, 3rd ed., 2002.
12. W. Bolton. *Mechatronics: Electronic Control Systems in Mechanical and Electrical Engineering.* Pearson Education, New York, 4th ed., 2015.
13. L. Michalski, K. Eckersdorf, J. Kucharski, J. McGhee. *Temperature Measurement.* Wiley Publications, New York, 2nd ed., 2002.
14. L. Michalski. *Temperature Measurement.* John Wiley & Sons, 2002.
15. D. Tandeske. *Pressure Sensors: Selection and Application.* CRC Press, New York, 1990.
16. D. Placko. *Fundamentals of Instrumentation and Measurement,* John Wiley & Sons, New York, 2007.
17. H. Eren. *Displacement Measurement.* Wiley Online Library, USA, 1999.

18. J. Delsing. *Flow Measurement and Instrumentation*. Elsevier, Oxford, UK, 2017.

19. W. Boyes. *Instrumentation Reference Book*. Butterworth-Heinemann, Boston, MA, 4th ed., 2008.

20. F. Franceschini, M. Galetto, D. Maisano. *Management by Measurement: Designing Key Indicators and Performance Measurement*. Springer, Berlin, Germany, 2007.

21. M.-H. Bao, S. Middelhoek. *Micro Mechanical Transducers, Volume 8: Pressure Sensors, Accelerometers and Gyroscopes*. Elsevier Science, USA, 2000.

22. T.-W. Lee, *Thermal and Flow Measurements*. CRC Press, Boca Raton, FL, 2008.

23. R. C. Baker. *Flow Measurement Handbook: Industrial Designs, Operating Principles, Performance, and Applications*. Cambridge University Press, New York, 2nd ed., 2016.

24. R. W. Miller. *Flow Measurement Engineering Handbook*. McGraw-Hill Education, New York, 3rd ed., 1996.

25. D. W. Spitzer. *Flow Measurement: Practical Guides for Measurement and Control (Practical Guides for Measurement and Control)*. ISA: The Instrumentation, Systems, and Automation Society, Research Triangle Park, NC, 2nd ed., 1991.

3

Programmable Logic Control Systems

LEARNING OBJECTIVES
• To understand the architecture of PLC
• To know the PLC programming concepts
• To recognize the input and output devices used in PLC
• To solve an industrial process problem using PLC programming

3.1 Introduction

Control engineering has evolved over time. In the past, systems were mainly controlled by humans. More recently electricity has been used for control, and early electrical control was based on relays. These relays allow power to be switched on and off without a mechanical switch. It is common to use relays to make simple logical control decisions. The development of low cost computers has brought the most recent revolution, the programmable logic controller (PLC). The advent of the PLC began in the 1970s, and it has since become the most common choice for manufacturing controls.

PLCs have been gaining popularity on the factory floor and will probably remain predominant for some time to come. Most of this is because of the advantages they offer.

- Cost effective for controlling complex systems.
- Flexible and can be reapplied to control other systems quickly and easily.
- Computational abilities allow more sophisticated control.
- Trouble shooting aids make programming easier and reduce downtime.
- Reliable components make these likely to operate for years before failure.

3.1.1 Ladder Logic

Ladder logic is the main programming method used for PLCs. As mentioned previously, ladder logic has been developed to mimic relay logic. The decision to use the relay logic diagrams was a strategic one. By selecting ladder logic as the main programming method,

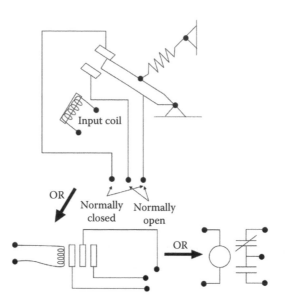

FIGURE 3.1
Simple relay layouts and schematics.

the amount of retraining needed for engineers and tradespeople was greatly reduced. Modern control systems still include relays, but these are rarely used for logic.

A relay is a simple device that uses a magnetic field to control a switch, as pictured in Figure 3.1. When a voltage is applied to the input coil, the resulting current creates a magnetic field. The magnetic field pulls a metal switch (or reed) toward it and the contacts touch, closing the switch. The contact that closes when the coil is energized is called normally open. The normally closed contacts touch when the input coil is not energized. Relays are normally drawn in schematic form using a circle to represent the input coil. The output contacts are shown with two parallel lines. Normally open contacts are shown as two lines and will be open (non-conducting) when the input is not energized. Normally closed contacts are shown with two lines with a diagonal line through them. When the input coil is not energized, the normally closed contacts will be closed (conducting).

Relays are used to let one power source close a switch for another (often high current) power source, while keeping them isolated. When we consider a PLC, there are inputs, outputs, and the logic. Figure 3.2 shows a more complete representation of the PLC. Here there are two inputs from push buttons. We can imagine the inputs as activating 24 Vdc relay coils in the PLC. This in turn drives an output relay that switches 115 Vac, which will turn on a light. In actuality, PLC inputs are never relays, but outputs are often relays. The ladder logic in the PLC is actually a computer program that the user can enter and change. Notice that both of the input push buttons are normally open, but the ladder logic inside the PLC has one normally open contact and one normally closed contact. Do not think that the ladder logic in the PLC needs to match the inputs or outputs. Many beginners will get caught trying to make the ladder logic match the input types.

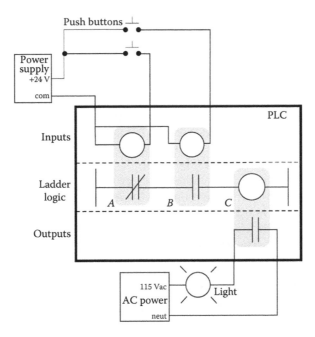

FIGURE 3.2
PLC illustrated with relays.

3.1.2 Programming

The first PLCs were programmed with a technique that was based on relay logic wiring schematics. This eliminated the need to teach the electricians, technicians, and engineers how to *program* a computer—but, this method has stuck, and it is the most common technique for programming PLCs today. An example of ladder logic can be seen in Figure 3.3. To interpret these diagrams, imagine that the power is on the vertical line on the left-hand

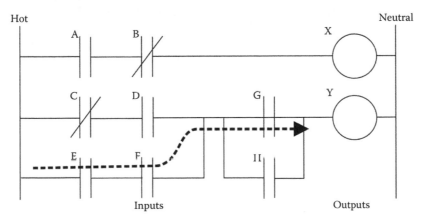

FIGURE 3.3
Simple ladder logic diagram. Power needs to flow through some combination of the inputs (A,B,C,D,E,F,G,H) to turn on outputs (X,Y).

side; we call this the hot rail. On the right-hand side is the neutral rail. In the figure there are two rungs, and on each rung there are combinations of inputs (two vertical lines) and outputs (circles).

If the inputs are opened or closed in the right combination, the power can flow from the hot rail, through the inputs, to power the outputs, and finally to the neutral rail. An input can come from a sensor, switch, or any other type of sensor. An output will be some device outside the PLC that is switched on or off, such as lights or motors. In the top rung, the contacts are normally open and normally closed. This means that if input A is on and input B is off, then power will flow through the output and activate it. Any other combination of input values will result in the output X being off.

The second rung is more complex—there are actually multiple combinations of inputs that will result in the output Y turning on. On the leftmost part of the rung, power could flow through the top if C is off and D is on. Power could also (and simultaneously) flow through the bottom if both E and F are true. This would get power half way across the rung, and then if G or H is true, the power will be delivered to output Y.

3.1.3 PLC Connections

When a process is controlled by a PLC, it uses inputs from sensors to make decisions and update outputs to drive actuators. The process is a real process that will change over time. Actuators will drive the system to new states (or modes of operation). This means that the controller is limited by the sensors available; if an input is not available, the controller will have no way to detect a condition.

The control loop is a continuous cycle of the PLC reading inputs, solving the ladder logic, and then changing the outputs. Like any computer, this does not happen instantly. Figure 3.4 shows the basic operation cycle of a PLC. When power is turned on initially, the PLC does a quick *sanity check* to ensure that the hardware is working properly. If there is a problem, the PLC will halt and indicate there is an error.

For example, if the PLC backup battery is low and power was lost, the memory will be corrupt and this will result in a fault. If the PLC passes the sanity check, it will then scan (read) all the inputs. After the inputs values are stored in memory, the ladder logic will be scanned (solved) using the stored values—not the current values. This is done to prevent logic problems when inputs change during the ladder logic scan. When the ladder logic scan is complete, the outputs will be scanned (the output values will be changed). After this, the system goes back to do a sanity check, and the loop continues indefinitely. Unlike

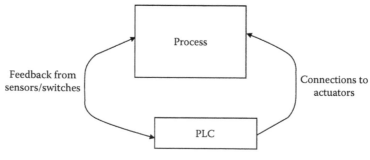

FIGURE 3.4
Separation of controller and process.

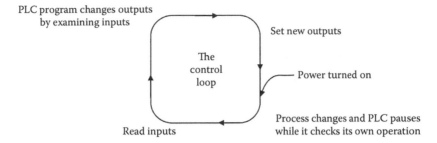

FIGURE 3.5
Scan cycle of a PLC.

normal computers, the entire program will be *run* every scan. Typical times for each of the stages is in the order of milliseconds (Figure 3.5).

3.1.4 Ladder Logic Inputs

PLC inputs are easily represented in ladder logic. In Figure 3.6 there are three types of inputs shown. The first two are normally open and normally closed inputs, discussed previously. The Immediate Input (IIT) function allows inputs to be read after the input scan while the ladder logic is being scanned. This allows ladder logic to examine input values more often than once every cycle.

3.1.5 Ladder Logic Outputs

In ladder logic there are multiple types of outputs, but these are not consistently available on all PLCs. Some of the outputs will be externally connected to devices outside the PLC, but it is also possible to use internal memory locations in the PLC. There are about six types of outputs. The first is a normal output—when energized the output will turn on and energize an output. The circle with a diagonal line through it is a normally on output. When energized, the output will turn off. This type of output is not available on all PLC types. When initially energized, the One Shot Relay (OSR) instruction will turn on for one scan,

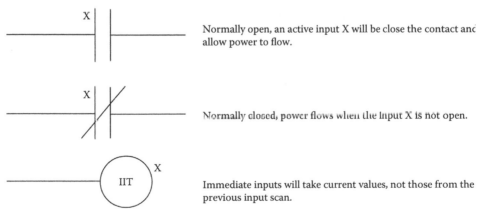

FIGURE 3.6
Ladder logic inputs.

but then be off for all scans after, until it is turned off. The L (latch) and U (unlatch) instructions can be used to lock outputs on. When an L output is energized, the output will turn on indefinitely, even when the output coil is de-energized. The output can only be turned off using a U output. The last instruction is the Immediate Output (IOT) that will allow outputs to be updated without having to wait for the ladder logic scan to be completed.

When power is applied (on), the output X is activated for the left output but turned off for the output on the right.

An input transition on will cause the output X to turn on for one scan (this is also known as a one shot relay). When the L coil is energized, X will be toggled on. It will stay on until the U coil is energized. This is like a flip-flop and stays set even when the PLC is turned off.

Some PLCs will allow immediate outputs that do not wait for the program scan to end before setting an output. (Note: This instruction will only update the outputs using the output table; other instructions must change the individual outputs.)

3.2 Programmable Logic Controller Hardware

Many PLC configurations are available, even from a single vendor. But, in each of these there are common components and concepts. The most essential components are:

Power supply: This can be built into the PLC or be an external unit. Common voltage levels required by the PLC (with and without the power supply) are 24 Vdc, 120 Vac, and 220 Vac.

CPU (*central processing unit*): This is a computer where ladder logic is stored and processed.

I/O (*input/output*): A number of input/output terminals must be provided so that the PLC can monitor the process and initiate actions.

Indicator lights: These indicate the status of the PLC including power on, program running, and a fault. These are essential when diagnosing problems. The configuration of the PLC refers to the packaging of the components.

Rack: A rack is often large (up to 18″ by 30″ by 10″) and can hold multiple cards. When necessary, multiple racks can be connected. These tend to be the highest cost, but also the most flexible and easy to maintain.

Mini: These are similar in function to PLC racks but about half the size.

Shoebox: A compact, all-in-one unit (about the size of a shoebox) that has limited expansion capabilities. Lower cost and compactness make these ideal for small applications.

Micro: These units can be as small as a deck of cards. They tend to have fixed quantities of I/O and limited abilities, but costs will be the lowest.

Software: A software-based PLC requires a computer with an interface card, but allows the PLC to be connected to sensors and other PLCs across a network.

3.2.1 Inputs and Outputs

Inputs to, and outputs from, a PLC are necessary to monitor and control a process. Both inputs and outputs can be categorized into two basic types: logical or continuous. Consider a light bulb. If it can only be turned on or off, it is logical control. If the light can be dimmed to different levels, it is continuous. Continuous values seem more intuitive, but logical values are preferred because they allow more certainty and simplify control. As a result, most controls applications (and PLCs) use logical inputs and outputs for most applications. Hence, we will discuss logical I/O and leave continuous I/O for later.

Outputs to actuators allow a PLC to cause something to happen in a process. A short list of popular actuators is given below in order of relative popularity.

Solenoid valves: Logical outputs that can switch a hydraulic or pneumatic flow.

Lights: Logical outputs that can often be powered directly from PLC output boards.

Motor starters: Motors often draw a large amount of current when started, so they require motor starters, which are basically large relays.

Servo motors: A continuous output from the PLC can command a variable speed or position.

Outputs from PLCs are often relays, but they can also be solid state electronics such as transistors for DC outputs or triacs for AC outputs. Continuous outputs require special output cards with digital to analog converters. Inputs come from sensors that translate physical phenomena into electrical signals. Typical examples of sensors are listed below in relative order of popularity.

Proximity switches: Use inductance, capacitance, or light to detect an object logically.

Switches: Mechanical mechanisms will open or close electrical contacts for a logical signal.

Potentiometer: Measures angular positions continuously using resistance.

LVDT (linear variable differential transformer): Measures linear displacement continuously using magnetic coupling.

Inputs for a PLC come in a few basic varieties. The simplest are AC and DC inputs. Sourcing and sinking inputs are also popular. The output method dictates that a device does not supply any power. Instead, the device only switches current on or off, like a simple switch.

Sinking: When active, the output allows current to flow to a common ground. This is best selected when different voltages are supplied.

Sourcing: When active, current flows from a supply, through the output device and to ground. This method is best used when all devices use a single supply voltage. This is also referred to as NPN (sinking) and PNP (sourcing). PNP is more popular. This will be covered in more detail in the chapter on sensors.

3.2.1.1 Inputs

In smaller PLCs, the inputs are normally built in and are specified when purchasing the PLC. For larger PLCs, the inputs are purchased as modules, or cards, with 8 or 16 inputs of the same type on each card. For discussion purposes, we will discuss all inputs as if they have been purchased as cards. The list below shows typical ranges for input voltages, and is roughly in order of popularity.

12–24 Vdc

100–120 Vacs

10–60 Vdc

12–24 Vac/dc

5 Vdc (TTL)

200–240 Vacs

48 Vdc

24 Vac

PLC input cards rarely supply power. This means that an external power supply is needed to supply power for the inputs and sensors. The example in Figure 3.7 shows how to connect an AC input card.

In the example there are two inputs; one is a normally open push button, and the second is a temperature switch, or thermal relay. Both switches are powered by the hot output of the 24 Vac power supply—this is like the positive terminal on a DC supply. Power is supplied to the left side of both switches. When the switches are open, there is no voltage passed to the input card. If either of the switches are closed, power will be supplied to the input card. In this case, inputs 1 and 3 are used—notice that the inputs start at 0.

The input card compares these voltages to the common. If the input voltage is within a given tolerance range, the inputs will switch on. Ladder logic is shown in the figure for the inputs. Here it uses Allen-Bradley notation for PLC-5 racks. At the top is the location of the input card I:013, which indicates that the card is an Input card in rack 01 in slot 3. The input number on the card is shown below the contact as 01 and 03 (Figure 3.7).

Many beginners become confused about where connections are needed in the circuit above. The key word to remember is *circuit*, which means that there is a full loop that the voltage must be able to follow. We can start following the circuit (loop) at the power supply. The path goes *through* the switches, *through* the input card, and back to the power supply where it flows back *through* to the start. In a full PLC implementation, there will be many circuits that must each be complete.

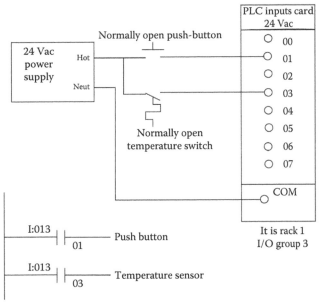

FIGURE 3.7
AC input card and ladder logic. Inputs are normally high impedance. This means that they will use very little current.

A second important concept is the common. Here the neutral on the power supply is the common, or reference voltage. In effect, we have chosen this to be our 0 V reference, and all other voltages are measured relative to it. If we had a second power supply, we would also need to connect the neutral so that both neutrals would be connected to the same common. Often, common and ground are confused. The common is a reference, or datum voltage that is used for 0 V, but the ground is used to prevent shocks and damage to equipment. The ground is connected under a building to a metal pipe or grid in the ground. This is connected to the electrical system of a building, to the power outlets, where the metal cases of electrical equipment are connected. When power flows through the ground, it is bad. Unfortunately, many engineers and manufacturers mix up ground and common. It is very common to find a power supply with the ground and common mislabeled.

One final concept that tends to trap beginners is that each input card is isolated. This means that if you have connected a common to only one card, then the other cards are not connected. When this happens, the other cards will not work properly. You must connect a common for each of the output cards.

There are many trade-offs when deciding which type of input cards to use. DC voltages are usually lower and therefore safer (i.e., 12–24 V). DC inputs are very fast; AC inputs require a longer on-time. For example, a 60 Hz wave may require up to 1/60 s for reasonable recognition. DC voltages can be connected to larger variety of electrical systems.

AC signals are more immune to noise than DC, so they are suited to long distances and noisy (magnetic) environments. AC power is easier and less expensive to supply to equipment. AC signals are very common in many existing automation devices.

3.2.1.2 Output Modules

As with input modules, output modules rarely supply any power, but instead act as switches. External power supplies are connected to the output card, and the card will switch the power on or off for each output. Typical output voltages are listed below and roughly ordered by popularity.

> 120 Vac
> 24 Vdc
> 12–48 Vac
> 5 Vdc (TTL)
> 230 Vac

These cards typically have 8–16 outputs of the same type and can be purchased with different current ratings. A common choice when purchasing output cards is relays, transistors, or triacs. Relays are the most flexible output devices. They are capable of switching both AC and DC outputs. But, they are slower (about 10 ms switching is typical), they are bulkier, they cost more, and they will wear out after millions of cycles. Relay outputs are often called dry contacts. Transistors are limited to DC outputs, and triacs are limited to AC outputs. Transistor and triac outputs are called switched outputs.

> *Dry contacts*: A separate relay is dedicated to each output. This allows mixed voltages (AC or DC and voltage levels up to the maximum), as well as isolated outputs to protect other outputs and the PLC. Response times are often greater than 10 ms. This method is the least sensitive to voltage variations and spikes.
>
> *Switched outputs*: A voltage is supplied to the PLC card, and the card switches it to different outputs using solid state circuitry (transistors, triacs, etc.). Triacs are well suited to AC devices requiring less than 1 A. Transistor outputs use NPN or PNP transistors up to 1 A typically. Their response time is well under 1 ms.

PLC outputs must convert the 5 Vdc logic levels on the PLC data bus to external voltage levels. This can be done with circuits similar to those shown in Figure 3.8. Basically, the circuits use an optocoupler to switch external circuitry. This electrically isolates the external electrical circuitry from the internal circuitry. Other circuit components are used to guard against excess or reversed voltage polarity.

Caution is required when building a system with both AC and DC outputs. If AC is accidentally connected to a DC transistor output, it will only be on for the positive half of the cycle and appear to be working with a diminished voltage. If DC is connected to an AC triac output, it will turn on and appear to work, but you will not be able to turn it off without turning off the entire PLC.

A major issue with outputs is mixed power sources. It is good practice to isolate all power supplies and keep their commons separate, but this is not always feasible. Some output modules, such as relays, allow each output to have its own common (Figure 3.8).

Other output cards require that multiple, or all, outputs on each card share the same common. Each output card will be isolated from the rest, so each common will have to be connected. It is common for beginners to only connect the common to one card and forget

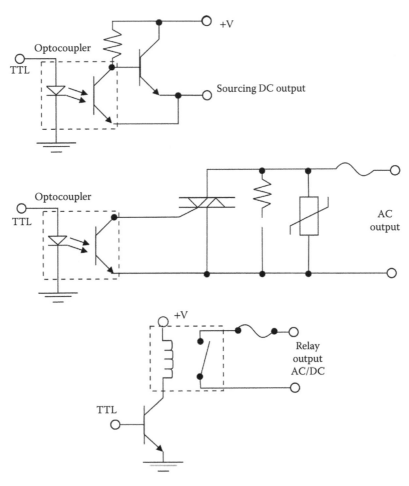

FIGURE 3.8
PLC output circuits.

the other cards. The output card shown in Figure 3.9 is an example of a 24 Vdc output card that has a shared common. This type of output card would typically use transistors for the outputs.

In this example, the outputs are connected to a low current light bulb (lamp) and a relay coil. Consider the circuit through the lamp, starting at the 24 Vdc supply. When the output 07 is on, current can flow in 07 to the COM, thus completing the circuit and allowing the light to turn on. If the output is off, the current cannot flow, and the light will not turn on. The output 03 for the relay is connected in a similar way. When the output 03 is on, current will flow through the relay coil to close the contacts and supply 120 Vac to the motor. Ladder logic for the outputs is shown in the bottom right of the figure. The notation is for an Allen-Bradley PLC-5. The value at the top left of the outputs, O: 012, indicates that the card is an output card, in rack 01, in slot 2 of the rack. To the bottom right of the outputs is the output number on the card 03 or 07. This card could have many different voltages applied from different sources, but all the power supplies would need a single shared common.

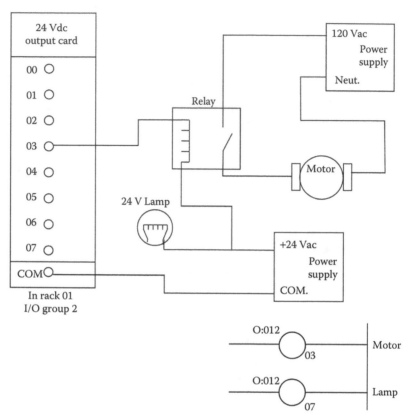

FIGURE 3.9
Example of a 24 Vdc output card (sinking).

3.2.2 Relays

Although relays are rarely used for control logic, they are still essential for switching large power loads. Some important terminology for relays is given below.

Contactor: Special relays for switching large current loads.

Motor starter: Basically, a contactor in series with an overload relay to cut off when too much current is drawn.

ARC suppression: When any relay is opened or closed, an arc will jump. This becomes a major problem with large relays. On relays switching AC, this problem can be overcome by opening the relay when the voltage goes to zero (while crossing between negative and positive). When switching DC loads, this problem can be minimized by blowing pressurized gas across during opening to suppress the arc formation.

AC coils: If a normal relay coil is driven by AC power, the contacts will vibrate open and closed at the frequency of the AC power. This problem is overcome by adding a shading pole to the relay. The most important consideration when selecting relays, or relay outputs on a PLC, is the rated current and voltage.

If the rated voltage is exceeded, the contacts will wear out prematurely, or if the voltage is too high, fire is possible. The rated current is the maximum current that should be used. When this is exceeded, the device will become too hot, and it will fail sooner. The rated values are typically given for both AC and DC, although DC ratings are lower than AC. If the actual loads used are below the rated values, the relays should work well indefinitely. If the values are exceeded, a small amount the life of the relay will be shortened accordingly. Exceeding the values significantly may lead to immediate failure and permanent damage.

Rated voltage: The suggested operation voltage for the coil. Lower levels can result in failure to operate; voltages above shorten life.

Rated current: The maximum current before contact damage occurs (welding or melting).

3.2.3 Electrical Wiring Diagrams

When control cabinet is designed and constructed ladder diagrams are used to document the wiring. In this example, the system would be supplied with AC power (120 or 220 Vac) on the left and right rails. The lines of these diagrams are numbered, and these numbers are typically used to number wires when building the electrical system. The switch before line 010 is a master disconnect for the power to the entire system. A fuse is used after the disconnect to limit the maximum current drawn by the system. Line 020 of the diagram is used to control power to the outputs of the system. The stop button is normally closed, while the start button is normally open.

The branch and output of the rung are CR1, which is a master control relay. The PLC receives power on line 30 of the diagram. The inputs to the PLC are all AC and are shown on lines 040–070. Notice that Input I:0/0 is a set of contacts on the MCR CR1. The three other inputs are a normally open push button (050), a limit switch (060), and a normally closed push button (070). After line 080, the MCR CR1 can apply power to the outputs. These power the relay outputs of the PLC to control a red indicator light (040), a green indicator light (050), a solenoid (060), and another relay (080). The relay on line 080 switches a relay that turns on another device *drill station*.

In the wiring diagram, the choice of a normally close stop button and a normally open start button are intentional. Consider line 020 in the wiring diagram. If the stop button is pushed, it will open the switch, power will not be able to flow to the control relay, and output power will shut off. If the stop button is damaged, say by a wire falling off, the power will also be lost and the system will shut down—safely. If the stop button used was normally open and this happened, the system would continue to operate while the stop button was unable to shut down the power. Now consider the start button. If the button was damaged, say a wire was disconnected, it would be unable to start the system, thus leaving the system unstarted and safe. In summary, all buttons that stop a system should be normally closed, while all buttons that start a system should be normally open.

3.2.3.1 *Joint International Committee Wiring Symbols*

To standardize electrical schematics, the Joint International Committee (JIC) symbols were developed (Figure 3.10).

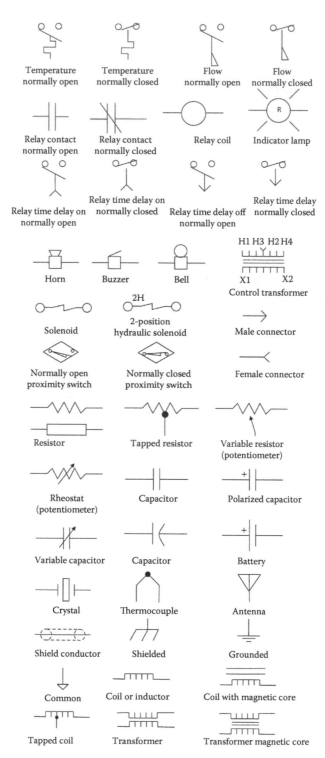

FIGURE 3.10
JIC schematic symbols.

3.3 Logical Sensors

Sensors allow a PLC to detect the state of a process. Logical sensors can only detect a state that is either true or false. Examples of physical phenomena that are typically detected are listed below.

- *Inductive proximity*: Is a metal object nearby?
- *Capacitive proximity*: Is a dielectric object nearby?
- *Optical presence*: Is an object breaking a light beam or reflecting light?
- *Mechanical contact*: Is an object touching a switch?

Recently, the cost of sensors has dropped and they have become commodity items, typically between $50 and $100. They are available in many forms from multiple vendors such as Allen-Bradley, Omron, Hyde Park, and Turk. In applications, sensors are interchangeable between PLC vendors, but each sensor will have specific interface requirements. This chapter will begin by examining the various electrical wiring techniques for sensors, and conclude with an examination of many popular sensor types.

3.3.1 Sensor Wiring

When a sensor detects a logical change, it must signal that change to the PLC. This is typically done by switching a voltage or current on or off. In some cases, the output of the sensor is used to switch a load directly, completely eliminating the PLC. Typical outputs from sensors (and inputs to PLCs) are listed below in relative popularity.

- *Sinking/Sourcing*: Switches current on or off.
- *Plain switches*: Switches voltage on or off.
- *Solid state relays*: These switch AC outputs.
- *Transistor–transistor logic (TTL)*: Uses 0 and 5 V to indicate logic levels.

3.3.1.1 Switches

The simplest example of sensor outputs are switches and relays. A simple example is shown in Figure 3.11. In the figure, a NO contact switch is connected to input 01. A sensor with a relay output is also shown. The sensor must be powered separately; therefore, the V+ and V– terminals are connected to the power supply. The output of the sensor will become active when a phenomenon has been detected. This means the internal switch (probably a relay) will be closed, allowing current to flow, and the positive voltage will be applied to input 06.

3.3.1.2 Transistor–Transistor Logic

Transistor–transistor logic (TTL) is based on two voltage levels, 0 V for false and 5 V for true. The voltages can actually be slightly larger than 0 V, or lower than 5 V and still be detected correctly. This method is very susceptible to electrical noise on the factory floor and should only be used when necessary. TTL outputs are common on electronic devices

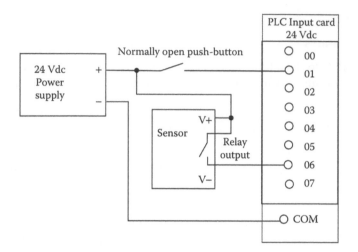

FIGURE 3.11
Circuit diagram of switches.

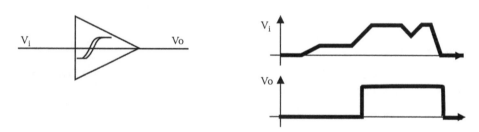

FIGURE 3.12
Schmitt trigger.

and computers, and will be necessary sometimes. When connecting to other devices, simple circuits can be used to improve the signal, such as the Schmitt trigger in Figure 3.12.

A Schmitt trigger will receive an input voltage between 0 and 5 V and convert it to 0 or 5 V. If the voltage is in an ambiguous range, about 1.5–3.5 V, it will be ignored. If a sensor has a TTL output, the PLC must use a TTL input card to read the values. If the TTL sensor is being used for other applications, it should be noted that the maximum current output is normally about 20 mA.

3.3.1.3 *Sinking/Sourcing*

Sinking sensors allow current to flow into the sensor to the voltage common, while sourcing sensors allow current to flow out of the sensor from a positive source. For both methods, the emphasis is on current flow, not voltage. By using current flow instead of voltage, many of the electrical noise problems are reduced. When discussing sourcing and sinking, we are referring to the *output* of the sensor that is acting like a switch. In fact, the output of the sensor is normally a transistor that will act like a switch (with some voltage loss). A PNP transistor is used for the sourcing output, and an NPN transistor is used for the sinking input. When discussing these sensors, the term sourcing is often interchanged with PNP, and sinking with NPN. A simplified example of a sinking output sensor is shown in Figure 3.13.

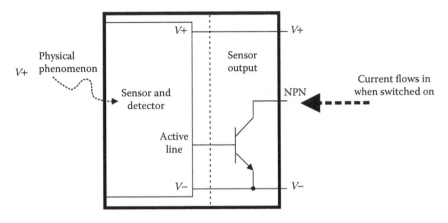

FIGURE 3.13
Simplified NPN/sinking sensor.

The sensor will have some part that deals with detection; this is on the left. The sensor needs a voltage supply to operate. If the sensor has detected some phenomenon, then it will trigger the active line. The active line is directly connected to an NPN transistor. If the voltage to the transistor on the *active line* is 0 V, then the transistor will not allow current to flow into the sensor. If the voltage on the active line becomes larger (say 12 V), then the transistor will switch on and allow current to flow into the sensor to the common.

Sourcing sensors are the complement to sinking sensors. The sourcing sensors use a PNP transistor, as shown in Figure 3.14. (Note: PNP transistors are always drawn with the arrow pointing to the center.) When the sensor is inactive, the active line stays at the V+ value, and the transistor stays switched off. When the sensor becomes active, the active line will be made 0 V, and the transistor will allow current to flow out of the sensor.

Most NPN/PNP sensors are capable of handling currents up to a few amps, and they can be used to switch loads directly. (Note: always check the documentation for

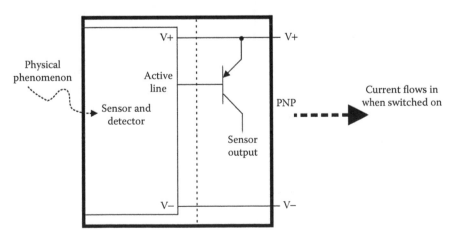

FIGURE 3.14
Simplified sourcing/PNP sensor.

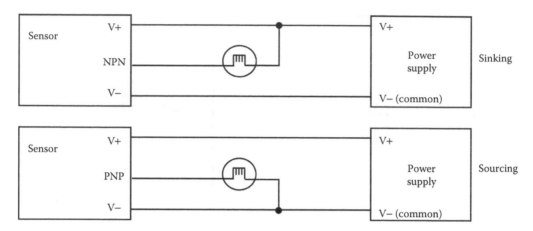

FIGURE 3.15
Direct control using NPN/PNP sensors.

rated voltages and currents.) An example using sourcing and sinking sensors to control lights is shown in Figure 3.15.

In the sinking system in Figure 3.16, the light has V+ applied to one side. The other side is connected to the NPN *output* of the sensor. When the sensor turns on, the current will be able to flow through the light, into the output to V– common. (Note: Yes, the current will be allowed to flow into the output for an NPN sensor.) In the sourcing arrangement, the light will turn on when the output becomes active, allowing current to flow from the V+, thought the sensor, the light and to V– (the common). At this point it is worth stating the obvious—the output of a sensor will be an input for a PLC. And, as we saw with the NPN sensor, this does not necessarily indicate where current is flowing. There are two viable approaches for connecting sensors to PLCs. The first is to always use PNP sensors

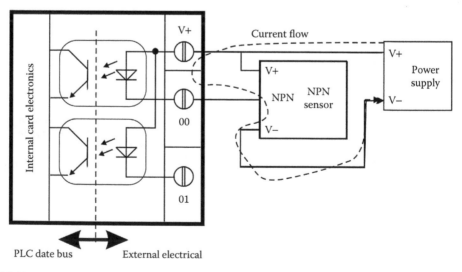

FIGURE 3.16
PLC input card for sinking sensors.

and normal voltage input cards. The second option is to purchase input cards specifically designed for sourcing or sinking sensors.

The dashed line in the figure represents the circuit or current flow path when the sensor is active. This path enters the PLC input card first at a V+ terminal (Note: there is no common on this card) and flows through an optocoupler. This current will use light to turn on a phototransistor to tell the computer in the PLC the input current is flowing. The current then leaves the card at input 00 and passes through the sensor to V–. When the sensor is inactive the current will not flow, and the light in the optocoupler will be off. The optocoupler is used to help protect the PLC from electrical problems outside the PLC.

Wiring is a major concern with PLC applications, so to reduce the total number of wires, two-wire sensors have become popular. But, by integrating three wires' worth of function into two, we now couple the power supply and sensing functions into one. Two-wire sensors are shown in Figure 3.17.

A two-wire sensor can be used as either a sourcing or sinking input. In both arrangements, the sensor will require a small amount of current to power the sensor, but when active it will allow more current to flow. This requires input cards that will allow a small amount of current to flow (called the leakage current), but also be able to detect when the current has exceeded a given value.

Most modern sensors have both PNP and NPN outputs, although if the choice is not available, PNP is the more popular choice. PLC cards can be confusing to buy, as each vendor refers to the cards differently. To avoid problems, look to see if the card is specifically for sinking or sourcing sensors, or look for a V+ (sinking) or COM (sourcing). Some vendors also sell cards that will allow you to have NPN and PNP inputs mixed on the same card.

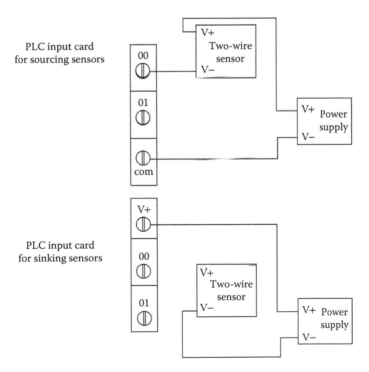

FIGURE 3.17
Two-wire sensors. These sensors require a certain leakage current to power the electronics.

3.3.1.4 Solid-State Relays

Solid-state relays switch AC currents. These are relatively inexpensive and are available for large loads. Some sensors and devices are available with these as outputs.

3.3.2 Presence Detection

There are two basic ways to detect object presence: contact and proximity. Contact implies that there is mechanical contact and a resulting force between the sensor and the object. Proximity indicates that the object is near, but contact is not required. The following sections examine different types of sensors for detecting object presence. These sensors account for a majority of the sensors used in applications.

3.3.2.1 Contact Switches

Contact switches are available as normally open and normally closed. Their housings are reinforced so that they can take repeated mechanical forces. These often have rollers and wear pads for the point of contact. Lightweight contact switches can be purchased for less than a dollar, but heavy duty contact switches will have much higher costs. Examples of applications include motion limit switches and part present detectors.

3.3.2.2 Reed Switches

Reed switches are very similar to relays, except a permanent magnet is used instead of a wire coil. When the magnet is far away, the switch is open, but when the magnet is brought near, the switch is closed as shown in Figure 3.18. These are very inexpensive and can be purchased for a few dollars. They are commonly used for safety screens and doors because they are harder to *trick* than other sensors.

3.3.2.3 Optical (Photoelectric) Sensors

Light sensors have been used for almost a century—originally photocells were used for applications such as reading audio tracks on motion pictures. But modern optical sensors are much more sophisticated. Optical sensors require both a light source (emitter) and detector. Emitters will produce light beams in the visible and invisible spectrums using LEDs and laser diodes. Detectors are typically built with photodiodes or phototransistors.

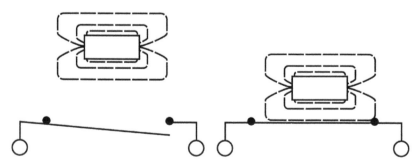

FIGURE 3.18
Reed switch.

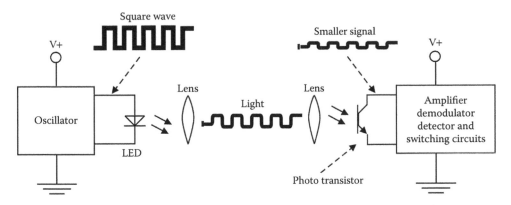

FIGURE 3.19
Basic optical sensor.

The emitter and detector are positioned so that an object will block or reflect a beam when present. A basic optical sensor is shown in Figure 3.19.

In the figure, the light beam is generated on the left, focused through a lens. At the detector side, the beam is focused on the detector with a second lens. If the beam is broken the detector will indicate an object is present. The oscillating light wave is used so that the sensor can filter out normal light in the room. The light from the emitter is turned on and off at a set frequency. When the detector receives the light, it checks to make sure that it is at the same frequency. If light is being received at the right frequency, then the beam is not broken. The frequency of oscillation is in the KHz range, and too fast to be noticed. A side effect of the frequency method is that the sensors can be used with lower power at longer distances.

An emitter can be set up to point directly at a detector; this is known as opposed mode. When the beam is broken, the part will be detected. This sensor needs two separate components, as shown in Figure 3.20.

Having the emitter and detector separate increases maintenance problems, and alignment is required. A preferred solution is to house the emitter and detector in one unit. But, this requires that light be reflected back as shown in Figure 3.21.

In the figure, the emitter sends out a beam of light. If the light is returned from the reflector, most of the light beam is returned to the detector. When an object interrupts the beam between the emitter and the reflector, the beam is no longer reflected back to the detector, and the sensor becomes active. A potential problem with this sensor is that reflective objects could return a good beam. This problem is overcome by polarizing the light at the emitter (with a filter), and then using a polarized filter at the detector. The reflector uses small cubic reflectors, and when the light is reflected the polarity is rotated by 90°. If the light is reflected off the object, the light will not be rotated by 90°.

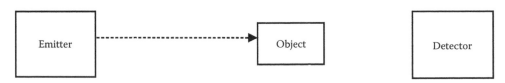

FIGURE 3.20
Opposed mode optical sensor.

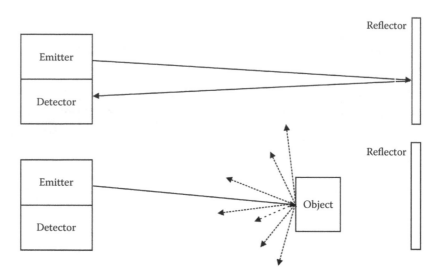

FIGURE 3.21
Retro-reflective optical sensor.

3.3.2.4 Capacitive Sensors

Capacitive sensors are able to detect most materials at distances up to a few centimeters. Recall the basic relationship for capacitance.

$$C = \frac{Ak}{d}$$

where:
  C is the capacitance (Farads)
  k is the dielectric constant
  A is the area of plates
  d is the distance between plates (electrodes)

In the sensor, the area of the plates and distance between them is fixed. But, the dielectric constant of the space around them will vary as different materials are brought near the sensor. An illustration of a capacitive sensor is shown in Figure 3.22. An oscillating field is used to determine the capacitance of the plates. When this changes beyond a selected sensitivity, the sensor output is activated.

These sensors work well for insulators (such as plastics) that tend to have high dielectric coefficients, thus increasing the capacitance. But, they also work well for metals because the conductive materials in the target appear as larger electrodes, thus increasing the capacitance as shown in Figure 3.23. In total, the capacitance changes are normally in the order of pF.

The sensors are normally made with rings (not plates) in the configuration shown in Figure 3.24. In the figure, the two inner metal rings are the capacitor electrodes, but a third outer ring is added to compensate for variations. Without the compensator ring, the sensor would be very sensitive to dirt, oil, and other contaminants that might stick to the sensor.

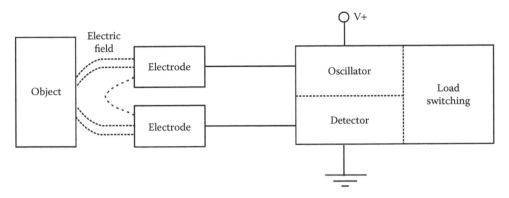

FIGURE 3.22
Capacitive sensor.

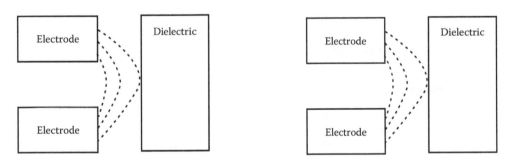

FIGURE 3.23
Dielectrics and metals increase the capacitance.

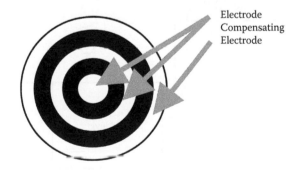

FIGURE 3.24
Sensors with ring.

3.3.2.5 *Inductive Sensors*

Inductive sensors use currents induced by magnetic fields to detect nearby metal objects. The inductive sensor uses a coil (an inductor) to generate a high frequency magnetic field as shown in Figure 3.25. If there is a metal object near the changing magnetic field, current will flow in the object. This resulting current flow sets up a new magnetic field that opposes the original magnetic field. The net effect is that it changes the inductance of the coil in the inductive sensor.

3.3.2.6 *Ultrasonic*

An ultrasonic sensor emits a sound above the normal hearing threshold of 16 KHz. The time that is required for the sound to travel to the target and reflect back is proportional to the distance to the target. The two common types of sensors are

Electrostatic: Uses capacitive effects. It has longer ranges and wider bandwidth, but is more sensitive to factors such as humidity.

Piezoelectric: Based on charge displacement during strain in crystal lattices. These are rugged and inexpensive. These sensors can be very effective for applications such as fluid levels in tanks and crude distance measurement.

3.3.2.7 *Hall Effect*

Hall effect switches are basically transistors that can be switched by magnetic fields. Their applications are very similar to reed switches, but because they are solid state, they tend to be more rugged and resist vibration. Automated machines often use these to do initial calibration and detect end stops.

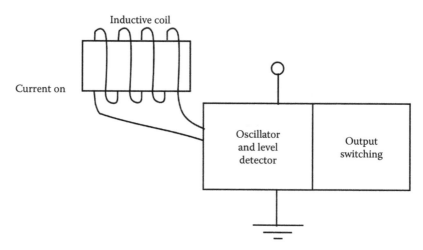

FIGURE 3.25
Inductive proximity sensor.

3.4 Logical Actuators

Actuators drive motions in mechanical systems. Most often this is by converting electrical energy into some form of mechanical motion.

3.4.1 Solenoids

Solenoids are the most common actuator components. The basic principle of operation is a moving ferrous core (a piston) that will move inside wire coil as shown in Figure 3.26. Normally, the piston is held outside the coil by a spring. When a voltage is applied to the coil and current flows, the coil builds up a magnetic field that attracts the piston and pulls it into the center of the coil. The piston can be used to supply a linear force. Well known applications of these include pneumatic values and car door openers.

As mentioned previously, inductive devices can create voltage spikes and may need snubbers, although most industrial applications have low enough voltage and current ratings they can be connected directly to the PLC outputs. Most industrial solenoids will be powered by 24 Vdc and draw a few hundred µA.

3.4.2 Valves

The flow of fluids and air can be controlled with solenoid controlled valves. An example of a solenoid controlled valve is shown in Figure 3.27. The solenoid is mounted on the side. When it is actuated, it will drive the central spool left. The top of the valve body has two ports that will be connected to a device such as a hydraulic cylinder. The bottom of the valve body has a single pressure line in the center with two exhausts to the side. In the top drawing, the power flows in through the center to the right-hand cylinder port. The left-hand cylinder port is allowed to exit through an exhaust port. In the bottom drawing, the solenoid is in a new position and the pressure is now applied to the left-hand port on the top, and the right-hand port can exhaust. The symbols to the left of the figure show the schematic equivalent of the actual valve positions. Valves are also available that allow the valves to be blocked when unused.

The solenoid has two positions and when actuated will change the direction that fluid flows to the device. The symbols shown here are commonly used to represent this type of valve.

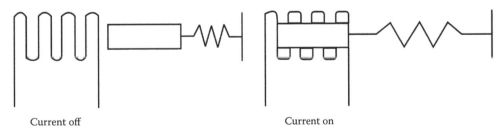

Current off   Current on

FIGURE 3.26
Solenoid.

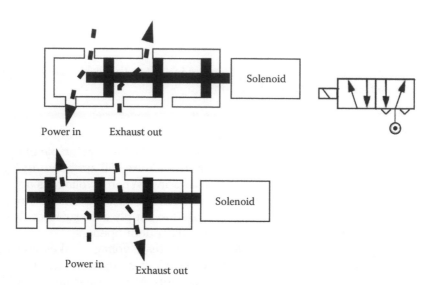

FIGURE 3.27
Solenoid controlled 5 ported, 4 way, 2 position valve.

Valve types are listed below. In the standard terminology, the "n-way," designates the number of connections for inlets and outlets. In some cases, there are redundant ports for exhausts. The normally open/closed designation indicates the valve condition when power is off. All of the valves listed are two-position valves, but three-position valves are also available.

2-way normally closed: These have one inlet and one outlet. When unenergized, the valve is closed. When energized, the valve will open, allowing flow. These are used to permit flows.

2-way normally open: These have one inlet and one outlet. When unenergized, the valve is open, allowing flow. When energized, the valve will close. These are used to stop flows. When system power is off, flow will be allowed.

3-way normally closed: These have inlet, outlet, and exhaust ports. When unenergized, the outlet port is connected to the exhaust port. When energized, the inlet is connected to the outlet port. These are used for single acting cylinders.

3-way normally open: These have inlet, outlet, and exhaust ports. When unenergized, the inlet is connected to the outlet. Energizing the valve connects the outlet to the exhaust. These are used for single acting cylinders.

3-way universal: These have three ports. One of the ports acts as an inlet or outlet, and is connected to one of the other two when energized/unenergized. These can be used to divert flows or select alternating sources.

4-way: These valves have four ports, two inlets, and two outlets. Energizing the valve causes connection between the inlets and outlets to be reversed. These are used for double acting cylinders.

Some of the ISO symbols for valves are shown in Figure 3.28. When using the symbols in drawings, the connections are shown for the un-energized state. The arrows show the flow paths in different positions. The small triangles indicate an exhaust port.

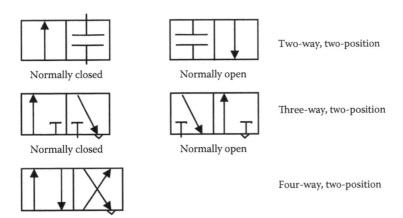

FIGURE 3.28
ISO valve symbols.

When selecting valves, there are a number of details that should be considered, as listed below:

Pipe size: Inlets and outlets are typically threaded to accept NPT (national pipe thread).

Flow rate: The maximum flow rate is often provided to hydraulic valves.

Operating pressure: A maximum operating pressure will be indicated. Some valves will also require a minimum pressure to operate.

Electrical: The solenoid coil will have a fixed supply voltage (AC or DC) and current.

Response time: This is the time for the valve to fully open/close. Typical times for valves range from 5 to 150 ms.

Enclosure: The housing for the valve will be rated as

Type 1 or 2: For indoor use, requires protection against splashes

Type 3: For outdoor use, will resist some dirt and weathering

Type 3R or 3S or 4: Water and dirt tight

Type 4X: Water and dirt tight, corrosion resistant

3.4.3 Cylinders

A cylinder uses pressurized fluid or air to create a linear force/motion as shown in Figure 3.29. In the figure, a fluid is pumped into one side of the cylinder under pressure, causing that side of the cylinder to expand, and advancing the piston. The fluid on the other side of the piston must be allowed to escape freely—if the incompressible fluid was trapped, the cylinder could not advance. The force the cylinder can exert is proportional to the cross-sectional area of the cylinder.

For force:

$$P = \frac{F}{A} \qquad F = PA$$

where:
P is the pressure of the hydraulic fluid
A is the area of the piston
F is the force available from the piston rod

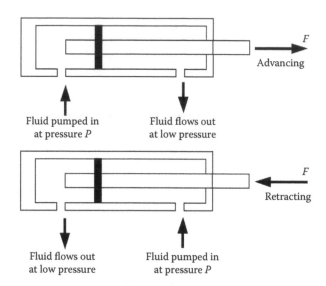

FIGURE 3.29
Cross section of a hydraulic cylinder.

Single-acting cylinders apply force when extending and typically use a spring to retract the cylinder. Double acting cylinders apply force in both directions.

3.4.4 Hydraulics

Hydraulics use incompressible fluids to supply very large forces at slower speeds and limited ranges of motion. If the fluid flow rate is kept low enough, many of the effects predicted by Bernoulli's equation can be avoided. The system uses hydraulic fluid (normally oil) pressurized by a pump and passed through hoses and valves to drive cylinders. At the heart of the system is a pump that will give pressures up to hundreds or thousands of psi. These are delivered to a cylinder that converts it to a linear force and displacement.

The hydraulic fluid is often noncorrosive oil chosen so that it lubricates the components. This is normally stored in a reservoir as shown in Figure 3.30. Fluid is drawn from the reservoir to a pump where it is pressurized. This is normally a geared pump so that it may deliver fluid at a high pressure at a constant flow rate. A flow regulator is normally placed at the high-pressure outlet from the pump. If fluid is not flowing in other parts of the system, this will allow fluid to recirculate back to the reservoir to reduce wear on the pump. The high-pressure fluid is delivered to solenoid controlled vales that can switch fluid flow on or off. From the vales, fluid will be delivered to the hydraulics at high pressure, or exhausted back to the reservoir.

Hydraulic systems can be very effective for high power applications, but the use of fluids and high pressures can make this method awkward, messy, and noisy for other applications.

3.4.5 Pneumatics

Pneumatic systems are very common, and have much in common with hydraulic systems with a few key differences. The reservoir is eliminated as there is no need to collect and store the air between uses in the system. Also, because air is a gas, it is compressible and

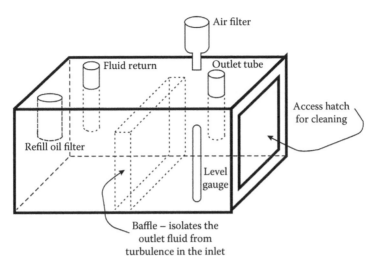

FIGURE 3.30
Hydraulic fluid reservoir.

regulators are not needed for recirculation. But, the compressibility also means that the systems are not as stiff or strong. Pneumatic systems respond very quickly and are commonly used for low force applications in many locations on the factory floor.

Some basic characteristics of pneumatic systems are:

- Stroke from a few millimeters to meters in length (longer strokes have more springiness).
- The actuators will give a bit; they are springy.
- Pressures are typically up to 85 psi above normal atmosphere.
- The weight of cylinders can be quite low.
- Additional equipment is required for a pressurized air supply—linear and rotatory actuators are available, and dampers can be used to cushion impact at ends of cylinder travel.

When designing pneumatic systems, care must be taken to verify the operating location. In particular, the elevation above sea level will result in a dramatically different air pressure. For example, at sea level the air pressure is about 14.7 psi, but at a height of 7800 ft (Mexico City), the air pressure is 11.1 psi. In other operating environments, such as in submersibles, the air pressure might be higher than at sea level.

Some symbols for pneumatic systems are shown in Figure 3.31. The flow control valve is used to restrict the flow, typically to slow motions. The shuttle valve allows flow in one direction, but blocks it in the other. The receiver tank allows pressurized air to be accumulated. The dryer and filter help remove dust and moisture from the air, prolonging the life of the valves and cylinders.

3.4.6 Motors

Motors are common actuators, but for logical control applications their properties are not that important. Typically, logical control of motors consists of switching low current

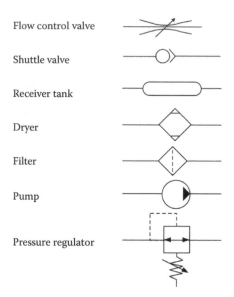

Flow control valve

Shuttle valve

Receiver tank

Dryer

Filter

Pump

Pressure regulator

FIGURE 3.31
Pneumatics components.

motors directly with a PLC, or for more powerful motors, using a relay or motor starter. Motors will be discussed in greater detail in the chapter on continuous actuators.

There are many other types of actuators, including heaters, lights, and sirens.

Heaters: They are often controlled with a relay and turned on and off to maintain a temperature within a range.

Lights: Lights are used on almost all machines to indicate the machine state and provide feedback to the operator. Most lights are low current and are connected directly to the PLC.

Sirens/Horns: Sirens or horns can be useful for unattended or dangerous machines to make conditions well known. These can often be connected directly to the PLC.

3.5 Boolean Logic Design

The process of converting control objectives into a ladder logic program requires structured thought. Boolean algebra provides the tools needed to analyze and design these systems.

3.5.1 Boolean Algebra

Boolean algebra was developed in the 1800s by James Bool, an Irish mathematician. It was found to be extremely useful for designing digital circuits, and it is still heavily used by electrical engineers and computer scientists. The techniques can model a logical system with a single equation. The equation can then be simplified and/or manipulated into new forms. The same techniques developed for circuit designers adapt very well to ladder logic programming.

Boolean equations consist of variables and operations and look very similar to normal algebraic equations. The three basic operators are AND, OR, and NOT; more complex operators include exclusive or (EOR), not and (NAND), not or (NOR). Small truth tables for these functions are given here. Each operator is shown in a simple equation with the variables A and B being used to calculate a value for X. Truth tables are a simple (but bulky) method for showing all of the possible combinations that will turn an output on or off.

By convention, a false state is also called off or 0 (zero). A true state is also called on or 1.

AND	OR	NOT
$X = A \cdot B$	$X = A + B$	$X = \overline{A}$

A	B	X
0	0	0
0	1	0
1	0	0
1	1	1

A	B	X
0	0	0
0	1	1
1	0	1
1	1	1

A	X
0	1
1	0

NAND	NOR	EOR
$X = \overline{A \cdot B}$	$X = \overline{A + B}$	$X = A \oplus B$

A	B	X
0	0	1
0	1	1
1	0	1
1	1	0

A	B	X
0	0	1
0	1	0
1	0	0
1	1	0

A	B	X
0	0	0
0	1	1
1	0	1
1	1	0

The symbols used in these equations, such as + for OR are not universal standards, and some authors will use different notations.

The EOR function is available in gate form, but it is more often converted to its equivalent, as shown below.

$$X = A \oplus B = A \cdot \overline{B} + \overline{A} \cdot B$$

3.5.2 Logic Design

Design ideas can be converted to Boolean equations directly, or with other techniques discussed later. The Boolean equation form can then be simplified or rearranged, and then converted into ladder logic, or a circuit. If we can describe how a controller should work in words, we can often convert it directly to a Boolean equation. In the example, a process description is given first. In actual applications, this is obtained by talking to the designer of the mechanical part of the system. In many cases the system does not exist yet, making this a challenging task. The next step is to determine how the controller should work. In this case it is written out in a sentence first, and then converted to a Boolean expression. The Boolean expression may then be converted to a desired form. The first equation contains an EOR, which is not available in ladder logic, so the next line converts this to an equivalent expression (2) using ANDs, ORs, and NOTs. The ladder logic developed is for the second equation. In the conversion, the terms that are ANDed are in series. The terms that are ORed are in parallel branches, and terms that are NOTed use normally closed contacts. This illustrates the same logical control function can be achieved with different, yet equivalent, ladder logic.

3.5.2.1 Process Description

A heating oven with two bays can heat one ingot in each bay. When the heater is on, it provides enough heat for two ingots. But if only one ingot is present, the oven may become too hot, so a fan is used to cool the oven when it passes a set temperature.

3.5.2.2 Control Description

If the temperature is too high and there is an ingot in only one bay, then turn on the fan.

3.5.2.3 Define Inputs and Outputs

B_1 is the bay 1 ingot present
B_2 is the bay 2 ingot present
F is the fan
T is the temperature overheat sensor

3.5.2.4 Boolean Equation

$$F = T \cdot (B_1 \oplus B_2) \tag{3.1}$$

$$F = T \cdot (B_1 \cdot \bar{B}_2 + \bar{B}_1 \cdot B_2) \tag{3.2}$$

$$F = B_1 \cdot \bar{B}_2 \cdot T + \bar{B}_1 \cdot B_2 \cdot T \tag{3.3}$$

Ladder logic for Equation 3.2 (Figure 3.32):

   To summarize, we will obtain Boolean equations from a verbal description or existing circuit or ladder diagram. The equation can be manipulated using the axioms of Boolean algebra. After simplification, the equation can be converted back into ladder logic or a circuit diagram. Ladder logic (and circuits) can behave the same even though they are in different forms. When simplifying Boolean equations that are to be implemented in ladder logic, there are a few basic rules.

1. Eliminate NOTs that are for more than one variable. This normally includes replacing NAND and NOR functions with simpler ones using De Morgan's theorem.
2. Eliminate complex functions such as EORs with their equivalent.

Assume that the Boolean equation that describes the controller is already known. This equation can be converted into both a circuit diagram and ladder logic. The circuit diagram contains about two dollars' worth of integrated circuits. If the design was mass produced,

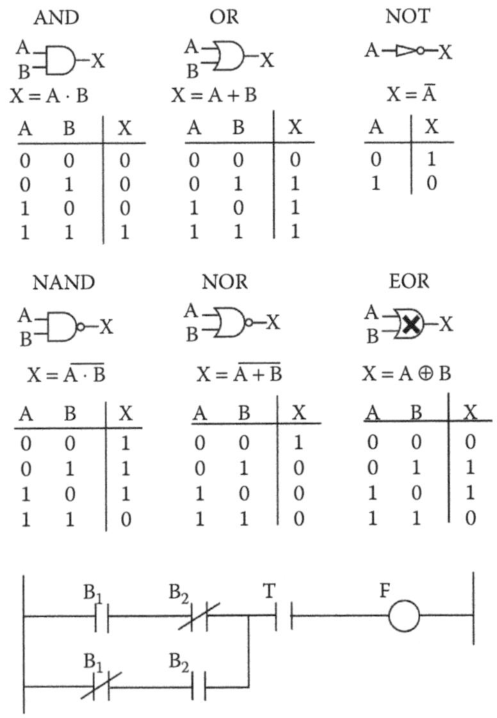

FIGURE 3.32
Ladder logic.

the final cost for the entire controller would be under $50. The prototype of the controller would cost thousands of dollars. If implemented in ladder logic, the cost for each controller would be approximately $500. Therefore, a large number of circuit-based controllers need to be produced before the breakeven occurs. This number is normally in the range of hundreds of units. There are some particular advantages of a PLC over digital circuits for the factory and some other applications. The PLC will be more rugged. The program can be changed easily. Less skill is needed to maintain the equipment.

3.5.3 Common Logic Forms

Knowing a simple set of logic forms will support a designer when categorizing control problems. The following forms are provided to be used directly, or provide ideas when designing.

3.5.3.1 Complex Gate Forms

In total, there are 16 different possible types of 2-input logic gates. The simplest are AND and OR; the other gates we will refer to as *complex* to differentiate. The three popular complex gates that have been discussed before are NAND, NOR, and EOR. All of these can be reduced to simpler forms with only ANDs and ORs that are suitable for ladder logic, as shown in Figure 3.33.

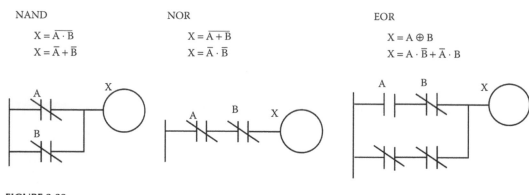

NAND

$$X = \overline{A \cdot B}$$
$$X = \overline{A} + \overline{B}$$

NOR

$$X = \overline{A + B}$$
$$X = \overline{A} \cdot \overline{B}$$

EOR

$$X = A \oplus B$$
$$X = A \cdot \overline{B} + \overline{A} \cdot B$$

FIGURE 3.33
Conversion of complex logic functions.

3.5.3.2 Multiplexers

Multiplexers allow multiple devices to be connected to a single device. These are very popular for telephone systems. A telephone *switch* is used to determine which telephone will be connected to a limited number of lines to other telephone switches. This allows telephone calls to be made to somebody far away without a dedicated wire to the other telephone. In older telephones, switch board operators physically connected wires by plugging them in. In modern computerized telephone switches the same thing is done, but to digital voice signals.

In Figure 3.34, a multiplexer is shown that will take one of four inputs bits—D_1, D_2, D_3, or D_4—and make it the output X, depending upon the values of the address bits, A_1 and A_2.

Ladder logic for the multiplexer can be seen in Figure 3.35.

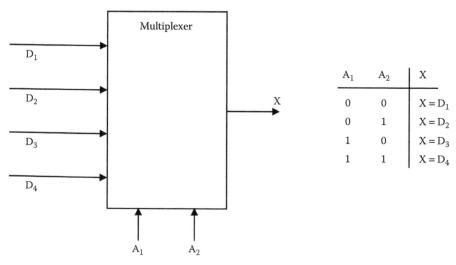

A_1	A_2	X
0	0	$X = D_1$
0	1	$X = D_2$
1	0	$X = D_3$
1	1	$X = D_4$

FIGURE 3.34
Multiplexer.

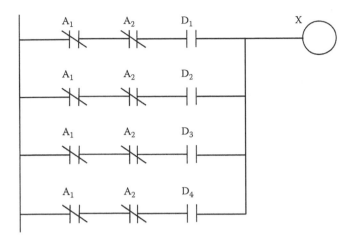

FIGURE 3.35
Multiplexer in ladder logic.

3.6 Programmable Logic Controller Operation

For simple programming, the relay model of the PLC is sufficient. As more complex functions are used, the more complex VonNeuman model of the PLC must be used. A VonNeuman computer processes one instruction at a time. Most computers operate this way, although they appear to be doing many things at once. Consider the computer components shown in Figure 3.36.

Input is obtained from the keyboard and mouse, output is sent to the screen, and the disk and memory are used for both input and output for storage. (Note: The directions of these arrows are very important to engineers; always pay attention to indicate where information is flowing.) This figure can be redrawn as in Figure 3.37 to clarify the role of inputs and outputs.

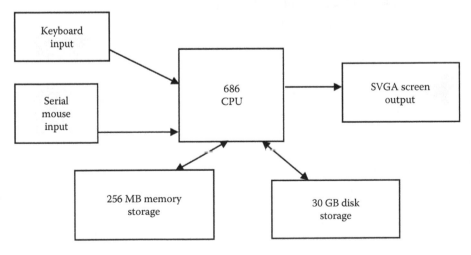

FIGURE 3.36
Simplified personal computer architecture.

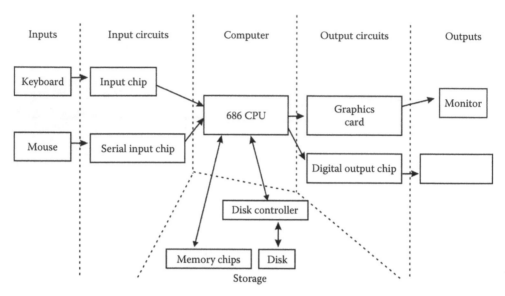

FIGURE 3.37
Input-output oriented architecture.

In this figure, the data enters the left side through the inputs. (Note: Most engineering diagrams have inputs on the left and outputs on the right.) It travels through buffering circuits before it enters the CPU. The CPU outputs data through other circuits. Memory and disks are used for storage of data that is not destined for output. If we look at a personal computer as a controller, it is controlling the user by outputting stimuli on the screen, and inputting responses from the mouse and the keyboard. A PLC is also a computer controlling a process. When fully integrated into an application, the analogies become:

Inputs: The keyboard is analogous to a proximity switch.

Input circuits: The serial input chip is like a 24 Vdc input card.

Computer: The 686 CPU is like a PLC CPU unit.

Output circuits: A graphics card is like a triac output card.

Outputs: A monitor is like a light.

Storage: Memory in PLCs is similar to memories in personal computers.

It is also possible to implement a PLC using a normal personal computer, although this is not advisable. In the case of a PLC, the inputs and outputs are designed to be more reliable and rugged for harsh production environments.

3.6.1 Operation Sequence

All PLCs have four basic stages of operations that are repeated many times per second. When turned on the first time, it will check its own hardware and software for faults. If there are no problems, it will copy all the input and copy their values into memory; this is called the input scan. Using only the memory copy of the inputs, the ladder logic program will be solved once; this is called the logic scan. While solving the ladder logic, the output values are only changed in temporary memory. When the ladder scan is done, the outputs

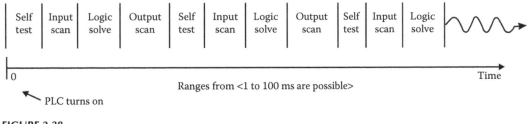

FIGURE 3.38
PLC scan cycle.

will update using the temporary values in memory; this is called the output scan. The PLC now restarts the process by starting a self-check for faults. This process typically repeats 10–100 times per second as is shown in Figure 3.38.

Self test: Checks to see if all cards error free, resets watch-dog timer, and so on. (A watchdog timer will cause an error and shut down the PLC if not reset within a short period of time—this would indicate that the ladder logic is not being scanned normally.)

Input scan: Reads input values from the chips in the input cards and copies their values to memory. This makes the PLC operation faster and avoids cases where an input changes from the start to the end of the program (e.g., an emergency stop). There are special PLC functions that read the inputs directly and avoid the input tables.

Logic solve/scan: Based on the input table in memory, the program is executed one step at a time, and outputs are updated. This is the focus of the later sections.

Output scan: The output table is copied from memory to the output chips. These chips then drive the output devices.

The input and output scans often confuse the beginner, but they are important. The input scan takes a *snapshot* of the inputs and solves the logic. This prevents potential problems that might occur if an input that is used in multiple places in the ladder logic program changed while halfway through a ladder scan, thus changing the behaviors of half of the ladder logic program. This problem could have severe effects on complex programs that are developed later in the book. One side effect of the input scan is that if a change in input is too short in duration, it might fall between input scans and be missed. When the PLC is initially turned on, the normal outputs will be turned off. This does not affect the values of the inputs.

3.6.1.1 The Input and Output Scans

When the inputs to the PLC are scanned, the physical input values are copied into memory. When the outputs to a PLC are scanned, they are copied from memory to the physical outputs. When the ladder logic is scanned, it uses the values in memory, not the actual input or output values. The primary reason for doing this is so that if a program uses an input value in multiple places, a change in the input value will not invalidate the logic. Also, if output bits were changed as each bit was changed, instead of all at once at the end of the scan, the PLC would operate much slower.

3.6.1.2 *The Logic Scan*

Ladder logic programs are modeled after relay logic. In relay logic, each element in the ladder will switch as quickly as possible. But in a program, elements can only be examined one at a time in a fixed sequence. Consider the ladder logic in Figure 3.39. The ladder logic will be interpreted left-to-right, top-to-bottom. In the figure, the ladder logic scan begins at the top rung. At the end of the rung it interprets the top output first, then the output branched below it. On the second rung it solves branches before moving along the ladder logic rung.

The logic scan sequence becomes important when solving ladder logic programs that use outputs as inputs. It also becomes important when considering output usage. Consider Figure 3.40, the first line of ladder logic will examine input A and set output X to have the same value. The second line will examine input B and set the output X to have the opposite value. So, the value of X was only equal to A until the second line of ladder logic was scanned. Recall that during the logic scan, the outputs are only changed in memory; the actual outputs are only updated when the ladder logic scan is complete. Therefore, the output scan would update the real outputs based upon the second line of ladder logic, and the first line of ladder logic would be ineffective.

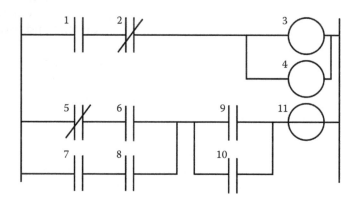

FIGURE 3.39
Ladder logic execution sequence.

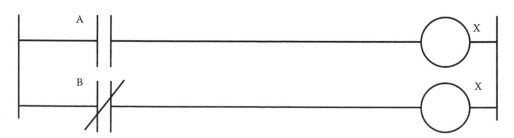

FIGURE 3.40
Duplicated output error.

3.6.2 Programmable Logic Controller Status

The lack of keyboard and other input-output devices is very noticeable on a PLC. On the front of the PLC there are normally limited status lights. Common lights indicate power on, which will be on whenever the PLC has power, program running, which will often indicate if a program is running or if no program is running, and fault, which will indicate when the PLC has experienced a major hardware or software problem. These lights are normally used for debugging. Limited buttons will also be provided for PLC hardware. The most common will be a run/program switch that will be switched to program when maintenance is being conducted, and back to run when in production. This switch normally requires a key to keep unauthorized personnel from altering the PLC program or stopping execution. A PLC will almost never have an on–off switch or reset button on the front. This needs to be designed into the remainder of the system. The status of the PLC can be detected by ladder logic also. It is common for programs to check to see if they are being executed for the first time.

The "first scan" input will be true the very first time the ladder logic is scanned, but false on every other scan. In this case, the address for "first scan" in a PLC-5 is "S2:1/14." With the logic in the example, the first scan will seal on "light" until "clear" is turned on. So, the light will turn on after the PLC has been turned on, but it will turn off and stay off after "clear" is turned on. The "first scan" bit is also referred to as the "first pass" bit.

3.6.3 Memory Types

There are a few basic types of computer memory that are in use today.

RAM (*Random Access Memory*): This memory is fast, but it will lose its contents when power is lost; this is known as volatile memory. Every PLC uses this memory for the central CPU when running the PLC.

ROM (*Read Only Memory*): This memory is permanent and cannot be erased. It is often used for storing the operating system for the PLC.

EPROM (*Erasable Programmable Read Only Memory*): This is memory that can be programmed to behave like ROM, but it can be erased with ultraviolet light and reprogrammed.

EEPROM (*Electronically Erasable Programmable Read Only Memory*): This memory can store programs like ROM. It can be programmed and erased using a voltage, so it is becoming more popular than EPROMs.

All PLCs use RAM for the CPU and ROM to store the basic operating system for the PLC. When the power is on, the contents of the RAM will be kept, but the issue is what happens when power to the memory is lost. Originally, PLC vendors used RAM with a battery so that the memory contents would not be lost if the power was lost. This method is still in use, but is losing favor. EPROMs have also been a popular choice for programming PLCs. The EPROM is programmed out of the PLC, and then placed in the PLC. When the PLC is turned on, the ladder logic program on the EPROM is loaded into the PLC and run. This method can be very reliable, but the erasing and programming technique can be time consuming. EEPROM memories are a permanent part of the PLC, and programs can be stored in them like EPROM. Memory costs continue to drop, and newer types (such as flash memory) are becoming available. These changes will continue to impact PLCs.

3.6.4 Software-Based Programmable Logic Controllers

The dropping cost of personal computers is increasing their use in control, including the replacement of PLCs. Software is installed that allows the personal computer to solve ladder logic, read inputs from sensors, and update outputs to actuators. These are important to mention here because they don't obey the previous timing model.

3.7 Latches, Timers, Counters, and More

More complex systems cannot be controlled with combinatorial logic alone. The main reason for this is that we cannot, or choose not to add sensors to detect all conditions. In these cases, we can use events to estimate the condition of the system. Typical events used by a PLC include:

First scan of the PLC: Indicating the PLC has just been turned on

Time since an input turned on/off: A delay

Count of events: To wait until set numbers of events have occurred

Latch on or unlatch: To lock something on or turn it off

The common theme for all of these events is that they are based upon one of two questions "How many?" or "How long?" An example of an event-based device is shown in Figure 3.41. The input to the device is a push button. When the push button is pushed, the input to the device turns on. If the push button is then released and the device turns off, it is a logical device. If when the push button is released the device stays on, it is one type of event-based device. To reiterate, the device is event based if it can respond to one or more things that have happened before. If the device responds only one way to the immediate set of inputs, it is logical.

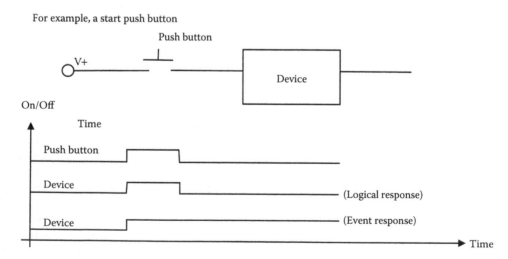

FIGURE 3.41
Event-driven device.

3.7.1 Latches

A latch is like a sticky switch—when pushed it will turn on but stick in place; it must be pulled to release it and turn it off. A latch in ladder logic uses one instruction to latch and a second instruction to unlatch, as shown in Figure 3.42. The output with an L inside will turn the output D on when the input A becomes true. D will stay on even if A turns off. Output D will turn off if input B becomes true and the output with a U inside becomes true. (Note: This will seem a little backwards at first.) If an output has been latched on, it will keep its value, even if the power has been turned off.

The operation of the ladder logic is shown in Figure 3.43. A timing diagram shows values of inputs and outputs over time. For example, the value of input A starts low (false) and becomes high (true) for a short while, and then goes low again. Here when input A

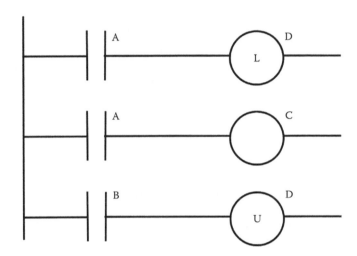

FIGURE 3.42
Ladder logic latch.

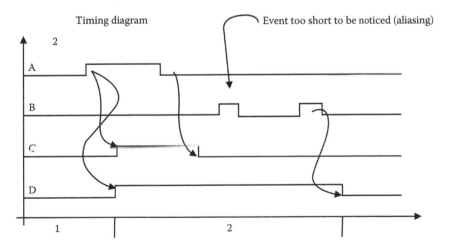

FIGURE 3.43
Timing diagram for the ladder logic in Figure 3.42.

turns on, both the outputs turn on. There is a slight delay between the change in inputs and the resulting changes in outputs, due to the program scan time. Here the dashed lines represent the output scan, sanity check, and input scan (assuming they are very short). The space between the dashed lines is the ladder logic scan. Consider that when A turns on initially, it is not detected until the first dashed line. There is then a delay to the next dashed line while the ladder is scanned, and then the output at the next dashed line. When A eventually turns off, the normal output C turns off, but the latched output D stays on. Input B will unlatch the output D. Input B turns on twice, but the first time it is on is not long enough to be detected by an input scan, so it is ignored. The second time it is on, it unlatches output D, and output D turns off.

1. The space between the lines is the scan time for the ladder logic. The spaces may vary if different parts of the ladder diagram are executed each time through the ladder (as with state space code). The space is a function of the speed of the PLC and the number of ladder logic elements in the program.

2. These lines indicate PLC input/output refresh times. At this time, all of the outputs are updated, and all of the inputs are read. The space between the lines is the scan time for the ladder logic. The spaces may vary if different parts of the ladder diagram are executed each time through the ladder (as with state space code).

3.7.2 Timers

There are four fundamental types of timers shown in Figure 3.44. An on-delay timer will wait for a set time after a line of ladder logic has been true before turning on, but it will turn off immediately. An off-delay timer will turn on immediately when a line of ladder logic is true, but it will delay before turning off. Consider the example of an old car. If you turn the key in the ignition and the car does not start immediately, that is an on-delay. If you turn the key to stop the engine but the engine doesn't stop for a few seconds, that is an off delay. An on-delay timer can be used to allow an oven to reach temperature before starting production. An off-delay timer can keep cooling fans on for a set time after the oven has been turned off.

A retentive timer will sum all of the on or off time for a timer, even if the timer never finished. A nonretentive timer will start timing the delay from zero each time. Typical applications for retentive timers include tracking the time before maintenance is needed. A nonretentive timer can be used for a start button to give a short delay before a conveyor begins moving. An example of an Allen-Bradley TON timer is shown in Figure 3.45.

	On-delay	Off-delay
Retentive	RTO	RTF
Nonretentive	TON	TOF

TON - Timer ON
TOF - Timer OFF
RTO - Retentive Timer ON
RTF - Retentive Timer Off

FIGURE 3.44
Four basic timer types.

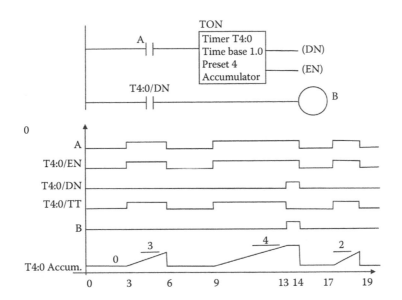

FIGURE 3.45
Allen-Bradley TON timer.

The rung has a single input A and a function block for the TON. (Note: This timer block will look different for different PLCs, but it will contain the same information.) The information inside the timer block describes the timing parameters.

The first item is the timer number T4:0. This is a location in the PLC memory that will store the timer information. The T4: indicates that it is timer memory, and the 0 indicates that it is in the first location. The time base is 1.0, indicating that the timer will work in 1.0 s intervals. Other time bases are available in fractions and multiples of seconds. The preset is the delay for the timer; in this case it is 4. To find the delay time, multiply the time base by the preset value: 4 * 1.0 s = 4.0 s. The accumulator value gives the current value of the timer as 0. While the timer is running, the accumulated value will increase until it reaches the preset value. Whenever the input A is true, the EN output will be true. The DN output will be false until the accumulator has reached the preset value. The EN and DN outputs cannot be changed when programming, but these are important when debugging a ladder logic program. The second line of ladder logic uses the timer DN output to control another output B.

The timing diagram operation of the TON timer with a 4 s on-delay is shown in Figure 3.45. A is the input to the timer, and whenever the timer input is true, the EN enabled bit for the timer will also be true. If the accumulator value is equal to the preset value, the DN bit will be set. Otherwise, the TT bit will be set and the accumulator value will begin increasing. The first time A is true, it is only true for 3 s before turning off; after this the value resets to zero. (Note: In a retentive time the value would remain at 3 s.) The second time A is true, it is on more than 4 s. After 4 s, the TT bit turns off, and the DN bit turns on. But when A is released, the accumulator resets to zero and the DN bit is turned off. A value can be entered for the accumulator while programming. When the program is downloaded, this value will be in the timer for the first scan. If the TON timer is not enabled, the value will be set back to zero. Normally zero will be entered for the preset value.

3.7.3 Counters

There are two basic counter types: count-up and count-down. When the input to a count-up counter goes true, the accumulator value will increase by 1 (no matter how long the input is true). If the accumulator value reaches the preset value, the counter DN bit will be set. A count-down counter will decrease the accumulator value until the preset value is reached. An Allen-Bradley count-up (CTU) instruction is shown in Figure 3.46. The instruction requires memory in the PLC to store values and status, in this case is C5:0. The C5: indicates that it is counter memory, and the 0 indicates that it is the first location. The preset value is 4 and the value in the accumulator is 2. If the input A were to go from false to true, the value in the accumulator would increase to 3. If A were to go off, then on again, the accumulator value would increase to 4, and the DN bit would go on. The count can continue above the preset value. If input B goes true, the value in the counter accumulator will become zero.

Count-down counters are very similar to count-up counters. And, they can actually both be used on the same counter memory location. Consider the example in Figure 3.46; the example input I/1 drives the count-up instruction for counter C5:1. Input I/2 drives the count-down instruction for the same counter location. The preset value for a counter is stored in memory location C5:1, so both the count-up and count-down instruction must have the same preset. Input I/3 will reset the counter.

The timing diagram in Figure 3.47 illustrates the operation of the counter. If we assume that the value in the accumulator starts at 0, then the I/1 inputs cause it to count up to 3, where it turns the counter C5:1 on. It is then reset by input I/3, and the accumulator value goes to zero. Input I/1 then pulses again and causes the accumulator value to increase again, until it reaches a maximum of 5. Input I/2 then causes the accumulator value to decrease to below 3, and the counter turns off again. Input I/1 then causes it to increase, but input I/3 resets the accumulator back to zero again, and the pulses continue until 3 is reached near the end. The program in Figure 3.48 is used to remove 5 out of every 10 parts from a conveyor with a pneumatic cylinder. When the part is detected, both counters will increase their values by 1. When the sixth part arrives, the first counter will then be done, thereby allowing the pneumatic cylinder to actuate for any part after the fifth. The second counter will continue until the eleventh part is detected, and then both of the counters will be reset.

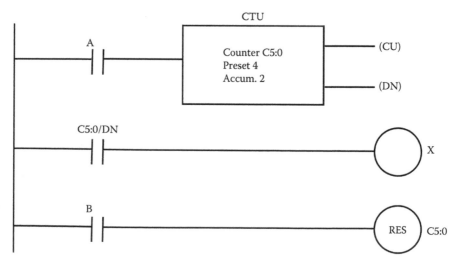

FIGURE 3.46
Allen-Bradley counter.

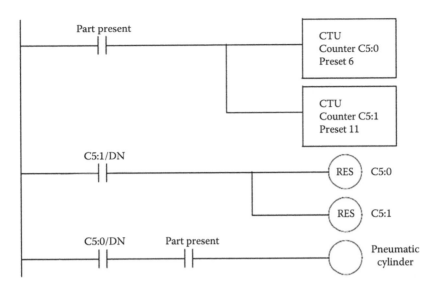

FIGURE 3.47
Counter example.

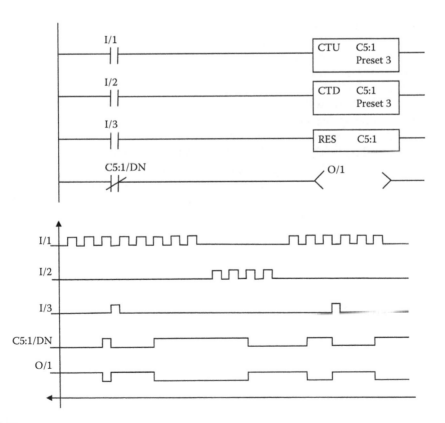

FIGURE 3.48
Counter example.

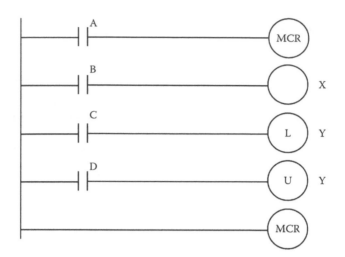

FIGURE 3.49
MCR instructions.

3.7.4 Master Control Relays (MCRs)

In an electrical control system, a master control relay (MCR) is used to shut down a section of an electrical system, as discussed in the earlier section of this chapter. This concept has been implemented in ladder logic also. A section of ladder logic can be put between two lines containing MCRs. When the first MCR coil is active, all of the intermediate ladder logic is executed up to the second line with an MCR coil. When the first MCR coil is inactive, the ladder logic is still examined, but all of the outputs are forced off. Consider the example in Figure 3.49. If A is true, then the ladder logic after will be executed as normal. If A is false, the following ladder logic will be examined, but all of the outputs will be forced off. The second MCR function appears on a line by itself and marks the end of the MCR block. After the second MCR, the program execution returns to normal. While A is true, X will equal B, and Y can be turned on by C, and off by D. But if A becomes false, X will be forced off, and Y will be left in its last state. Using MCR blocks to remove sections of programs will not increase the speed of program execution significantly because the logic is still examined.

If the MCR block contained another function, such as a TON timer, turning off the MCR block would force the timer off. As a general rule, normal outputs should be outside MCR blocks unless they must be forced off when the MCR block is off.

3.8 Structured Logic Design

Traditionally ladder logic programs have been written by thinking about the process and then beginning to write the program. This always leads to programs that require debugging. And, the final program is always the subject of some doubt. Structured design techniques, such as Boolean algebra, lead to programs that are predictable and reliable. The structured design techniques in this and the following chapters are provided to make ladder logic design routine and predictable for simple sequential systems.

Most control systems are sequential in nature. Sequential systems are often described with words such as "mode" and "behavior." During normal operation, these systems will

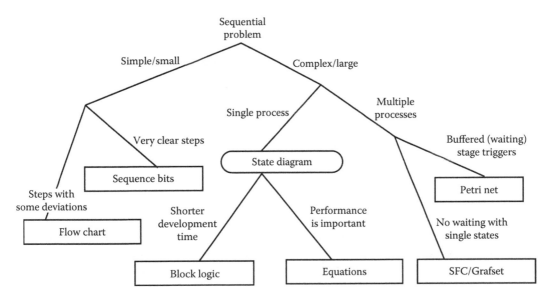

FIGURE 3.50
Sequential design techniques.

have multiple steps or states of operation. In each operational state, the system will behave differently. Typical states include start-up, shut-down, and normal operation. Consider a set of traffic lights—each light pattern constitutes a state. Lights may be green or yellow in one direction and red in the other. The lights change in a predictable sequence. Sometimes traffic lights are equipped with special features such as cross walk buttons that alter the behavior of the lights to give pedestrians time to cross busy roads.

Sequential systems are complex and difficult to design. In the previous section, timing charts and process sequence bits were discussed as basic design techniques. But, more complex systems require more mature techniques, such as those shown in Figure 3.50. For simpler controllers, we can use limited design techniques such as process sequence bits and flow charts. More complex processes, such as traffic lights, will have many states of operation, and controllers can be designed using state diagrams. If the control problem involves multiple states of operation, such as one controller for two independent traffic lights, then Petri net or SFC based designs are preferred.

3.8.1 Process Sequence Bits

A typical machine will use a sequence of repetitive steps that can be clearly identified. Ladder logic can be written that follows this sequence. The steps for this design method are:

1. Understand the process.
2. Write the steps of operation in sequence and give each step a number.
3. For each step, assign a bit.
4. Write the ladder logic to turn the bits on/off as the process moves through its states.
5. Write the ladder logic to perform machine functions for each step.
6. If the process is repetitive, have the last step go back to the first.

3.8.2 Timing Diagrams

Timing diagrams can be valuable when designing ladder logic for processes that are only dependent on time. The timing diagram is drawn with clear start and stop times. Ladder logic is constructed with timers that are used to turn outputs on and off at appropriate times. The basic method is:

1. Understand the process.
2. Identify the outputs that are time dependent.
3. Draw a timing diagram for the outputs.
4. Assign a timer for each time when an output turns on or off.
5. Write the ladder logic to examine the timer values and turn outputs on or off.

Consider the handicap door opener design in Figure 3.51 that begins with a verbal description. The verbal description is converted to a timing diagram, with t = 0 being when the door open button is pushed. On the timing diagram, the critical times are 2, 10, and 14 s.

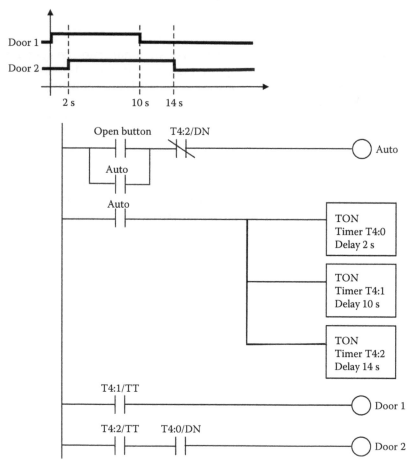

FIGURE 3.51
Design with a timing diagram.

The ladder logic is constructed in a careful order. The first item is the latch to seal-in the open button, but shut off after the last door closes. *Auto* is used to turn on the three timers for the critical times. The logic for opening the doors is then written to use the timers.

> *Description*: A handicap door opener has a button that will open two doors. When the button is pushed (momentarily), the first door will start to open immediately; the second door will start to open 2 s later. The first door power will stay open for a total of 10 s, and the second door power will stay on for 14 s. Use a timing diagram to design the ladder logic.

3.9 Flowchart-Based Design

A flowchart is ideal for a process that has sequential process steps. The steps will be executed in a simple order that may change as the result of some simple decisions. The symbols used for flowcharts are shown in Figure 3.52. These blocks are connected using arrows to indicate the sequence of the steps. The different blocks imply different types of program actions. Programs always need a *start* block, but PLC programs rarely stop so the *stop* block is rarely used. Other important blocks include *operations* and *decisions*. The other functions may be used but are not necessary for most PLC applications.

A flowchart is shown in Figure 3.53 for a control system for a large water tank. When a start button is pushed, the tank will start to fill, and the flow out will be stopped. When full, or the stop button, is pushed, the outlet will open up, and the flow in will be stopped. In the flowchart, the general flow of execution starts at the top. The first operation is to open the outlet valve and close the inlet valve. Next, a single decision block is used to wait for a button to be pushed. When the button is pushed, the *yes* branch is followed and the inlet valve is opened, and the outlet valve is closed. Then the flow chart goes into a loop that uses two decision blocks to wait until the tank is full or the stop button is pushed.

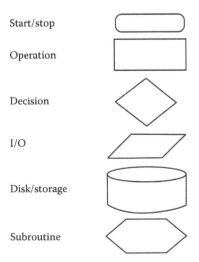

Start/stop

Operation

Decision

I/O

Disk/storage

Subroutine

FIGURE 3.52
Flowchart symbols.

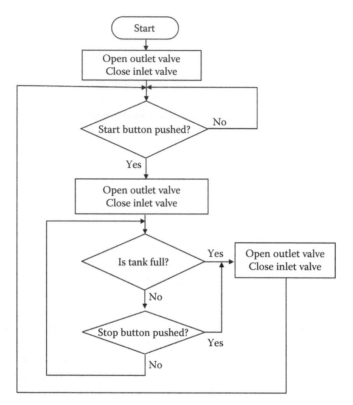

FIGURE 3.53
Flowchart for tank filler.

If either case occurs, the inlet valve is closed and the outlet valve is opened. The system then goes back to wait for the start button to be pushed again. When the controller is on, the program should always be running, so only a start block is needed. Many beginners will neglect to put in checks for stop buttons.

The general method for constructing flowcharts is:

1. Understand the process.
2. Determine the major actions; these are drawn as blocks.
3. Determine the sequences of operations; these are drawn with arrows.
4. When the sequence may change, use decision blocks for branching.

Once a flowchart has been created, ladder logic can be written. There are two basic techniques that can be used. The first presented uses blocks of ladder logic code. The second uses normal ladder logic (Figure 3.54).

STEP 1: Add labels to each block in the flowchart

Each block in the flowchart will be converted to a block of ladder logic. To do this, we will use the MCR (master control relay) instruction (it will be discussed in more detail later). The instruction is shown in the figure and will appear as a matched pair of outputs labelled MCR. If the first MCR line is true, then the ladder logic on the following lines will be scanned as normal to the second MCR. If the

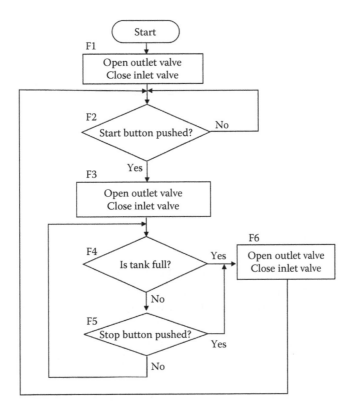

FIGURE 3.54
Labeling blocks in the flowchart.

first line is false, the lines to the next MCR block will all be forced off. If a normal output is used inside an MCR block, it may be forced off. Therefore, latches will be used in this method. The first part of the ladder logic required will reset the logic to an initial condition. The line will only be true for the first scan of the PLC, and at that time it will turn on the flowchart block F1 which is the *reset all values off* operation. All other operations will be turned off (Figure 3.55).

STEP 2: Write ladder logic to force the PLC into the first state

The ladder logic for the first state is shown in Figure 3.56. When F1 is true, the logic between the MCR lines will be scanned; if F1 is false, the logic will be ignored. This logic turns on the outlet valve and turns off the inlet valve. It then turns off operation F1 and turns on the next operation F2.

FIGURE 3.55
Initial reset of states.

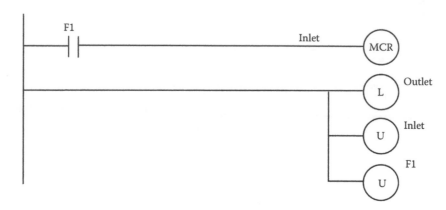

FIGURE 3.56
Ladder logic for the operation F1.

STEP 3: Write ladder logic for each function in the flowchart

The ladder logic for operation F2 is simple, and when the start button is pushed, it will turn off F2 and turn on F3. The ladder logic for operation F3 opens the inlet valve and moves to operation F4 (Figure 3.57).

The ladder logic for operation F4 turns off F4, and if the tank is full, it turns on F6; otherwise F5 is turned on. The ladder logic for operation F5 is very similar (Figure 3.58).

The ladder logic for operation F6 turns the outlet valve on and turns off the inlet valve. It then ends operation F6 and returns to operation F2.

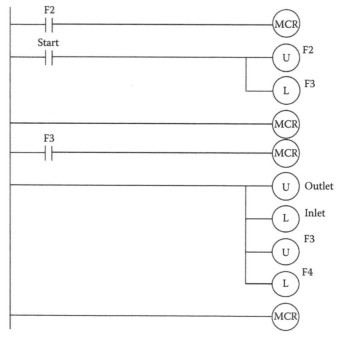

FIGURE 3.57
Ladder logic for flowchart operations F2 and F3.

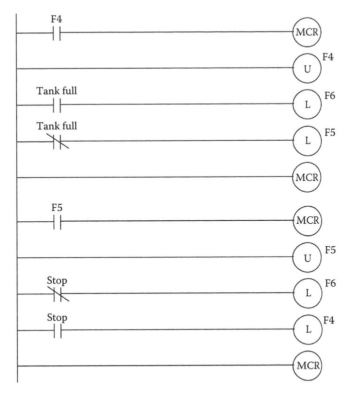

FIGURE 3.58
Ladder logic for operations F4 and F5.

3.10 Programmable Logic Controller Memory

Advanced ladder logic functions allow controllers to perform calculations, make decisions, and do other complex tasks. Timers and counters are examples of ladder logic functions. They are more complex than basic input contacts and output coils, and they rely upon data stored in the memory of the PLC. The memory of the PLC is organized to hold different types of programs and data.

3.10.1 Memory Addresses

The memory in a PLC is organized by data type as shown in Figure 3.59. There are two fundamental types of memory used in Allen-Bradley PLCs—program and data memory. Memory is organized into blocks of up to 1000 elements in an array called a file. The program file holds programs, such as ladder logic. There are eight data files defined by default, but additional data files can be added if they are needed (Table 3.1).

1. These are a collection of up to 1000 slots to store up to 1000 programs. The main program will be stored in program file 2. SFC programs must be in file 1, and file 0 is used for program and password information. All other program files from 3 to 999 can be used for *subroutines*.

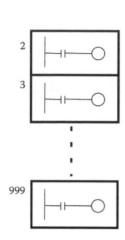

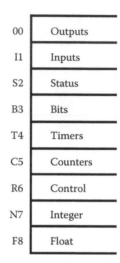

FIGURE 3.59
PLC memory.

TABLE 3.1

Allen-Bradley Data Types

Type	Length
A: ASCII	1/2
B: Bit	1/16
BT: Block transfer	6
C: Counter	3
D: BCD	1
F: Floating point	2
MG: Message	56
N: Integer (signed, unsigned, 2 s compliment, BCD)	1
PD: PID controller	82
R: Control	3
SC: SFC status	3
ST: ASCII string	42
T: Timer	3

2. This is where the variable data is stored on which the PLC programs operate. This is quite complicated, so a detailed explanation follows.

3.10.2 Program Files

In a PLC-5, the first three program files, from 0 to 2, are defined by default. File 0 contains system information and should not be changed, and file 1 is reserved for SFCs. File 2 is available for user programs, and the PLC will run the program in file 2 by default. Other program files can be added from file 3–999. Typical reasons for creating other programs are for subroutines. When a user creates a ladder logic program with programming software, it is converted to a mnemonic-like form and then transferred to the PLC, where it is stored in a program file. The contents of the program memory cannot be changed while the PLC is running. If, while a program was running, it was overwritten with a new program, serious problems could arise.

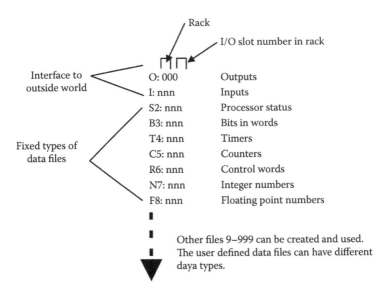

FIGURE 3.60
Data files for an Allen-Bradley PLC-5.

3.10.3 Data Files

Data files are used for storing different information types, as shown in Figure 3.60. These locations are numbered from 0 to 999. The letter in front of the number indicates the data type. For example, F8: is read as *floating point numbers* in *data file* 8. Numbers are not given for O: and I:, but they are implied to be O0: and I1:. The number that follows the colon is the location number. Each file may contain from 0 to 999 locations that may store values. For the input, I: and output O: files the locations are converted to physical locations on the PLC using rack and slot numbers. The addresses that can be used will depend upon the hardware configuration. The status S2: file is more complex and is discussed later. The other memory locations are simply slots to store data in. For example, F8:35 would indicate the 36th value in the 8th data file, which is floating point numbers.

Only the first three data files are fixed: O:, I:, and S2:; all of the other data files can be moved. It is also reasonable to have multiple data files with the same data type. For example, there could be two files for integer numbers N7: and N10:. The length of the data files can be from 0 up to 999 as shown in Figure 3.60. But, these files are often made smaller to save memory.

Figure 3.61 shows the default data files for a PLC-5. There are many additional data types. A full list is shown in Table 3.1. Some of the data types are complex and contain multiple data values, including BT, C, MG, PD, R, SC, and T. Recall that timers require integers for the accumulator and preset, and TT, DN, and EN bits are required. Other data types are based on single bits, 8 bit bytes, and 16 bit words.

When using data files and functions, we need to ask for information with an address. The simplest data addresses are data bits (we have used these for basic inputs and outputs already). Memory bits are normally indicated with a forward slash followed by a bit number: /n. The first example is from an input card I:000; the third input is indicated with the bit address /02. The second example is for a counter C5: done bit/DN. This could also be replaced with C5:4/15 to get equivalent results. The DN notation and others like it are

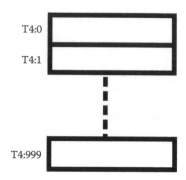

FIGURE 3.61
Locations in a data file.

used to simplify the task of programming. The example B3/4 will get the fourth bit in bit memory B3. For bit memory, the slash is not needed because the data type is already represented in bytes.

3.10.4 Ladder Logic Functions

Ladder logic input contacts and output coils allow simple logical decisions. Functions extend basic ladder logic to allow other types of control. For example, the addition of timers and counters allowed event based control. A longer list of functions includes:

Combinatorial logic
- Relay contacts and coils

Events
- Timer instructions
- Counter instructions

Data handling
- Moves
- Mathematics
- Conversions

Numerical logic
- Boolean operations
- Comparisons

Lists
- Shift registers/stacks
- Sequencers

Program control
- Branching/looping
- Immediate inputs/outputs
- Fault/interrupt detection

Input and output

- PID
- Communications
- High speed counters
- ASCII string functions

Most of the functions will use PLC memory locations to get values, store values, and track function status. Most function will normally become active when the input is true. But, some functions, such as TOF timers, can remain active when the input is off. Other functions will only operate when the input goes from false to true; this is known as positive edge triggered. Consider a counter that only counts when the input goes from false to true; the length of time the input is true does not change the function behavior. A negative edge triggered function would be triggered when the input goes from true to false. Most functions are not edge triggered; unless stated, assume functions are not edge triggered.

3.11 Analog Inputs and Outputs

An analog value is continuous, not discrete, as shown in Figure 3.62. In the previous chapters, techniques were discussed for designing logical control systems that had inputs and outputs that could only be on or off. These systems are less common than the logical control systems, but they are very important.

Typical analog inputs and outputs for PLCs are listed below.

Inputs:

- Oven temperature
- Fluid pressure
- Fluid flow rate

Outputs:

- Fluid valve position
- Motor position
- Motor velocity

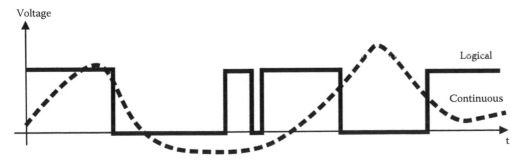

FIGURE 3.62
Logical and continuous values.

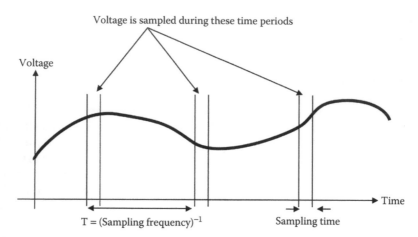

FIGURE 3.63
Sampling an analog voltage.

3.11.1 Analog Inputs

To input an analog voltage (into a PLC or any other computer), the continuous voltage value must be *sampled* and then converted to a numerical value by an A/D converter. Figure 3.63 shows a continuous voltage changing over time. There are three samples shown on the figure. The process of sampling the data is not instantaneous, so each sample has a start and stop time. The time required to acquire the sample is called the *sampling time*. A/D converters can only acquire a limited number of samples per second. The time between samples is called the sampling period T, and the inverse of the sampling period is the sampling frequency (also called sampling rate). The sampling time is often much smaller than the sampling period. The sampling frequency is specified when buying hardware, but for a PLC, a maximum sampling rate might be 20 Hz.

This data is noisier, and even between the start and end of the data sample there is a significant change in the voltage value. The data value sampled will be somewhere between the voltage at the start and end of the sample. The maximum (V_{max}) and minimum (V_{min}) voltages are a function of the control hardware. These are often specified when purchasing hardware, but reasonable ranges are

0–5 V

0–10 V

–5–5 V

–10–10 V

The number of bits of the A/D converter is the number of bits in the result word. If the A/D converter is 8-bit, then the result can read up to 256 different voltage levels. Most A/D converters have 12 bits; 16 bit converters are used for precision measurements.

There are other practical details that should be considered when designing applications with analog inputs:

- *Noise*: Since the sampling window for a signal is short, noise will have added effect on the signal read. For example, a momentary voltage spike might result in a higher than normal reading. Shielded data cables are commonly used to reduce the noise levels.

- *Delay*: When the sample is requested, a short period of time passes before the final sample value is obtained.
- *Multiplexing*: Most analog input cards allow multiple inputs. These may share the A/D converter using a technique called multiplexing. If there are 4 channels using an A/D converter with a maximum sampling rate of 100 Hz, the maximum sampling rate per channel is 25 Hz.
- *Signal conditioners*: Signal conditioners are used to amplify or filter signals coming from transducers before they are read by the A/D converter.
- *Resistance*: A/D converters normally have high input impedance (resistance), so they affect circuits they are measuring.
- *Single ended inputs*: Voltage inputs to a PLC can use a single common for multiple inputs; these types of inputs are called *single* ended inputs. These tend to be more prone to noise.
- *Double ended inputs*: Each double ended input has its own common. This reduces problems with electrical noise, but also tends to reduce the number of inputs by half.

This device is an 8-bit A/D converter. The main concept behind this is the successive approximation logic. Once the reset is toggled, the converter will start by setting the most significant bit of the 8-bit number. This will be converted to a voltage V_e that is a function of the $+/-V_{ref}$ values. The value of V_e is compared to V_{in}, and a simple logic check determines which is larger. If the value of V_e is larger, the bit is turned off. The logic then repeats similar steps from the most to least significant bits. Once the last bit has been set on/off and checked, the conversion will be complete, and a done bit can be set to indicate a valid conversion value.

Quite often an A/D converter will multiplex between various inputs. As it switches, the voltage will be sampled by a *sample and hold circuit*. This will then be converted to a digital value. The sample and hold circuits can be used before the multiplexer to collect data values at the same instant in time (Figure 3.64).

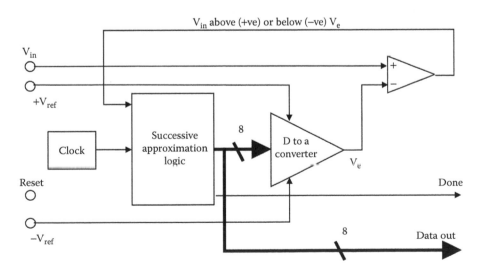

FIGURE 3.64
Successive approximation A/D converter.

3.11.1.1 Analog Inputs with a PLC

The PLC 5 ladder logic in Figure 3.65 will control an analog input card. The block transfer write (BTW) statement will send configuration data from integer memory to the analog card in rack 0, slot 0. The data from N7:30 to N7:66 describes the configuration for different input channels. Once the analog input card receives this, it will start doing analog conversions. The instruction is edge triggered, so it is run with the first scan, but the input is turned off while it is active, BT10:0/EN.

This instruction will require multiple scans before all of the data has been written to the card. The *update* input is only needed if the configuration for the input changes, but this would be unusual. The block transfer read (BTR) will retrieve data from the card and store it in memory, N7:10 to N7:29. This data will contain the analog input values. The function is edge triggered, so the enable bits prevent it from trying to read data before the card is configured, BT10:0/EN. The BT10:1/EN bit will prevent it from starting another read until the previous one is complete. Without these the instructions experience continuous errors. The *MOV* instruction will move the data value from one analog input to another memory location when the BTR instruction is done (Figure 3.65).

3.11.2 Analog Outputs

Analog outputs are much simpler than analog inputs. To set an analog output, an integer is converted to a voltage. This process is very fast and does not experience the timing problems with analog inputs.

3.11.2.1 Pulse Width Modulation Outputs

An equivalent analog output voltage can be generated using pulse width modulation (PWM), as shown in Figure 3.66. In this method, the output circuitry is only capable of

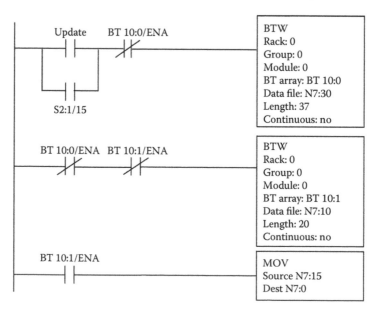

FIGURE 3.65
Ladder logic to control an analog input card.

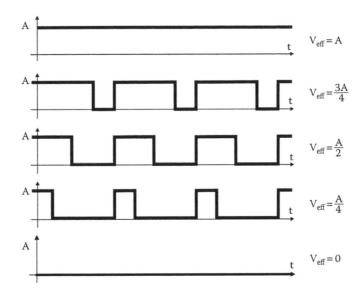

FIGURE 3.66
Pulse width modulated (PWM) signals.

outputting a fixed voltage (in the figure "A") or 0 V. To obtain an analog voltage between the maximum and minimum, the voltage is turned on and off quickly to reduce the effective voltage. The output is a square wave voltage at a high frequency, typically over 20 KHz, above the hearing range. The duty cycle of the wave determines the effective voltage of the output. It is the percentage of time the output is on relative to the time it is off. If the duty cycle is 100%, the output is always on. If the wave is on for the same time it is off, the duty cycle is 50%. If the wave is always off, the duty cycle is 0% (Figure 3.66).

PWM is commonly used in power electronics, such as servo motor control systems. In this case, the response time of the motor is slow enough that the motor effectively filters the high frequency of the signal. The PWM signal can also be put through a low pass filter to produce an analog DC voltage.

3.12 Continuous Actuators

Continuous actuators allow a system to position or adjust outputs over a wide range of values. Even in their simplest form, continuous actuators tend to be mechanically complex devices. For example, a linear slide system might be composed of a motor with an electronic controller driving a mechanical slide with a ball screw. The cost for such actuators can easily be higher than for the control system itself. These actuators also require sophisticated control techniques that will be discussed in later chapters. In general, when there is a choice, it is better to use discrete actuators to reduce costs and complexity.

3.12.1 Electric Motors

An electric motor is composed of a rotating center, called the rotor, and a stationary outside, called the stator. These motors use the attraction and repulsion of magnetic

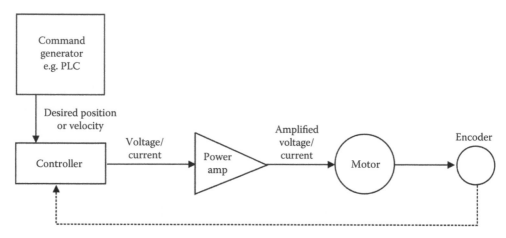

FIGURE 3.67
Typical feedback motor controller.

fields to induce forces, and hence motion. Typical electric motors use at least one electromagnetic coil, and sometimes permanent magnets to set up opposing fields. When
a voltage is applied to these coils, the result is a torque and rotation of an output shaft.
There are a variety of motor configurations that yield motors suitable for different
applications. The speed and torque of the motors will be varied by varying the voltage
applied to the motor.

A control system is required when a motor is used for an application that requires
continuous position or velocity. A typical controller is shown in Figure 3.67. In any controlled system, a command generator is required to specify a desired position. The controller will compare the feedback from the encoder to the desired position or velocity
to determine the system error. The controller will then generate an output based on the
system error. The output is then passed through a power amplifier, which in turn drives
the motor. The encoder is connected directly to the motor shaft to provide feedback of
position.

3.12.1.1 Basic Brushed DC Motors

In a DC motor, there is normally a set of coils on the rotor that turn inside a stator popu-
lated with permanent magnets. Figure 3.68 shows a simplified model of a motor. The mag-
netics provide a permanent magnetic field for the rotor to push against. When current is
run through the wire loop, it creates a magnetic field.

The power is delivered to the rotor using a commutator and brushes, as shown in
Figure 3.69. In the figure, the power is supplied to the rotor through graphite brushes
rubbing against the commutator. The commutator is split so that every half revolution the
polarity of the voltage on the rotor, and the induced magnetic field reverses to push against
the permanent magnets.

The direction of rotation will be determined by the polarity of the applied voltage, and
the speed is proportional to the voltage. A feedback controller is used with these motors
to provide motor positioning and velocity control. These motors are losing popularity to
brushless motors. The brushes are subject to wear, which increases maintenance costs. In
addition, the use of brushes increases resistance and lowers the motor's efficiency.

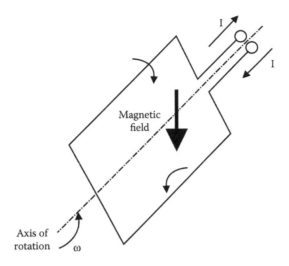

FIGURE 3.68
Simplified rotor.

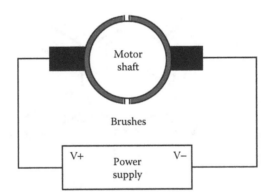

FIGURE 3.69
Split ring commutator.

3.12.1.2 AC Motors

Power is normally generated as 3-phase AC, so using this increases the efficiency of electrical drives. In AC motors, the AC current is used to create changing fields in the motor. Typically, AC motors have windings on the stator with multiple poles. Each pole is a pair of windings. As the AC current reverses, the magnetic field in the rotor appears to rotate (Figures 3.70 and 3.71).

- The number of windings (poles) can be an integer multiple of the number of phases of power. More poles results in a lower rotation of the motor.
- Rotor types for induction motors are listed below. Their function is to intersect changing magnetic fields from the stator. The changing field induces currents in the rotor.

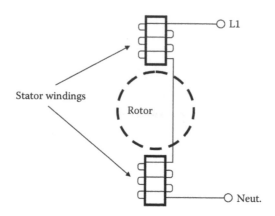

FIGURE 3.70
2-pole single phase AC motor.

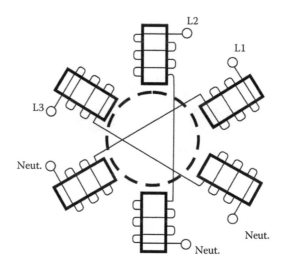

FIGURE 3.71
6-pole 3-phase AC motor.

These currents in turn set up magnetic fields that oppose fields from the stator, generating a torque.

Squirrel cage: Has the shape of a wheel with end caps and bars.

Wound rotor: The rotor has coils wound. These may be connected to external contacts via commutator.

Induction motors require slip. If the motor turns at the precise speed of the stator field, it will not see a changing magnetic field. The result would be a collapse of the rotor magnetic field. As a result, an induction motor always turns slightly slower than the stator field. The difference is called the slip. This is typically a few percent. As the motor is loaded, the slip will increase until the motor stalls. An induction motor has the windings on the stator. The rotor is normally a squirrel cage design. The squirrel cage is a cast aluminum core that when exposed to a changing magnetic field will set up an opposing field. When an AC voltage is applied to the stator coils, an AC magnetic field is created, the squirrel cage sets up an opposing magnetic field, and the resulting torque causes the motor to turn. The motor will turn at a frequency close to that of the applied voltage, but there is always some slip. It is possible to control the speed of the motor by controlling the frequency of the AC voltage.

3.12.1.3 Brushless DC Motors

Brushless motors use a permanent magnet on the rotor and use windings on the stator. Therefore, there is no need to use brushes and a commutator to switch the polarity of the voltage on the coil. The lack of brushes means that these motors require less maintenance than the brushed DC motors. A typical brushless DC motor could have three poles, each corresponding to one power input, as shown in Figure 3.72. Each of coils is separately controlled. The coils are switched on to attract or repel the permanent magnet rotor.

To continuously rotate these motors, the current in the stator coils must alternate continuously. If the power supplied to the coils is a 3-phase AC sinusoidal waveform, the motor will rotate continuously. The applied voltage can also be trapezoidal, which will give a similar effect. The changing waveforms are controlled using position feedback from the motor to select switching times. The speed of the motor is proportional to the frequency of the signal. A typical torque speed curve for a brushless motor is shown in Figure 3.73.

3.12.1.4 Stepper Motors

Stepper motors are designed for positioning. They move one step at a time with a typical step size of 1.8°, giving 200 steps per revolution. Other motors are designed for step sizes of 2°, 2.5°, 5°, 15°, and 30°. There are two basic types of stepper motors, unipolar and bipolar, as shown in Figure 3.74. The unipolar uses center tapped windings and can use a single power supply. The bipolar motor is simpler but requires a positive and negative supply and more complex switching circuitry.

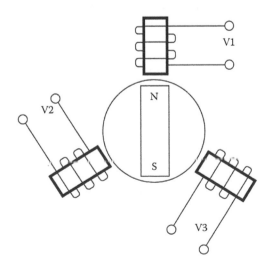

FIGURE 3.72
Brushless DC motor.

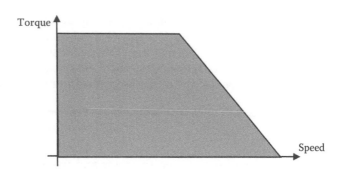

FIGURE 3.73
Torque speed curve for a brushless DC motor.

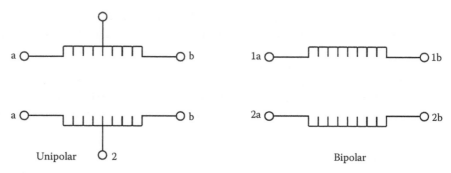

FIGURE 3.74
Unipolar and bipolar stepper motor windings.

The motors are turned by applying different voltages at the motor terminals. The voltage change patterns for a unipolar motor are shown in Figure 3.75. For example, when the motor is turned on we might apply the voltages as shown in line 1. To rotate the motor, we would then output the voltages on line 2, then 3, then 4, then 1, and so on. Reversing the sequence causes the motor to turn in the opposite direction. The dynamics of the motor and load limit the maximum speed of switching; this is normally a few thousand steps per second. When not turning, the output voltages are held to keep the motor in position.

To turn the motor, the phases are stepped through 1, 2, 3, 4, and then back to 1. To reverse the direction of the motor, the sequence of steps can be reversed, 4, 3, 2, 1, 4…. If a set of outputs is kept on constantly, the motor will be held in position.

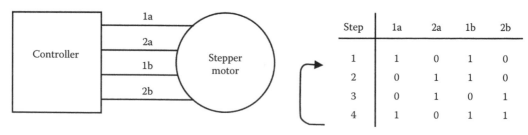

FIGURE 3.75
Stepper motor control sequence for a unipolar motor.

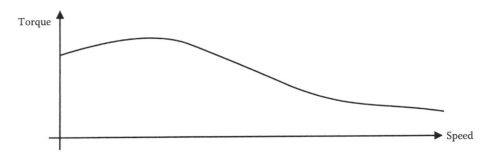

FIGURE 3.76
Stepper motor torque speed curve.

Stepper motors do not require feedback except when used in high-reliability applications and when the dynamic conditions could lead to slip. A stepper motor slips when the holding torque is overcome, or it is accelerated too fast. When the motor slips, it will move a number of degrees from the current position. The slip cannot be detected without position feedback. Stepper motors are relatively weak compared to other motor types. The torque speed curve for the motors is shown in Figure 3.76. In addition, they have different static and dynamic holding torques. These motors are also prone to resonant conditions because of the stepped motion control.

The motors are used with controllers that perform many of the basic control functions. At the minimum, a *translator* controller will take care of switching the coil voltages. A more sophisticated *indexing* controller will accept motion parameters, such as distance, and convert them to individual steps. Other types of controllers also provide finer step resolutions with a process known as *microstepping*. This effectively divides the logical steps and converts them to sinusoidal steps.

Translators: The user indicates maximum velocity and acceleration and a distance to move.

Indexer: The user indicates direction and number of steps to take.

Microstepping: Each step is subdivided into smaller steps to give more resolution.

3.12.1.5 Wound Field Motors

- Uses DC power on the rotor and stator to generate the magnetic field (i.e., no permanent magnets)
- Shunt motors
  - Have the rotor and stator coils connected in parallel.
  - When the load on these motors is reduced, the current flow increases slightly, increasing the field and slowing the motor.
  - These motors have a relatively small variation in speed as they are varied, and are considered to have a relatively constant speed.
  - The speed of the motor can be controlled by changing the supply voltage, or by putting a rheostat/resistor in series with the stator windings.

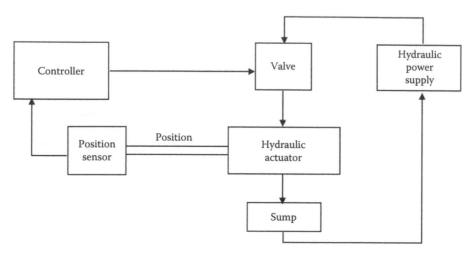

FIGURE 3.77
Hydraulic servo system.

3.12.2 Hydraulics

Hydraulic systems are used in applications requiring a large amount of force and slow speeds. When used for continuous actuation, they are mainly used with position feedback. An example system is shown in Figure 3.77. The controller examines the position of the hydraulic system, and drivers a servo valve. This controls the flow of fluid to the actuator and also it provides the hydraulic power to drive the system.

The valve used in a hydraulic system is typically a solenoid controlled valve that is simply opened or closed. Newer, more expensive, valve designs use a scheme such as PWM, which opens/closes the valve quickly to adjust the flow rate.

3.13 Continuous Control

Continuous processes require continuous sensors and/or actuators. For example, an oven temperature can be measured with a thermocouple. Simple decision-based control schemes can use continuous sensor values to control logical outputs, such as a heating element. Linear control equations can be used to examine continuous sensor values and set outputs for continuous actuators, such as a variable position gas valve. Two continuous control systems are shown in Figure 3.78. The water tank can be controlled by valves. In a simple control scheme, one of the valves is set by the process, but we control the other to maximize some control object. If the water tank was actually a water tank, the outlet valve would be the domestic and industrial water users. The inlet valve would be set to keep the tank level at maximum. If the level drops, there will be a reduced water pressure at the outlet, and if the tank becomes too full, it could overflow. The conveyor will move boxes between stations. Two common choices are to have it move continuously, or to move the boxes between positions, and then stop. When starting and stopping, the boxes should be accelerated quickly, but not so quickly that they slip. And, the conveyor should stop at precise positions. In both of these systems, a good control system design will result in better performance.

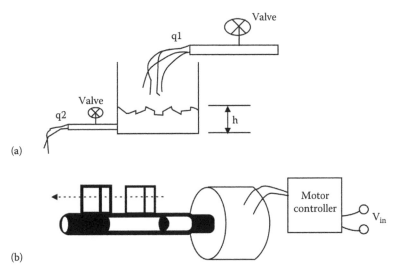

FIGURE 3.78
Continuous controlling units. (a) Water tank. (b) Motor driven conveyor.

Continuous control systems typically need a target value; this is called a *set point*. The controller should be designed with some objective in mind. Typical objectives are listed below.

Fastest response: Reach the setpoint as fast as possible (e.g., hard drive speed)

Smooth response: Reduce acceleration and jerks (e.g., elevators)

Energy efficient: Minimize energy usage (e.g., industrial oven)

Noise immunity: Ignores disturbances in the system (e.g., variable wind gusts)

An engineer can design a controller mathematically when performance and stability are important issues. A common industrial practice is to purchase a *PID* unit, connect it to a process, and tune it through trial and error. This is suitable for simpler systems, but these systems are less efficient and prone to instability. In other words, it is quick and easy, but these systems can go out-of-control.

3.13.1 Control of Logical Actuator Systems

Many continuous systems will be controlled with logical actuators. Common examples include building heating, ventilation, and air-conditioning (HVAC) systems. The system setpoint is entered on a *thermostat*. The controller will then attempt to keep the temperature within a few degrees as shown in Figure 3.79. If the temperature is below the bottom limit, the heater is turned on. When it passes the upper limit, it is turned off, and it will stay off until it passes the lower limit. If the gap between the upper and lower the boundaries is larger, the heater will turn on less often but be on for longer, and the temperature will vary more. This technique is not exact, and time lags will often lead to overshoot above and below the temperature limits.

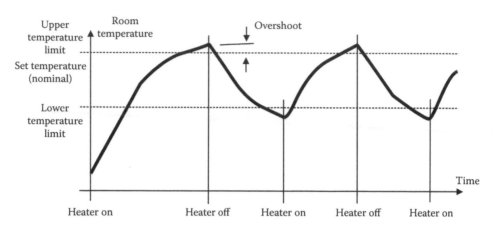

FIGURE 3.79
Continuous control with a logical actuator.

3.13.2 Control of Continuous Actuator Systems

3.13.2.1 *Block Diagrams*

Figure 3.80 shows a simple block diagram for controlling arm position. The system set-point, or input, is the desired position for the arm. The arm position is expressed with the joint angles. The input enters a summation block, shown as a circle, where the actual joint angles are subtracted from the desired joint angles. The resulting difference is called the *error*. The *error* is transformed to joint torques by the first block labeled *neural system and muscles*. The next block, *arm structure and dynamics*, converts the torques to new arm positions. The new arm positions are converted back to joint angles by the *eyes*.

The blocks in block diagrams represent real systems that have inputs and outputs. The inputs and outputs can be real quantities, such as fluid flow rates, voltages, or pressures.

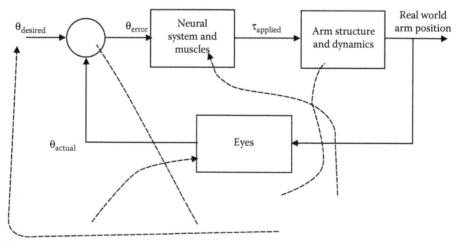

FIGURE 3.80
Block diagram.

The inputs and outputs can also be calculated as values in computer programs. In continuous systems, the blocks can be described using differential equations. Laplace transforms and transfer functions are often used for linear systems.

3.13.2.2 Proportional Controllers

Figure 3.81 shows a block diagram for a common servo motor controlled positioning system. The input is a numerical position for the motor, designated as C. (Note: The relationship between the motor shaft angle and C is determined by the encoder.) The difference between the desired and actual C values is the system error. The controller then converts the error to a control voltage V. The current amplifier keeps the voltage V the same, but increases the current (and power) to drive the servomotor. The servomotor will turn in response to a voltage, and drive an encoder and a ball screw. The encoder is part of the negative feedback loop. The ball screw converts the rotation into a linear displacement X. In this system, the position X is not measured directly, but it is estimated using the motor shaft angle.

3.13.3 PID Control Systems

Proportional-integral-derivative (PID) controllers are the most common controller choice. The basic controller equation is shown in Figure 3.82. The equation uses the

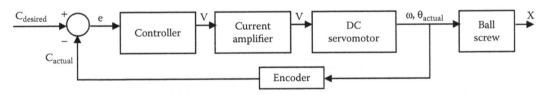

FIGURE 3.81
Servomotor feedback controller.

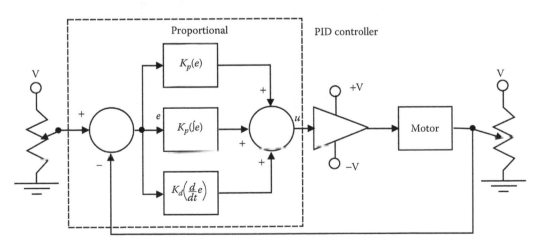

FIGURE 3.82
PID control system.

system error e to calculate a control variable u. The equation uses three terms. The proportional term, K_p, will push the system in the right direction. The derivative term, K_d, will respond quickly to changes. The integral term, K_i, will respond to long-term errors. The values of K_c, K_i, and K_p can be selected, or tuned, to get a desired system response.

Figure 3.82 shows a (partial) block diagram for a system that includes a PID controller. The desired set point for the system is a potentiometer set up as a voltage divider. A summer block will subtract the input and feedback voltages. The error then passes through terms for the proportional, integral, and derivative terms; the results are summed together. An amplifier increases the power of the control variable u to drive a motor. The motor then turns the shaft of another potentiometer, which will produce a feedback voltage proportional to shaft position.

3.13.3.1 Water Tank Level Control

Problem: The system in Figure 3.83 will control the height of the water in a tank. The input from the pressure transducer, Vp, will vary between 0 V (empty tank) and 5 V (full tank). A voltage output, Vo, will position a valve to change the tank fill rate. Vo varies between 0 V (no water flow) and 5 V (maximum flow).

The system will always be on: the emergency stop is connected electrically. The desired height of a tank is specified by another voltage, Vd. The output voltage is calculated using Vo = 0.5 (Vd − Vp). If the output voltage is greater than 5 V it will be made 5 V, and below 0 V it will be made 0 V.

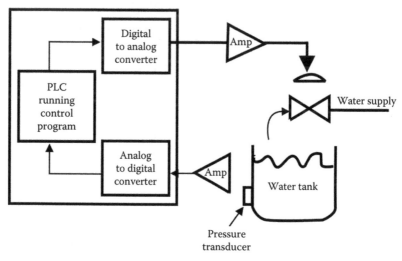

FIGURE 3.83
Water tank level controller.

3.14 Summary

The outcome of the chapter is to understand the working of PLC and to give a clear view of programming methods of PLC. The types of inputs and outputs used in PLC are explained in detail. The user will understand the procedure to program a PLC for an industrial concept.

References

1. F. Petruzella. *Programmable Logic Controllers (English)*. Tata McGraw-Hill, Education Pvt. Ltd., India, 3rd ed., 2014.
2. G. Bolton. *Programmable Logic Controllers (English)*. Elsevier India, 5th ed., 2015.
3. J. W. Webb, R. A. Reis. *Programmable Logic Controllers*. Phi learning, New Delhi, India, 2009.
4. Hackworth. *Programmable Logic Controllers: Programming Methods and Applications (With CD) (English)*. Pearson India, 1st ed., 2011.
5. M. Gottardo. *Advanced PLC Programming*. Lulu.com, Italy, 2015.
6. W. Bolton. *Programmable Logic Controllers*. Newnes, Oxford, UK, 5th ed., 2009.
7. G. Olsson. *Programmable Logic Controllers*. Springer, USA, 2005.
8. M. Rabiee. *Programmable Logic Controllers: Hardware and Programming*. Goodheart-Willcox, Tinley Park, IL, 3rd ed., 2012.
9. J. A. Rehg, G. J. Sartori. *Programmable Logic Controllers*. Pearson Prentice Hall, Upper Saddle River, NJ, 2nd ed., 2008.
10. G. A. Mazur, W. J. Weindorf. *Introduction to Programmable Logic Controllers*. American Technical Publication, Orland Park, IL, 2nd ed., 2009.
11. S. C. Jonathon Lin. *Programmable Logic Controllers*. Industrial Press, 1st ed., 2016. https://www.bookdepository.com/publishers/Industrial-Press-Inc-U-S. Industrial Press Inc., U.S.A.
12. K. Kamel, E. Kamel. *Programmable Logic Controllers: Industrial Control*. McGraw-Hill Education, New York, 1st ed., 2013.
13. J. R. Hackworth, F. D. Hackworth. *Programmable Logic Controllers: Programming Methods and Applications*. Prentice Hall, Upper Saddle River, NJ, 2003.
14. K. T. Erickson. *Programmable Logic Controllers: An Emphasis on Design and Application*. Dogwood Valley Press, LLC, Rolla, MO, 2nd ed., 2011.
15. J. Stenerson. *Fundamentals of Programmable Logic Controllers, Sensors, and Communications*. Prentice Hall, Englewood Cliffs, NJ, 2004.
16. J. W. Webb. R. A. Reis. *Programmable Logic Controllers: Principles and Applications*. Prentice Hall, Englewood Cliffs, NJ, 5th ed., 2002.
17. G. Dunning. *Introduction to Programmable Logic Controllers*, Thomson/Delmar Learning, Clifton Park, NY, 3rd ed., 2005.
18. E. P. Adrover. *Introduction to PLCs: A Beginner's Guide to Programmable Logic Controllers*. Elvin Perez Adrover, USA, 2012.
19. T. Borden, R. A. Cox. *Technician's Guide to Programmable Controllers*. Delmar Cengage Learning, Clifton Park, NY, 6th ed., 2012.
20. R. Sartori. *Programmable Logic Controllers*. Pearson International, Upper Saddle River, NJ, 2nd ed., 2008.
21. S. Brian Morriss. *Programmable Logic Controllers*. Prentice Hall, Upper Saddle River, NJ, 1st ed., 1999.
22. F. Petruzella. *Activities Manual to Accompany Programmable Logic Controllers*. McGraw-Hill Science/Engineering/Math, New York, 3rd ed., 2005.
23. C. T. Jones. *Programmable Logic Controllers the Complete Guide to the Technology*. Patrick Turner Publishing Company, Atlanta, GA, 1996.

4

Instrumentation and Control Systems on Blowroom Sequence

LEARNING OBJECTIVES

- To explain the blowroom preparatory process
- To identify the working of instruments used in the blowroom sequence
- To comprehend the method of measuring systems used in the blowroom process
- To show expertise in the control system process in the blowroom sequence

4.1 Introduction

It the cotton received is from different ginners, it is better to maintain the percentage of cotton from different ginners through the lot, even though the type of cotton is the same. If an automatic bale opening machine is used, the bales should be arranged as follows. Let us assume that there are five different micronaire and five different colors in the mixing, and 50 bales are used. Five to 10 groups should be made by grouping the bales in a mixing so that each group will have the same average micronaire and average color as that of the overall mixing. The position of a bale for micronaire and color should be fixed for the group, and it should repeat in the same order for all the groups.

4.2 Bale Management

It is always advisable to use a mixing with very low micronaire range preferably 0.6–1.0. Because it is easy to optimize the process parameters in the blowroom and cards drafting faults will be less, the dyed cloth appearance will be better because of uniform dye pickup, and so on. It is advisable to use single cotton in a mixing, provided the length, strength micronaire, maturity coefficient, and trash content of the cotton will be suitable for producing the required counts. An automatic bale opener is a must if more than two cottons are used in the mixing to avoid barre or shade variation problems.

The cotton fiber is an industrial raw material that is utilized primarily by the textile manufacturing industry. Therefore, in order to reach and serve the ultimate source of demand for cotton, the consumers, it is necessary to serve the needs of textile manufacturers well enough to displace other available fibers—whether they be foreign cotton, other natural fibers, or manmade fibers. Success for the cotton production sector requires understanding and accommodation of the major trends impacting the textile industry.

It is impossible to explain the accelerating trend toward globalization apart from the information revolution. Too many people still limit their concept of globalization to increased trade, which is only one component. Globalization is also related to the ease and the speed of obtaining and communicating knowledge and information. Globalization advanced at a moderately accelerating, evolutionary pace throughout much of the twentieth century.

Throughout the majority of the twentieth century, most newly developed capital equipment for the textile industry was used to produce *things*. Now, however, much of the new equipment is used to produce *information*, which is then used to monitor and control production processes, to facilitate buying and selling activities, to predict outcomes and control risk, and so on. For example, the ongoing development of robotics to perform tasks necessary for production is largely based on the capability to generate information (on a real-time basis) and use it to guide the activities of the robotic machinery.

4.2.1 Objective Measurement and Quality Control

Developments within the last 25 years have brought within our reach the ability to objectively measure the properties of fibers going into textile manufacturing processes and the quality of products being produced. Therefore, we can reach new levels of exactness in quality control, which translates quickly into higher levels of manufacturing efficiency. This is critically important for cotton, which is a natural fiber with complex, biological distributions of properties that could not be adequately measured and manipulated until computers became very powerful and fast. The capability for measurement and statistical process control translates into an improvement in cotton's comparative advantage relative to the manmade fibers.

One of the most visible manifestations of ongoing developments related to the information revolution is the focus on "quick response" (QR) by the textile industry. QR was formally conceived about a decade ago as a business strategy for all participants in textile channels (textile manufacturers, apparel manufacturers, and retailers) to exploit technology and collaboration in order to shorten the response times between junctures in the system. In effect, investments in the technology and organization required for QR have been more efficient than investments in warehouses and inventories.

Opening in the blowroom means opening into small flocks. Technological operation of opening means the volume of the flock is increased while the number of fibers remains constant, that is, the specific density of the material is reduced. The larger the dirt particle, the better they can be removed. Since almost every blowroom machine can shatter particles, as far as possible a lot of impurities should be eliminated at the start of the process. Opening should be followed immediately by cleaning, if possible in the same machine. The higher the degree of opening, the higher the degree of cleaning. A very high cleaning effect is almost always purchased at the cost of a high fiber loss. Higher roller speeds give a better cleaning effect but also more stress on the fiber (Figure 4.1).

Cleaning is made more difficult if the impurities of dirty cotton are distributed through a larger quantity of material by mixing with clean cotton. The cleaning efficiency is strongly dependent on the trash percentage. It is also affected by the size of the particle and stickiness of cotton. Therefore, cleaning efficiency can be different for different cottons with the same trash percentage.

There are three types of feeding apparatus in the blowroom opening machines:

1. Two feed rollers (clamped)
2. Feed roller and feed table
3. Feed roller and pedals

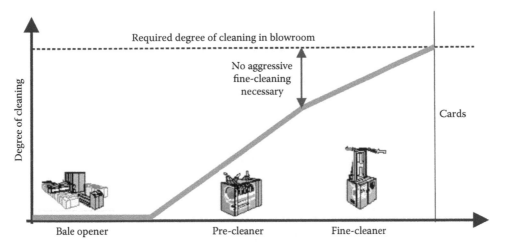

Required degree of cleaning in blowroom

No aggressive
fine-cleaning
necessary

Cards

Degree of cleaning

Bale opener   Pre-cleaner   Fine-cleaner

FIGURE 4.1
Blowroom process.

Two feed roller arrangements gives the best forwarding motion, but unfortunately results in greatest clamping distance between the cylinders and the beating element. Feed roller and pedal arrangement gives secure clamping throughout the width and a small clamping distance, which is critical for an opening machine. In a feed roller and table arrangement, the clamping distance can be made very small. This gives intensive opening, but clamping over the whole width is poor because the roller presses only on the highest points of the web. Thin places in the web can be dragged out of the web as a clump by the beaters.

4.3 Mixing Bale Opener

The mixing bale opener is provided with infinitely variable speed drive for required production rates. It has a pair of feed rollers with coarse toothed cleaning rollers controlled with a PLC driving system. Knife girds in the machine are used for efficient cleaning of trashy cotton. Wider width stainless steel material passage attached with the automatic flow controlling devices is used for uniform material feed, which is provided with an adjustable filling trunk operated with a photoelectric sensing unit.

4.4 Bale Pluckers

The bale plucker has a flexible programming system, to allocate one to four bale blocks per side and one to four assortments. It has track length of 10–50 m in steps of 2.5 m and bale lay down up to 350 bales. It is suitable for all kinds of spinning applications. A PLC controlled system is used for machine parts movements and positioning.

4.5 Compact Blowroom

High in performance, yet compact in design, a complete blowroom line is composed of only four multi-function machines. A lower current consumption and lower exhaust air volumes improve the economic efficiency. Nevertheless, there are no compromises regarding cleaning, opening, and blending quality, even at a production rate of 800 kg/h.

The compact blowroom consists of the following four machines:

- Automatic bale opener
- Multi-function separator
- Mixer–cleaner combination with high production cleaner
- Foreign part separator

4.5.1 Automatic Bale Opener

Three qualities of cotton can be fed into three separate cleaning lines. When processing blended materials, the machine can proves the components adjacent to each other. This can take place on one side of the track or on both, in which case the tower of the machine automatically moves to the other side when it reaches the end of a line of bales. On completion of its work here, it moves again to the first side and so on. It adapts its traversing speed automatically to the production requirement. A frequency-controlled drive motor allows traverse speeds of between 6 and 13 mts/min (Figure 4.2).

The computer system Blend commander controls the operation of the machine and monitors all its functions. A color monitor visualizes all the production operations and shows up any malfunctions. Operation and monitoring are greatly simplified due to this control system. Micro Computer Control with Color Monitor is equipped with the Blend

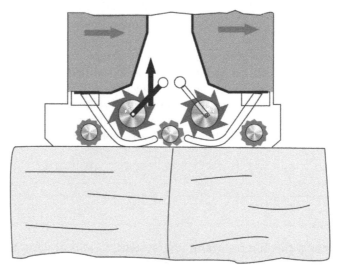

FIGURE 4.2
Inversion mechanism in the detacher.

commander as a standard feature. This computer controls the machine and illustrates the operating conditions on a monitor. Operation of the machine is thus very simple.

Flow control drives: For installations with continuous material flow, electronically controlled drives are used. The feed table monitor gives a warning signal before the material supply runs out.

Network: A network connects the different machine controls to the center plant control system, Line Commander (LC) Software. A comprehensive software system is an integral part of each control. Faults are located automatically and indicated to the operator.

4.5.2 The Multi-Function Separator

The next machine, the multi-function separator, is a completely new development. Fan for blowing off the Blendomat is integrated in this machine. The fan speed is controlled fully automatically and is always only as high as is needed for the suction of the Blendomat. This saves electric power. A heavy part separator in the multifunction separator protects the cleaning line against damage. Heavy parts or burning tufts detected by a spark sensor are separated into containers and extinguished, if required (Figure 4.3).

4.5.3 Mixer–Cleaner Combination with High Production Cleaner

Fine cleaning is done with different types of machines. Some fine cleaners use single opening rollers and some use multiple opening rollers. If single roller cleaning machines are used, depending upon the amount and type of trash in the cotton, the number of fine cleaning points can be either one or two. Saw tooth beaters can be used if trash particles are more and the machine is not using suction and deflector blades, that is, beater and regular grid bar arrangements. The cleaning points in CVT1, CVT3, CVT4, and so on, consist of opening roller, deflector blades, mote knives, and suction hood. Trash particles released due to centrifugal forces are separated at the mote knives and continuously taken away by the suction. This gives

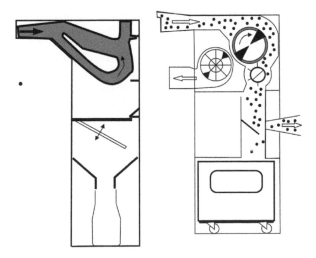

FIGURE 4.3
Heavy part separator.

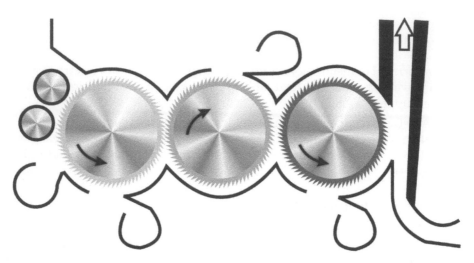

FIGURE 4.4
Mixer–cleaner combination with high production cleaner.

better cleaning. Suction plays a major role in these machines. If suction is not consistent, the performance will be adversely affected. Very high suction will result in more white fiber loss, and less suction will result in low cleaning efficiency. The minimum recommended pressure in the waste chamber (P2) is 700 Pa. It can be up to 1000 Pa. Material suction (P1) should be around 500 Pa. Whenever the suction pressure is changed, the deflector blade settings should be checked. Deflector blade settings cannot be same for all three or four rollers. The setting for deflector blades in the panel should be 3, 12, and 30 for first, second, and third deflector blades. The deflector blade setting should be done in such a way that the setting should be opened until the fibers start slipping on the deflector blade. The wider the deflector blade setting, the higher will be the waste. If the setting is too wide, white fiber loss will be very high. For saw ginned cottons, the above concepts are helpful because constant suction concentrated directly at the mote knives ensures much removal of dust from the cotton (Figure 4.4).

4.5.3.1 Computer Controlled Cleaning Efficiency

Every Cleano mat cleaner, whether it has 1 or ¾ rolls, is controlled and monitored by the integral microprocessor system clean commander. In the Cleano mat cleaner, three cleaning rolls are used, as shown in Figure 4.5. The density of clothing, that is, points per square inch, increases gradually from the first to the third roll. Thus, opening and cleaning of cotton is based on the principle of stepwise cleaning. The cleaning rolls clean the material on both sides as it passes over the carding segments, openings, and mote knives. The mote knives are equipped with direct suction units that collect the released dirt. Like the clothing wire density, the rotational speed also increases from the first to the third roll, which results in proportional increase of the peripheral speed and the centrifugal forces acting on the material.

In other words, the material is subjected to comparatively lower centrifugal and pneumatic forces at the intake, increasing to very high magnitude at the outlet. The Cleanomat as well as other Cleanomat cleaners operate with priority on gentle treatment of the material without sacrificing high efficiency cleaning.

The micro-computer control system, that is, the Clean commander, facilitates control of the amount of waste being extracted and the degree of cleaning by motorized adjustment of the deflector blades in front of the mote knives of the cleaning rolls.

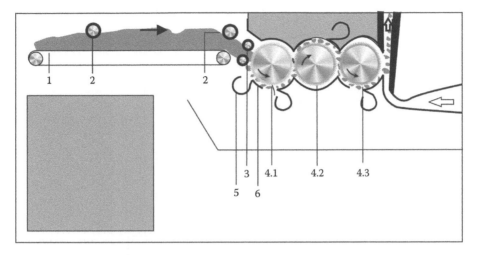

FIGURE 4.5
Cleanomat cleaner. 1—Feed lattice, 2—Pressure Rolls, 3—Feed Rolls. 4.1—Fully spiked roll, 4.2—Coarse saw-toothed roll, 4.3—Medium saw toothed roll, 5—Mote knives, and 6—Carding segment.

For example, the deflector blades in front of the mote knife, for controlling the amount of waste, are individually adjusted by moors via a keyboard. A rapid adoption to a new raw material or to an altered degree of cleaning is therefore possible. Stored experienced values help with the settings. The amount of suction for fibers and waste is also permanently monitored by the system.

4.5.3.2 Servo Motors

With the cleaners, the amount of waste extracted is adjusted by means of a servo motor. This gives the capability of optimizing the degree of cleaning for each cotton quality.

4.5.4 Foreign Part Separator

The end of the compact blowroom line, the foreign part separator, also is a multi-function machine. When a fine dedusting has taken place in the upper part of the machine, the cotton is scanned by two color cameras. The separation of detected foreign parts is effected in a selective way by compressed air nozzles. This avoids unnecessary fiber losses (Figure 4.6).

4.6 Detection and Removal of Contamination

4.6.1 Premier Fiber Eye

The new generation Fiber Eye identifies foreign material with an array of 160 high speeds and high sensitive photo sensors, and ejects the material with high speed air nozzles. The technology is very efficient and at the same time cost effective. Premier Fiber Eye reliably detects all types of foreign materials including jute, hair, paper, polypropylene, feather,

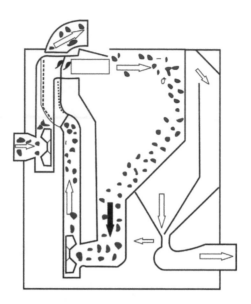

FIGURE 4.6
Dedusting machine.

leaves, colored threads, dyed fibers, cloth bits, and sticks. The signals from the individual photo sensors are processed through an intelligent custom-designed software that ensures foolproof detection and elimination of the foreign material.

After detection of the contaminants, the contaminated material is ejected into the waste chamber using 32 high speed air nozzles. Three or five nozzles are activated at a time, resulting in minimum good fiber loss while at the same time effectively ejecting the detected contaminants.

Premier Fiber Eye has an automatic calibration mechanism for adjustment of the detection threshold during regular operation based on the base cotton material. This ensures effective removal of contaminants at all times irrespective of the natural variations in the cotton color. The efficiency of detection depends to a great extent on the proper upkeep and maintenance of the equipment.

Premier Fiber Eye is uniquely designed to facilitate easy maintenance. Modular lighting sections enable easy cleaning of the lights. Other dirt prone areas are also easily accessible for quick cleaning. Location and number of detectors—There was a school of thought suggesting placement of "the detector" in the front of the blowroom so that the contaminations are not broken down into multiple smaller fragments. But with more installations and actual experience, almost all the manufacturers are in favor of locating it at the end of the line. Smaller tuft size and lower production rate results in better detection efficiency.

4.6.1.1 Double Value with Premier Fiber Eye

In order to ensure the maximum removal of contamination, Premier recommends two Fiber Eye systems in each line, one in the beginning and the other after the fine beater. This configuration enables the highest level of detection and yet assures the most economical solution for foreign fiber cleaning.

4.6.2 Textile Cotton Eye—Contaminant Removal in Cotton

Cotton contamination can lead to serious product quality problems. Detection of contaminants is a laborious and tedious process. The raw material inspectors cannot be expected to be vigilant at all times. Cotton Eye is a real-time system that detects and removes contamination from cotton. Cotton contamination that escapes detection can break up into tiny pieces and spread over a long length of yarn. This leads to defects in the finished product resulting in substantial losses.

Cotton moving through the blowroom line enters into Cotton Eye's rectangular section. The section comprises an illuminated chamber where the passing cotton is inspected by four ultra-high–speed charge coupled device (CCD) cameras. These CCD cameras are strategically positioned to enhance detection capabilities. The solution's powerful image processing techniques detects cotton contaminants and activates the high-speed air gun ejection mechanism. The contaminated tuft is separated from the main line and is sent into the central waste collecting systems.

Cotton Eye also comprises a touch screen interface to enable online graphical reporting and easy Cotton Eye setup. The solution combines robust software architecture and industrial strength hardware to scale to any requirement. The solution is easy to configure, deploy, administer, and use.

4.6.3 Detection and Removal of Contamination by Securomat

BARCO is proud to introduce the Cotton Sorter, a real-time system for the detection and removal of contamination from raw cotton in the blow line. All unwanted material that differs from the cotton throughput is detected by ultra-fast CCD cameras and removed by means of high-speed air guns.

It is very easy to install the Cotton Sorter in an existing opening line. The system is installed in the pipeline that transports the cotton tufts from the bale opener. The cotton flow through the machine is straight. No need to add fan capacity to elevate the cotton and drop it through the inspection zone. The Cotton Sorter provides the means to safeguard a high and consistent quality level, regardless of operator practices. The early removal of contamination results in an improved and consistent quality level, less customer complaints, and a higher efficiency on OE machines and winders due to a lower number of foreign fiber cuts by the yarn clearer.

4.6.3.1 *Operating Principle*

In the Cotton Sorter, the material passes through a transparent tunnel. In the inspection zone, the cotton is diffusely illuminated. The individual tufts are observed by means of four ultra-fast, high-resolution CCD line scan cameras. Simultaneous image acquisition from both sides enhances the detection capability. Image processing includes correction for non-uniform lightening, contrast enhancement, and feature extraction. Defects that are of a different color than the raw cotton and are of a certain size are detected in the cotton flow.

In order to allow exact timing between contaminant detection and activation of the downstream ejection mechanism, the speed of the cotton flow is continuously measured. After the corresponding delay, the appropriate high-speed air guns are activated downstream.

4.6.3.2 Operator Interface

The Cotton Sorter comes with a touch screen-based operator's console, allowing comfortable and easy machine setup, operation, and follow up. Contamination to be ejected includes deviation in color and size. The contaminant size is defined as a combination of contaminant width (number of pixels) and contaminant length (number of scan lines). Tolerances in color are defined by means of threshold levels (limits). Setting a low level results in objects with a color value below this limit being identified as a contaminant.

The result of the scanning by the cameras is shown on the camera display menu. On the *x*-axis, the number of camera pixels is indicated, whereas the *y*-axis displays the color level. A value of zero corresponds with black, whereas 255 means white. It is obvious that the flow of normal cotton is recognized as close to the white level.

4.6.3.2.1 Securomat: A New System for the Effective Removal of Foreign Matter in Spinning Preparation

There are only very few processing stages from cotton production to garment manufacture that to date have not been confronted with the issue of cotton contamination. Since an ultimate solution to the problem at the cotton production end seems unlikely in the foreseeable future, most of the burden still remains with the spinner and fabric producer. A partial solution to the foreign matter dilemma in the spinning process was initially provided by electronic yarn monitoring systems with foreign fiber detection installed on spinning or winding frames.

The effectiveness of foreign matter removal in opening and cleaning is a function of tuft size or the degree of opening. Basically, the ratio between tuft size and the size of a typical foreign object is extremely disadvantageous. Numerous contaminants are fully enclosed and concealed by large and compact fiber tufts, preventing any possibility of optical detection. A homogeneous web consisting of individualized fibers would provide ideal conditions for detecting foreign matter. This condition would be achieved at the card, but it is too late then to extract contaminants because of heavy fibrillation occurring in the main carding zone between the cylinder and the revolving flats. However, from a technological point of view, near-optimum conditions can be found immediately before the card since mean tuft size continuously decreases with each processing step.

The logical conclusion of these fundamental considerations is a system that is characterized by the basic features such as (1) detection of foreign matter via digital cameras, monitoring the fibers present on the surface of a rotating spiked cylinder, or (2) the system is positioned right before the cards.

These elementary thoughts have eventually led to the development of the securomat in the opening and cleaning line. The securomat replaces the dust extraction machine or a separate fine opener with chute feed, if there is one. Therefore, as stated earlier, foreign fiber detection takes place at the end of the cleaning line and immediately before carding, where the maximum degree of fiber opening has been achieved, while contaminants are still in coherent form (Figure 4.7).

The separation of the fibrous material from the transport air flow is performed by a modified material separator—a dome-like design consisting of a curved perforated sheet metal surface and pivoting distributor flaps (1). This is also where considerable amounts of dust are extracted. The exhaust air and dust emission is not conveyed to the filter—as would normally be the case—but is used as the transport air flow for removing foreign matter (2). This saves on filter capacity and all associated costs. The material separator itself has no moving parts apart from the distributor flaps and is therefore characterized by low energy consumption and low maintenance requirements.

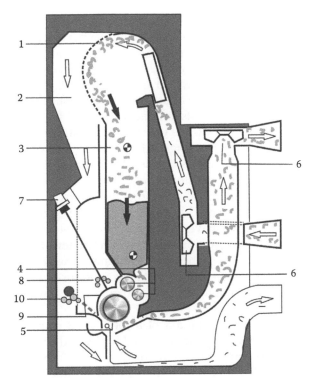

FIGURE 4.7
System for the effective removal of foreign matter.

From the material separator, the dedusted cotton is dropped into a chute (3). The feed roll arrangement (4) conveys the fiber into the working zone of a spiked cylinder (5), which ensures uniform cotton tuft opening. The fans to supply the fiber to the material separator and to doff the cylinder and feed the cards (6) are driven by frequency converter variable speed motors. They are both on board the machine, providing for a compact layout of the opening and cleaning line.

The surface of the rotating spiked cylinder is partially covered with fibers and permanently monitored by two trilinear RGB CCD color line scan cameras (7). RGB means red, green, and blue. These are the basic colors or wavelengths to which the CCD chip responds. Each camera provides 1024 effective pixels per line. Due to the two-camera arrangement, optical image errors in terms of focus, brightness, and color convergence are minimized.

Consequently, uniform detection sensitivity can be achieved across the width of the spiked cylinder. The cameras and the illumination system (8) are installed in dustproof housing. The cylinder cover (9) can easily be removed for cleaning the viewing window. As an integral component of the machine control unit, a powerful computer system transforms and analyzes the unprocessed camera data. With the aid of an RGB color line scan sensor, each color can be broken down into the three basic colors: red, green, and blue. Brightness is also determined as the fourth criterion. The actual color components and the brightness value of the specific material processed are established as reference values.

During normal operation, the system recognizes objects that deviate from the color and/ or brightness of the metallic surface of the spiked cylinder partially covered with the fibrous material. By comparing the actual signal with the reference values and upon exceeding certain thresholds, which can be adjusted manually, objects are identified as contaminants and removed by compressed air nozzles. This also applies to pastel shades, for example, yellow contaminants, which would not provide a satisfactory contrast against the background of a compact cotton tuft. Among the tiny tufts and individualized fibers present on the cylinder surface, even the smallest contaminants are exposed and distinctly presented to the camera system. Hence, high camera resolution, high tuft openness, and optimum defect presentation on the cylinder permit the reliable detection of very small objects. The basic principle of detecting foreign matter with the securomat requires a color difference between the reference background and the foreign object. Therefore, by theory, fully transparent and white materials cannot be detected. In practice, however, it is observed that fully transparent foils, for instance, can be detected if they have been stained or discolored, which is often the case. Furthermore, many transparent objects provide bright reflections from the surface. This will also trigger detection in the brightness channel. In summary, transparent materials can sometimes be detected and ejected successfully if certain conditions are fulfilled. However, these conditions are subjected to the random principle. No guarantees are therefore made concerning transparent materials.

Contaminants are removed with the aid of a total of 32 compressed air nozzles (10) arranged across the working width of the cylinder. However, the compressed air impulse is confined to a few nozzles covering the actual position of the foreign object on the cylinder surface. The nozzles are individually controlled by pneumatic valves and activated precisely at the moment when a foreign object passes the nozzle array. The compressed air flow is tangentially directed toward the cylinder and activated for a few milliseconds only. Compressed air consumption is therefore negligible even at high extraction rates. This form of selective removal produces a minimum loss of usable lint of only 1–2 g of fiber per intervention, which corresponds to a total waste figure of 0.1%–0.2% under practical conditions. Low loss of usable lint is an essential prerequisite for highly sensitive settings and the removal of very small pieces of foreign matter.

The waste removal concept is also very innovative. Instead of conveying the extracted foreign matter into a separate waste container, which has to be dumped manually at regular intervals, the extracted matter can be fed directly to the filter. Where this option is either prohibitive (i.e., recycling, waste processing) or is not desired (e.g., because the extracted objects are to be inspected), a separate condenser with a waste container underneath can be supplied.

4.6.4 Cotton Contamination Cleaning Machine

The most effective way of removing contaminants and other foreign matter from cotton in the blowroom of spinning mills is with the use of high speed CCD cameras. Sizes of contaminant detection is programmable in length or width independently or both. It has easily programmable adjustments for color variations and sensitivity adjustments for contaminants.

Nozzles are programmable to enable them to select those nearer to contamination, that is, you can select either one on each side or two and in any fashion to effect the ejection (Figure 4.8).

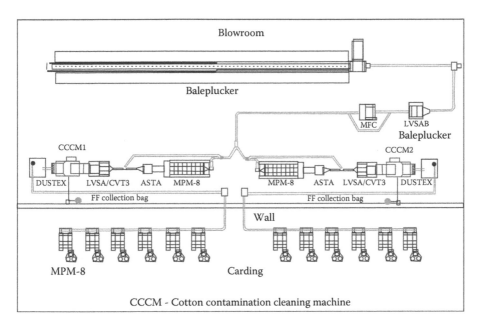

FIGURE 4.8
Material flow to CCCM.

4.7 Dust and Metal Extraction Machine

The dust and metal extraction machine (DMEM) consists of a detector to detect even the smallest of metal pieces like ring travelers and a high-speed diverter to drop fibers mixed up with these particles to a collection chamber. The system can be mounted "ON LINE" with a small increase in piping length. Unlike magnetic elbow, even a small metal inside a big ball of fiber can be separated. It minimizes the card wire damage and reduces fire generation.

4.8 Automatic Waste Evacuation System (Intermittent)

Highly efficient and reliable, the automatic waste evacuation system has a centralized electronic control and is equipped with highly sophisticated programmable logic controllers (PLCs), starters, pressure/electrical interlocks, and hooter arrangements. The compressed air requirement of the entire system is a mere 0.4 m³/h at 6 kg/cm². The system employs pneumatically operated, non-clog type flap valves of proven design to suck waste from individual suction points. The waste is then transported to the waste collection room by means of a powerful material handling ventilator.

At the waste collection room, the cotton dust/waste collected from various suction points is segregated and fed into compactors. Here the waste is separated from the transport air. These wastes are compacted and filled into gunny bags or containers and compacted to be baled directly. The residual air from the compactors is filtered through cellar/rotary filters.

4.9 Chute Feeding Systems

The chute feed operates on the two trunk system with continuous regulation of the web of tufts. In the feed duct, the material passes over the chute feeds where the tufts are separated by purely aerodynamic means to fill the upper material reserve trunk. The movable feed roll transfers the material from the upper trunk via a feed pan to the opening roll. The material is conveyed to the lower trunk and condensed by airstreams generated by the fan. A pressure transducer adjusts the speed of the feed roll in accordance with the varying pressure of the bottom reserve trunk. The exhaust air from the upper trunk is conveyed into a dust extraction duct whereas the exhaust air from the lower trunk is retrieved and recycled by the fan.

4.10 Summary

At the end of this chapter, user will understand the process involved in blowroom sequence. Also, the instruments and measuring devices used in the blowroom process were explained. Explanation of the different control systems used in the process gives a clear picture of working blowroom sequence.

References

1. https://www.slideshare.net/Farhanullahbaig/blowroom-process-in-spinning.
2. http://www.rieter.com/en/rikipedia/articles/fibre-preparation/the-blowroom/summary-of-the-process/the-blowroom-installation-as-asequence-of-machines/.
3. https://textlnfo.wordpress.com/2011/10/26/process-parameter-in-blow-room/.
4. https://www.lakshmimach.com/textile-machinery/range-of-products/card-sliver-system/blowroom-line/.
5. https://www.lakshmimach.com/textile-machinery/range-of-products/card-sliver-system/baleplucker-la23s/.
6. http://www.synchronics.co.in/items/blow-room-plc-eprom.aspx.
7. http://www.indiantextilejournal.com/articles/FAdetails.asp?id=2010.
8. https://www.truetzschler-spinning.de/fileadmin/user_upload/truetzschler_spinning/downloads/broschuere/Putzerei/Putzerei_EN.pdf.
9. http://ategroup.com/textile-engineering/spinning/product-family/blowroom/.
10. http://www.rieter.com/en/rikipedia/articles/fibre-preparation/the-blowroom/the-machines-comprising-ablowroom-installation/machines-for-fine-cleaning/rieter-uniflex-b60-fine-cleaner/.
11. http://ptj.com.pk/Web%202003/9-2003/art-sheikh.htm.
12. https://www.truetzschler-spinning.de/en/products/blow-room/detailed-information/cleaner-opener/.
13. https://textlnfo.wordpress.com/tag/blow-room/.
14. http://www.rieter.com/en/rikipedia/articles/fibre-preparation/the-blowroom/summary-of-the-process/basic-operations-in-the-blowroom/dust-removal/.

15. http://www.premier-1.com/news_content_4.html.
16. https://www.cotton.org/tech/physiology/cpt/fiberquality/upload/CPT-Oct90-REPOP.pdf.
17. http://www.cottoninfo.com.au/sites/default/files/img/Contamination.PDF.
18. http://www.stitextile.net/ProcessDesc.aspx.
19. http://www.indiantextilejournal.com/articles/FAdetails.asp?id=3622.
20. https://www.scribd.com/doc/75290560/Trutzschler-Blowroom.
21. http://pdf.directindustry.com/pdf/trutzschler/blow-room/133297-703879.html.
22. http://vetal.com/category/cotton-contamination-cleaning-machine/.
23. http://www.rieter.com/en/rikipedia/articles/fibre-preparation/the-blowroom/damage-prevention-and-fire-protection/metal-detection/electronic-metal-extractors/.
24. http://www.humidificationplant.com/automaticwaste-evacuation-systems.html.
25. http://textilecentre.blogspot.com/2013/07/chute-feeding-or-aero-feed-system-how_28.html.
26. http://www.rieter.com/en/rikipedia/articles/fibre-preparation/the-card/the-operating-zones-of-the-card/material-feed/the-two-piece-chute-system/.

5

Instrumentation and Control System in Carding

LEARNING OBJECTIVES

- To know the working of carding machine
- To identify the instruments and measuring devices used in carding machine
- To comprehend the concept of automation used in carding
- To understand the concept of sensors in carding

5.1 Introduction

Fundamentally, textile manufacturing is a labor-intensive operation. An improvement in the textile technology is mainly confined to increase in productivity, cost reduction, quality upgradation, and product development. One of the main upgradations in the textile industry is the introduction of various control systems in almost all the processes.

Advantages of control systems are:

- Smooth running, increased production, and improved quality
- Savings in plant cost
- Savings in raw material
- Increased safety
- Improved working conditions
- Saving in manpower
- Improved plant management and supervision

The proverbs, "The card is the heart of the spinning mill" and "Well carded is half spun," demonstrate the immense significance of carding for the final result of the spinning operation. In the carding machine (as shown in the Figure 5.1), the raw material is supplied to the feed arrangement consisting of feed roller and a feed plate. It pushes the material slowly into the lickerin. The portion of sheet projecting from the feed roller must be combed and opened to flocks by the taker in. These flocks are passed over the grid equipment and transferred to the main cylinder. In the moving parts like mote knives, grids, and carding segments, the material loses most of its impurities. Suction ducts carry away the waste. Then along the cylinder they penetrate into the flats and open up into individual fibers. The actual carding action occurs here. The underside of the carding cylinder is enclosed with grids. After the carding operation has been completed, the main cylinder carries the material to the doffer. The doffer combines the fiber into the web. A stripping device draws

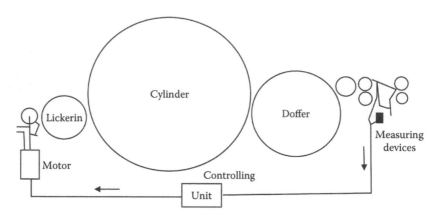

FIGURE 5.1
Long-term autoleveler.

the web from the doffer after the calendar roller has compressed the sliver to some extent, then the coiler deposits it in the can. The working roller, cylinder, and flats are provided with clothing.

5.1.1 Function of Carding

- Opening of flocks into individual fiber state
- Disentanglement of neps
- Fiber blending
- Fiber orientation
- Sliver formation
- Elimination of dust and short fiber

5.1.2 Autoleveling

A relatively high degree of evenness is required in the sliver. For various reasons, the card cannot always operate absolutely even. So, spinning mills are accordingly forced to use autoleveling equipment under highly varying circumstances. For export quality yarn, count variation is expected to meet 1.0%–2.5%. An extremely good count consistency is the most important condition for an excellent appearance of the woven or knitted end product. In order to achieve this, it is a must to control the evenness and coefficient of variation of count both at sliver as well as at yarn stage. Control of these cumulative factors can be taken care only by autolevelers. Autoleveling is usually performed by adjusting the feed rollers' speed.

5.1.2.1 Different Types of Autolevelers

5.1.2.1.1 Long-Term Autoleveling

Only the mean count value is autoleveled. A sensor in the delivery performs measuring. The pulses derived in this way are processed electronically so that the speed of the feed roller can be adapted to the delivered sliver weight via mechanical or electronic regulating devices (see Figure 5.1).

5.1.2.1.2 *Medium-Term Autoleveling*

In Zellweger equipment, a medium-term autoleveler is provided as an addition to auto levelers. Both mean value, and to a large extent the complete variance length spectrum is autoleveled. The measuring device is built into the protective cover above the doffer. The device measures reflection of infrared light from the fibers. After comparison with the set value, a different signal is generated and passed to an electronic regulating unit. This operates via a regulating drive to adjust the infeed speed of the card (Figure 5.2).

5.1.2.1.3 *Short-Term Autoleveling*

The regulating lengths of the short-term autoleveler are 10–25 cm. Its advantage is the correction of periodicities and other short disturbances (Figure 5.3).

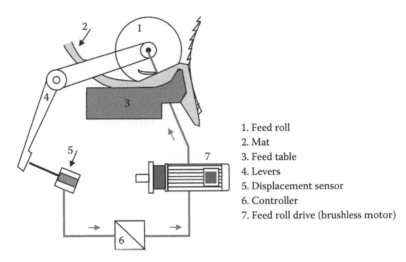

1. Feed roll
2. Mat
3. Feed table
4. Levers
5. Displacement sensor
6. Controller
7. Feed roll drive (brushless motor)

FIGURE 5.2
Medium-term autoleveler.

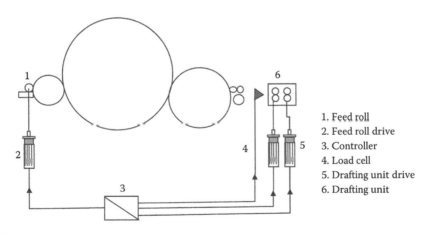

1. Feed roll
2. Feed roll drive
3. Controller
4. Load cell
5. Drafting unit drive
6. Drafting unit

FIGURE 5.3
Short-term autoleveler.

Regulation at the delivery: A measuring point is provided upstream from the drafting arrangement to sense the volume of the incoming sliver and transmit corresponding pulse signals to the electronic control unit. The control signal is passed to a regulating device, which adopts the speed of the delivery roller to the measured sliver volume.

Autoleveling in the infeed: The volume of the incoming batt is detected by the reference to the movement of feed plate relative to the feed roller. A signal corresponding to this volume is fed to the microcomputer that operates via a regulating gate transmission to continuously adapt to the speed of the feed roller to a predetermined set infeed volume. This is a feed plate measuring device used in Rieter C4 cards. Depending on the system of operations, autolevelers can also be classified as follows:

Open-loop system: In this system, sliver hank deviations are detected at the feed end itself, and the corrections are made near the delivery end. No monitoring is possible after the deviation has been corrected.

Closed-loop system: The sliver coming out at the delivery end is constantly monitored for any deviation from the standard value. Any deviation detected is fed to a correction system at the feed end, which corrects the draft. This corrected sliver is again monitored at the delivery and any deviation detected is fed to the feed correction system.

Combination system: This is the combination of long-term stability and accuracy of the closed-loop system and speed of the open-loop system. In this system, the detection and correction of faults are as that of open loop, and the corrected sliver is checked by a monitor. If the deviation exceeds the preset limit, the system stops the machine.

5.2 Measuring Devices

Active measuring device: The sliver to be measured is drawn through a specially designed measuring trumpet by a calendar roller. The movement of the material through the trumpet compresses the sliver, which in turn displaces the air trapped between the fibers. Hence as the thickness of the sliver varies, the air pressure inside the trumpet also varies proportionately, and this is measured by a pressure transducer. But it varies with the fiber fineness.

Strain gauge measuring device: The trumpet is used in coordination with the strain gauge mounted suitably at the entry point to measure the variation of its fineness.

Sensor measuring device: A transducer measures the sound waves along with the sliver and accordingly changes are made in the draft.

Tongue and groove measuring device: The sliver is drawn between a pair of rollers, one fixed and one movable. The material when passing through displaces the movable roller depending upon the thickness. This movement is converted into signals by the displacement transducer and is used to correct the variations of its fineness.

Feed roller measuring device: The volume of incoming material displaces the feed roller, and feed variations are sensed by measuring this movement in relation to the feed plate. The signal is sent to a microcomputer and necessary corrections are made.

Feed plate measuring device: Similar to the feed roller measuring device; the only difference is the feed plate is movable instead of a feed roller.

Optical measuring device: An optical device detects the variations in the cross section of the fiber layer on the cylinder in carding. Thus, micro controllers are used for closed-loop control, constant sliver count production, and shift counting. The doffer speed is continuously adjusted to the delivery by a frequency counter. A starting and breaking control prevents the sliver breaks.

5.3 Sensofeed

With the integral feed plate, sensofeed, the web is guided to the transfer point feed plate, feed roller, and lickerin via 10 spring elements. Each spring element exactly adjusts itself to the momentary mass of the web to be fed. The deflections of all the spring elements are processed to become one signal for short-wave regulation. Thus, it is possible to avoid thickness variation and to feed an even web to the lickerin system (Figures 5.4 and 5.5).

As an additional function, the integral feed plate sensofeed is equipped with a feed monitoring device for the card. Metal particles or thick places in the web are detected and result in an immediate stop of the machine. Through an electrical reversing motion, these foreign particles can be taken away before causing damage to the card.

The web feed consists of three opening and cleaning rolls in series arrangement. The tufts are open in a gentle way and to a higher degree than in the case of a conventional lickerin. The first roller runs at a slower speed than the conventional one and is equipped with needles. The second and third clothing rolls open the tufts further and form the web.

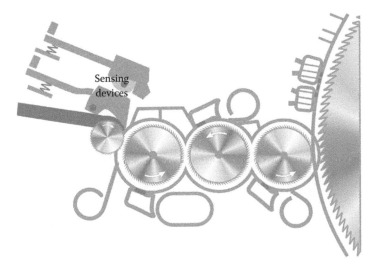

Sensing devices

FIGURE 5.4
Sensofeed and webfeed.

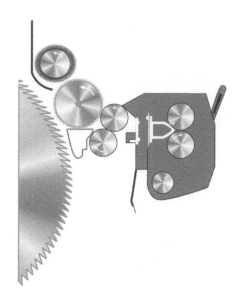

FIGURE 5.5
Doffer area with webfeed.

5.4 Flat Control

For an exact distance measuring, independent of the person carrying it out, the flat measuring system has been developed. An electronic measuring flat exactly measures the distance to the cylinder. The measurement can be made with the cylinder at rest or under running conditions (Figure 5.6).

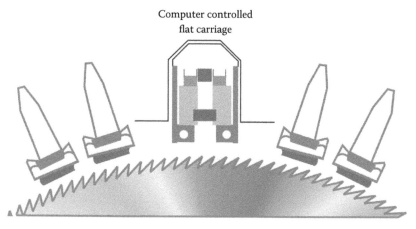

Computer controlled
flat carriage

FIGURE 5.6
Flat control (FCT).

5.5 Precision Flat Setting

The precision flat setting (PFS) is a new system; the flat setting is made centrally. At each side of the card, an actuator is turned manually or by a motor. That way, the distance of all the flat bars in working position toward the cylinder is widened or reduced. In the case of manual adjustment, the scale shows the actual setting in relation to the basic setting. In the case of adjustment by the motor, the position is chosen and shown on the display of the machine control. A procedure that formerly required the full attention of an experienced mechanic can now be done in a precise, reproducible way within a few seconds by pushing a button. New flexible bend disposes of six adjustments instead of four adjustment points. The deviation in the conventional system is 5/100; in the new system it is 1/100.

Various drive motors and transmission elements together form the drive solution for high speed card. It consists of three-phase AC motor drive cylinder and lickerin, feed rolls, and doffer. There are no gears or brushes, which makes the system maintenance free. The web doffing unit is equipped with a separate variable drive. Thus, it is possible to choose an ideal draft for any speed. Even during the time of start-up and slow down, the right draft is adjusted to the actual speed. This results in more even slivers, from the first to last meter in the can.

5.6 IGS Top Automatic Grinding System

This is installed permanently over the returning flats after the flat cleaning unit. In the flat return zone, the flats are pressed against the rotating grinding brush bristles containing an abrasive substance that partially sharpens the wire points. An electronic control system ensures that all the flats are grounded in one cycle, and then the mechanism goes out of action until the next cycle commences.

IGS classic: This contains a grinding stone, which moves automatically across the cylinder clothing during working. This process is performed 400 times during the planned life cycle of the clothing, and therefore the carding is always carried out with sharpened wire points.

5.7 Carding Drives

An inventor driven AC motor, drives lickerin, cylinder, and flats. The doffer and the web doffing system are equipped with a separate variable drive in order to operate with the ideal draft at any speed, ensuring the highest degree of sliver regularity. The motors for the feed roll and the drafting unit have no carbon brushes, and therefore maintenance is less.

5.8 Coiler Sliver Stop Motion

Whenever sliver breaks, the stop motion stops the machine, which is mounted on the coiler stand for web detection. The performance of the stop motion will be affected by fluff accumulation.

5.9 Thermistor Protection Unit

The thermistor protection unit is used for the protection of motors and is built in the motor coil. It trips the motor experiencing excessive heat due to loading by sensing the changes in resistance, thereby preventing the motor from burn out.

5.10 Safety Switches

The safety system positively monitors the cylinder speed to avoid cylinder chocking, belt slippage, or breakage. It observes the presence of the drive belt using an optical sensor and stops the feed when the belt slips or fails to safeguard the carding elements. In the carding unit, switches are provided for safety purposes.

5.11 On Card Filter

Crosrol on the card filter unit provides individual dust and waste extraction from the card. Dust and trash from the lickerin passes to one waste collection chamber. Cleaner waste from flat strips, the card delivery roller, and the coiler head passes to a second waste collection chamber. Two-way valves at each on card filter and at the central filter separate clean from dirty waste. Dirty waste is extracted from each of up to 10 cards in sequence. The central control unit then activates all the two-way valves simultaneously, and clean waste is drawn from each card in sequence. Limit switches prevent operation of card or extraction fan if on card filter is not in position.

5.12 Continuous Quality and Production Monitoring

The card is controlled by a microcomputer control card commander, which collects and monitors all the relevant data for the machine including can coiler, can changer, the regulation device, CCD and ICFD, and the integrated draw frame. All the data for production are entered at the operator's console of the card commander. The operator guidance is written in plain text in many languages. All data for a particular quality must be entered only once and can be stored in the set value memory. They can be called up anytime, can be changed according to the respective requirements, and can be stored again. The input and display units can be made depending upon the customer requirements.

The card commander collects all data and represents them on the display, such as delivery speed, total draft, can filling volume, production rate, sliver count, web thickness, pressure/bottom trunk, speeds (DKF opening roll, feed roll, lickerin, cylinder, doffer), flat speed, temperatures/electronics, network addresses, spectrograms, length variation diagrams, flat distance diagrams, nep evaluation, overload drives, metal intake detectors, thick place monitoring, negative pressure suction, safety devices, and sliver monitoring.

5.13 Card Manager

The card manager incorporates the electronic, closed loop, long-term autoleveler as standard equipment. Sliver variations are detected by ultra-reliable tongue and groove rollers, using the precision transducer at the taker in and adjusting the feed roller speed accordingly. This transducer is insensitive to the delivery speed and unaffected by fiber type or atmospheric changes. Sensing after correction point gives the closed-loop control stability. An optional closed-loop mid-term autoleveler augments the long-term system, attenuating sliver irregularities at wavelength as short as one meter. A vast range of card performance, control, and service data is instantly accessed through a simple key pad with only seven touch sensitive keys. It is very simple and has a pictorial display.

The card manager displays the updated graphical reports such as machine status; graphical indication of progress in the filling can; sliver linear density; CV% at selected wavelength; production rate and delivery speed; automatic reset at start of shift; clear pictorial indication of location of faults, plus suggested remedial action; and card clothing monitor. A serial connection link gives multiple choices of connection between machines, facility for remote programming, and direct interface to computer mill management systems.

5.14 Summary

The user will understand the working of carding machine and the working principle of different sensors and transducers used in the carding machine. The importance of automation is explained in the carding process.

References

1. http://w3.siemens.com/mcms/mc-solutions/en/mechanical-engineering/textile-machine/staple-fiber-spinning/carding-machine/pages/carding-machine.aspx.
2. http://www.rieter.com/en/rikipedia/articles/fibre-preparation/the-card/autoleveling-equipment/the-principle-of-short-term-autoleveling/autoleveling-in-the-infeed/.
3. http://textilecentre.blogspot.com/2016/11/autoleveller.html.
4. https://encrypted.google.com/patents/EP0708849B1?cl=en.
5. http://yarnmanufacturing.blogspot.in/2009/10/auto-leveler-autoleveller-in-carding.html.
6. http://www.cottonyarnmarket.net/OASMTP/Carding%20Machine%20-%20Operating%20Principle.pdf.
7. https://www.scribd.com/document/92646859/Carding-Machine.

8. http://www.rieter.com/en/rikipedia/articles/technology-ofshort-staplespinning/opening/carding/the-most-important-working-regions-in-carding/transfer-zone-at-the-doffer/.

9. https://www.omicsgroup.org/journals/card-setting-a-factor-for-controlling-sliver-quality-and-yarn-2165-8064-1000246.php?aid=71235.

10. http://www.microepsilon.com/applications/areas/Weg/Karde_Deckeleinstellung/?sLang=en.

11. http://www.directindustry.com/prod/trutzschler/product-133297-1865509.html.

12. https://www.scribd.com/document/218598989/TC11-EN-carding.

13. http://www.rieter.com/cz/rikipedia/articles/fibre-preparation/the-card/card-clothing/metallic-clothing/the-most-important-operating-parameters-of-the-clothing/.

14. https://www.truetzschler-spinning.de/fileadmin/user_upload/truetzschler_spinning/downloads/broschuere/Karde/Karde_EN.pdf.

15. http://www.rieter.com/en/rikipedia/articles/technology-ofshort-staple-spinning/opening/carding/the-most-important-working-regions-in-carding/carding-between-main-cylinder-and-flats/.

16. https://www.lakshmimach.com/textile-machinery/range-of-products/card-sliver-system/card-lc-300a-v3/.

17. http://www.rieter.com/en/rikipedia/articles/spinning-preparation/the-drawframe/monitoring-and-autoleveling/leveling-drawframes-with-closed-loop-control/.

18. http://vaibavsri.in/pdfdocs/1447501935.pdf.

19. http://textilelearner.blogspot.in/2012/01/carding-process-setting-of-carding.html.

20. http://www.rieter.com/en/machines-systems/products/fibre-preparation/c-70-card/.

6

Instrumentation and Control Systems in Draw Frame and Speed Frame

LEARNING OBJECTIVES

- To understand the working principle of the draw frame machine
- To identify the sensors and transducers used in the draw frame machine
- To understand the working of speed frame machine
- To comprehend the working of measuring devices used in speed frame
- To explain the importance of monitoring devices used in textile machinery

6.1 Control Systems in Draw Frame

6.1.1 Introduction

Drafting is the process of elongating a strand of fibers, with the intention of orienting the fibers in the direction of the strand and reducing its linear density. In a roller drafting system, the strand is passed through a series of sets of rollers, each successive set rotating at a surface velocity greater than that of the previous set. During drafting, the fibers must be moved relative to each other as uniformly as possible by overcoming the cohesive friction. Uniformity implies in this context that all fibers are controllably rearranged with a shift relative to each other equal to the degree of draft.

The parallel arrangement of fibers in the sliver is accomplished by the draw frame to which 4–8 card slivers are presented. The feed roller pair takes care of the slivers. The slivers then run into the drafting arrangement consisting of three pairs of rollers forming the back and the main draft zone. The drafted sliver coming out of the drafting zone is condensed by a trumpet and is placed in a can.

Parallelization of fibers, drafting and doubling, formation of sliver, hooks created in the card are straightened, and the autoleveler maintains absolute sliver fineness.

6.1.2 Autoleveler

The principle of open-loop control is used in this autoleveling. The thickness of the incoming sliver is sensed by a pair of tongue and groove rollers called scanning rollers. The displacement transducer converts this angular movement into voltages. A plate is connected to the rollers and is moved into the electromagnetic field of the transducer. This movement of the plates cuts the flux, and a voltage is induced. This signal is transferred to the electronic memory, which then transmits it to the set point stage with a certain delay. The correction delay is determined by the pulses, which can be set by first in first out (FIFO). The distance between the measuring rollers and the front draft zone is divided into 177–192 pulses.

FIFO is a register with the first measured variation stored in the first register, the second measured variation stored in the second register and so on. This system ensures the change in the draft takes place exactly when the corresponding deviating length of the sliver passes the main draft zone. The set point stage uses the measuring voltage and the machine speed, which are measured by two tachogenerators to calculate the speed of the servo drive. The middle roller is driven by a differential gear arrangement, which has a constant drive from the front roller and variable speed from the servo motor. The correction length is further reduced now for effective autoleveling. Speed of the variable speed motor is continuously measured, and if it does not correspond with the intended speed, the machine is switched off. Accurate leveling is ensured by the high dynamic servo drive, so the correction times are of the order of a few milliseconds and the correction lengths a few millimeters.

This system is comprised of an electronic unit, control drive using servo motor with control Tacho and differential gear with constant Tacho transducer, measuring trumpet, transformer, a power unit with brake resistance, and a capacitive measuring unit (Figure 6.1).

6.1.2.1 Principle of Measurement and Auto Levelling

In short-term variations, a control circuit with dead time and P-controller with delay and a capacitive measuring unit are capable of doing the job. In medium- and long-term variations, a closed control loop with dead time and PI- controller can be used. A measuring unit can be used on request.

The draw frame autoleveler was the first to achieve universal acceptance. In these systems, the measurement is made either at the input or the output sliver. The sliver count is automatically corrected with reference to the required value by variation of the draft. In this system, the sliver count at the output stage is measured. The autoleveler detects deviations, if any, from the pre-set nominal value. If deviations are determined, the actuator changes the input speed of the drafting rollers, changing the draft in proportion to the control signal. The Tacho-generator at the input roller and the actuator controller ensure

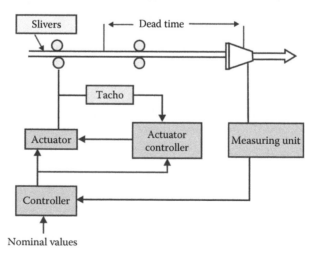

FIGURE 6.1
Basic principle—The ADC system for autoleveling of sliver input.

that the actuator corrects the count in proportion to the deviation. The degree of correction is automatically and continuously checked. This is a closed-loop system, which controls the mean sliver count and guarantees the best possible accuracy.

The length of the sliver over which correction of the count takes place is the "correction length." This corresponds to approximately 6 times the distance between the measuring point and the autoleveling correction point. The time the sliver takes to travel between these two points is called the "dead time."

6.1.2.2 Application Concept

The following Figure 6.2 shows the correct position of placing the autolevelers.

6.1.2.3 Position and Range of Correction

The main draft zone, pre-draft or auxiliary draw box should be maintained the speed range of maximum ±25% with respect to feed material variation. The principle of the open-loop control system is used. The accurate measurement of the mass variation is achieved by high frequency scanning at constant intervals. The signal reception is independent of the running in speed, and not affected by the ambient conditions and type of the fiber processed (Figure 6.2).

6.1.2.4 Storage of the Measured Values

The measured values are transmitted in phase with the sliver movement to an electronic memory. The correction value is immediately transmitted via a highly dynamic servo drive to the feed system. An epicycle gear superimposes the constant speed of the main motor with the variable speed of the servo drive. The signal in the correction computer actuates a change in the autoleveler draft as soon as the measured part of the sliver enters the drafting zone. Even with high delivery speed of 900 m/m, the correction speed is twice that of the quickest cross-sectional changes normally occurring in the sliver.

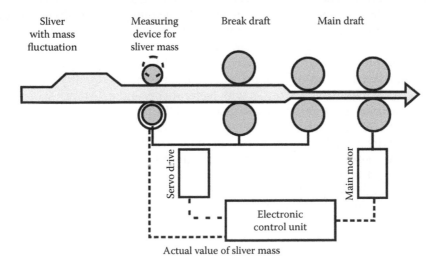

FIGURE 6.2
Open-loop autoleveler.

6.1.3 Sliver Data

Functions of sliver data are quality data monitoring, production monitoring, reports, reports and diagrams, and inputs and displays.

Zellweger Uster describe sliver data as an extremely useful online monitoring device that detects quality disturbances in the sliver with reference to the pre-set limits for sliver count, evenness, periodic faults, drafting wave faults, and number of thick places per 100 m. If these values are exceeded, the machine is either stopped or an alarm is actuated depending upon the degree of deviation. Sliver data also detects bottlenecks and disturbances in the production processes well in time and provides details of the causes of machine stoppages, which are removed by adopting a rapid elimination procedure.

Quality data monitoring shows the sliver count, evenness (CV%), spectrogram of drafting waves, and periodic variations and stops the machine if the set limit value is exceeded. Production monitoring monitors the production, efficiency of the machine, and the number of machine stoppages per cause. All these data are per machine, article, and group.

Count diagram, CV% diagram, spectrograms, stop reports, disturbance reports, summaries, and long-term reports for trend recognition are available in print or on video. They display the most important quality and production parameters at the machine as well as the central input of limit values and settings of measuring unit and input of control spectrograms. They also provide causes of machine stoppages at the machine terminal.

6.1.4 Sliver Alarm

The monitoring system is comprised of a machine station, operating terminal, production sensor, sliver measuring unit, and alarm lamp. Main functions are monitoring of sliver count and evenness, optical indication at the operating terminal and at the alarm lamp, and stopping the machine if the limit value is exceeded.

6.1.4.1 Principle of Measurement

The sliver alarm is a fiber press measuring system along with automatic, intermittent measuring of unit cleaning. The sliver alarm can be used at cards, draw frame, and comber.

An automatic draw frame autoleveler is used for leveling out the sliver weight variations. This system consists of a micro terminal, control unit, transformer, ballast resistor, control drive using servo motor, Tacho transducer distance sensor, T and G measuring unit for the sliver, autoleveler, and FP measuring trumpet for sliver monitoring. Uster sliver control is conceived with an open loop and additional online quality monitoring. The tongue and groove measuring the organ with feeler rollers and a distance sensor serves for the precise measurement of the mass variation of all the feed slivers. The measured value is compared with the predetermined set value in the USC unit. The difference in the value is amplified and delayed. The input speed and other speeds are determined by the Tachos. Through this the control to the speed is set and adjusted.

The controlling element is comprised of differential gears, a servo motor comprised of a Tacho-generator and rotor position sensor, and a brake and brushless servo motor.

The power brake unit consists of a brake, ballast control, and four quadrant servo amplifiers. The fiber press measuring trumpet, together with the pre-amplifier, serves the purpose of a rapid and precise measurement of the cross-section of slivers before the calendar rollers. The FP measuring trumpet delivers the actual signal for the online quality monitoring of the drawn slivers.

Collection and monitoring of values on the micro terminals are production in meters and kilograms, stopped times, efficiency, production speed, can changes, count deviation A%, coefficient of variation CV%, CV% at various lengths such as 1, 3, 5 m, and so on, and CVL%.

The sliver alarm is connected to the sliver data, which is shown in the Figure 6.2.

It is possible to connect the sliver data to the polylink or to the sliver alarm at any time and without much investment. There is a possibility of testing for leaks in measuring connections at Ucc-L, Ucc-S, Uster Adc-e, and uster tex alarm using uster leak tester.

6.1.5 Sliver Watch

6.1.5.1 Contamination Detection on Draw Frames and Lappers

Sliver Watch is used to detect the contamination present in the fiber. This detection system is installed in the creel of draw frames handling the first step of drawing or in the creel of the lapper. The purpose is to stop the draw frame or the lapper if a contaminated sliver is detected so that the operator can eliminate the contamination, thus avoiding that second step drawing and spinning spreading the contamination over many meters of yarn.

6.1.5.2 Yarning Principle of Sliver Watch

The sliver passes through a diffuse transparent guide. Transmitters or LEDs illuminate the sealed inside of the detector. The receivers sense the amount of light in the spherical inside of the sensor. Foreign material absorbs a fraction of the light emitted by the LEDs; this increases the signal and allows for detection. Then light is absorbed from all angles, allowing an uninterrupted view of the contamination, regardless of its position in the sliver. The shape of the sliver guide does not compress the sliver. The sliver with difference in color from the color of the fiber material can be considered as foreign fiber in the sliver. The absence of a sliver, or a non-moving sliver in the detector, means that the detector replaces the sliver breakage stop motion in the creel of the draw frame.

The length of the contamination is the main setting and it is set to minimum length setting to avoid irrelevant stops. Filter settings are available to detect short, intensive contaminants. Sensitivity settings are based on a scale from 5–20.

All sensors are wired to a central machine unit, which has to be present on each machine. Sliver Watch has been tested for speeds up to 1200 m/min. A central unit with full color display and sealed key board allows user-friendly and menu-driven settings to define the contaminants that have to be removed. A lamp tree serves as a warning system. The stop position is such that the contamination is always at the same location, ready for the operator to take it out.

6.1.5.3 Sliver Watch on Heather Yarns

Heather yarns (blends of cotton with 1%, 3%, 5%, or more black polyester) can be monitored by Sliver Watch. Not enough black or too much black polyester in the sliver makes Sliver Watch stop the draw frame. As the position of the PE/CO blend cans versus the pure cotton cans in the creel is important, Sliver Watch will stop the machine if a can position is not respected by the operator.

6.1.5.4 *Production and Quality Data*

Production data, such as machine efficiency, produced kg, produced sliver length, number of can changes, number of production stops, as well as quality stops, are available in real time and on shift base. Statistic evaluation includes histogram of defects, number of contaminants per shift, number of sliver breaks, and so on.

Monitoring on cards: The installation of a Sliver Watch on cards allows the mill to keep track of the level of contamination in the production. The purpose is to monitor the contamination index of the sliver can. The contamination index of the can is an expression that relates to the number of contaminants per 1000 m of sliver.

6.1.6 Sliver Monitoring

This system supervises the sliver thickness at the output of the draw frame. This compares the value with the nominal value and stops the machine when the set limit is overstepped. The following parameters are supervised: count deviation A% limits, CV% limits, and spectogram. data output includes stops and stop times covering the preceding 8 h, efficiency values and updated values of CV%, and CVL at 1, 3, and 5 m.

6.1.7 Automation Material Transport

With today's handling equipment, robots and transport devices designed with microprocessor controlled noncontacting sensors and detection systems are used. Material transport between two machines has become fully mechanized. In the drawing section, breaker and finisher draw frames are connected by CAN link or a can-connect system.

6.1.7.1 *Cubican*

This automatic can changing mechanism is used for rectangular cans. It can hold about 50% more slivers than regular round cans. The can filling volume is also sensed by the bottom controller, which is set based on the length time basis. As the sliver is delivered into the can, the bottom controller, which is in raised position initially, is lowered according to the length of the sliver falling.

> *Cubican sliver deposit*: Filling of the Cubican through a special coiler, back and forth movement of the Cubican under the coiler plate during the filling procedure.
>
> *Cubican can changer*: The empty cans are taken from the can magazine and conveyed to the coiler of the RSB.

6.1.7.2 *CANlog—Handling System for Cans on Trolleys*

6.1.7.2.1 *Automatic Can Feeding from Trolley*

The features of the system include automatic trolley traverse, can supply on trolley, waiting position for empty can, and full can trolley.

The benefits of the system are:

- Labor saving about 10%
- No manual can lifting

- Reliable can feeding
- Systematic can buffer directly at draw frame

6.1.7.3 CAN Link—Draw Frame Interlinking System

Following are the functions and benefits of using automatic CAN link-draw frame interlinking system:

Functions

- Filling of cans on first passage
- Automatic pushing of full cans to spare row aside
- After feed cans run empty, full spare cans are pushed into feed position
- Empty cans are pushed to empty can row
- Automatic transport of empty cans to can change of first passage

Benefits

- Eliminates manual can transport
- Increases efficiency of draw frame set
- Reduces down time for change of feed cans
- Reduces number of required cans
- Avoids mix up of raw materials
- Provides clear production lines

6.1.7.4 Cannyone

Cannyone is an automatic can transport system individually designed for draw frame to draw frame and draw frame to speed frame. With this system, several process stages can be flexibly connected and independently programmed. The assignment takes place via a central transport computer. This computer monitors and controls the entire system and clearly displays the respective status on the screen. Changes in the machinery assignments take place within seconds via the computer.

6.1.8 Sliver Expert System

Rieter has developed the Sliver Professional offline expert system to provide easy, rapid technological support. The unique software offers two valuable functions: machine setting recommendations and assistance with fault finding.

6.1.8.1 Machine Setting Recommendations

The Sliver Professional is easy to operate following the input of raw material data and feed/delivery sliver weights. Sliver Professional provides a complete recommendation for setting draw frame parameters. The recommended settings together with the operator's own optimization can be saved in the PC. This enables the user to manage his application—specific draw frame setting data. The great benefit of Sliver professional is the prevention of widely incorrect settings.

6.1.8.2 Rapid Elimination of Faults

A further valuable function provided by Sliver Professional is an analytical program for periodic spectrogram faults, or so-called stacks. After manual input of the wavelength of the spectrogram fault, the software defines all the machine components that could be responsible for stacks. The results are displayed in plain text and by a drive diagram with color-coded machine components.

Solid 3D displays also provide information on the position of the components defined. This is not to say that all the marked components need to be replaced, but the potential sources of error should be checked thoroughly and the components replaced if necessary. Using Sliver Professional means that locating the causes of periodic faults is not confined to specialists. It speeds up the elimination of faults and thus increases the availability of the draw frame.

6.1.9 Integrated Draw Frame

The integrated draw frame is driven by three servo motors. Motor 1 drives feed and center cylinder of drafting system. Motor 2 drives the delivery cylinder, delivery roll, and sliver coiling ring. Motor 3 drives the rotary can plate. All three drives are synchronized with the card. The card sliver is scanned before entering the drafting system by the measuring funnel. The second measuring funnel after the drafting system permanently checks the sliver.

6.2 Control Systems in Comber

Within the overall spinning process, the comber serves to improve the raw material. It is mainly used to get fine yarns. The lap from the previous machine is fed into the feed roller to enter the nippers. The clamped fibers of the nippers are swept by the combing segment mounted on the combing cylinder and the opening of the nipper allows the fibers to move toward the detaching rollers. The detaching rollers return part of the previously drawn off stock by means of a reverse rotation, so that the web protrudes from the back of the detaching device. The new fibers coming toward the roller are placed on the returned web and the top comb entry combs the trailing part of the comb. It is then pulled out of the previous nipper, and piecing occurs. In this way, the combing cycle continues. The web is then collected by a trumpet, and the calendar roller condenses the sliver. The slivers then enter the drafting arrangement. Six or eight slivers are drawn in the drafting system together to give single or double delivery. The functions of the comber are to:

- Elimination of the short fibers
- Elimination of remaining impurities
- Elimination of a large proportion of neps
- Formations of sliver having maximum possible evenness

6.2.1 Sliver Lap Machine

Sliver stop motion model: This model consists of separate units on either side of the table for sensing sliver break, immediate stoppage after an end break, and easy piecing as sliver is stopped well before the rollers.

Extended features are:

- Recording and processing of production data
- Uster sliver data
- Data analyzing
- Languages selection

Regulated lap loading

- The regulated automatic lap loading safeguards the ideal pressure of the lap on the lap rollers during the complete build up cycle

6.2.2 Ribbon Lap Machine

6.2.2.1 Ribbon Breakage Photo Cell

The ribbon breakage photo cell is used to stop the machine whenever there is a break of lap on the feed end or breakage of fleece on the table end. It consists of one control unit and two sensor units: one for the feed end and other for the lap end and empty beam stop motion. Top roller lapping is avoided, as the machine stops immediately.

The raw material from the card is unsuitable for combing, as it needs further disposition of fibers within the sheet. The suitable preparatory machines are sliver lap and ribbon lap machines or draw frames, or a combination of the two.

6.2.2.2 Comber Photo Control for Entry Lamp

This unit consists of one control unit and eight sensors for sensing the empty lap. By fixing this unit, the continuity of the lap is maintained. One of the advantages is that the lap is stopped well before the contact rollers, so piecing is easier. The unit also includes non-contact detectors, a dual timer, and a cyclic timer, where the machine is switched to slow speed for the pre-selected time interval. Machine run from slow speed to full speed is also controlled through this unit. The machine need not be switched off for cleaning. Timely switching to slow speed maintains the cleanliness of the machine. Waste removal can be done in programmed method.

6.2.3 Computer Aided Top Performance

- An important step in the direction of the total fiber control in the combing process has been taken with CAPD.
- A thorough knowledge of the kinematics and engineering interactions and relationships combined with computer aided process development were prerequisites for achieving nip rates of up to 400 per min.
- *Auto lap piecer*: The run out lap spool falls automatically beneath the lap rollers and the lap is torn off. The remaining lap end is pieced automatically with the full lap supplied from SLM. Thus, the piecing is ensured and machine efficiency is improved.
- *Spool cleaner:* Spool cleaner automatically strips the remaining fleece on spools of comber preparatory machines and combers. The stripped spools are then automatically returned to the magazine of SLM or RLM.

6.2.4 Automatic Lap Transport

Automated processes call for gentle but reliable lap transport. Manual lap transport is not feasible for the comber. The automatic lap transport system SERVOlap meets the requirements of lap positioning best. The overhead SERVO lap consists of system transport 8 laps at the time from the combing preparation UNILAP to the 3 UNIlap and 18 combers set up in line. Comber and ROBOlap are controlled and monitored by a state of the art control system, designed for easy operation, as well as by unskilled labor.

In modern combers, a modular system based on overhead transport with suspended rail does the package handling automatically. The transport system is capable of conveying 8 laps from the UNIlap to the comber automatically. The empty tubes are automatically returned to the UNIlap. Transportation of roving packages to the ring frame and the return of empty bobbins to the speed frame are carried out by snake or Electro jet system.

6.2.4.1 SERVOlap

Operator effort is reduced to supervisory functions by the SERVOlap automated lap transport system and the ROBOlap fully automatic lap changer and batt piecer. Machine efficiency is thus boosted by a further 2%–3%.

Detach batt and attach machine-side end. Pneumatic tube cleaning and removal of batt remains via a separate suction system. The step by step operation of the SERVOlap is:

- Convey empty tubes to tube store and tilt full laps into working position
- Prepare batt start (on lap side) for the automatic piecing operation
- Pneumatic release of the lap end by means of the Aero-pic system and mechanical joining of the two batt ends

6.3 Speed Frame Controls

6.3.1 Introduction

The sliver from the draw frame or from the comber is present to the roving frame. When the sliver first enters the drafting arrangement, the drafting arrangement attenuates the sliver with a draft of between 5 and 20. The delivered strand is passed through a flyer, which imparts twist to it. The turns are created by the rotating flyer and are transmitted into the unsupported length between the flyer and the delivery from the drafting zone. The flyers form a part of the driven spindle and are rotated with the spindle. The up and down movement of the flyer helps in building up the roving strand on the bobbin, which is placed on the spindle.

Thus winding of roving on the bobbin occurs, which is the end product of roving frame after that the process will be moved on to the ring frame. The process for the ring frame is:

- Attenuation of the sliver
- Twist insertion
- Winding of roving in a package suitable for frame

The requirements placed on spinning mills, and also the degree of automation most suitable for their needs, essentially depend on their product mix and the type of local economy in which they operate. In countries where labor costs are low, the amortization period required for fully automated systems is often very long.

The basic machine with fully automatic doffer can be linked with the bobbin transfer system and then with subsequent ring spinning machines.

Higher productivity and efficiency means less number of speed frames for particular production requirement, less time loss in doffing, and less requirement of labor, less requirement of spares and lubricants, and minimum down time for maintenance. Ten lots of data can be stored and retrieved at any time, which means less down time for lot changing and no count deviations between the front and rear flyer row as the roving enters the heads of both flyer rows at exactly the same angle.

The main feature of this machine is the automatic flyer speed control device. The computer automatically establishes the correct speed progression pattern and adjusts the flyer speed via an inverter drive system to maintain a constant centrifugal force on the roving. A cone drum less mechanism is used for building the bobbins to reduce vibrations while running at a high speed.

A microprocessor control system enables the bobbin rail to return to the set position once the limit length has been reached. This reduces the average surplus winding on each bobbin from 40 mm to almost nil, eliminating material wastage from bobbin overfills and thus improving roving yield and machine efficiency. Two separate motors are used to drive roving and winding operations. This allows the winding to be halted slightly before the roving bobbins are full to provide extra length of roving between the front roller and flyer top. This prevents excess tension during restarting of the machine. For easy restarting, the machine also has an automatic roving end positioning device. When the bobbin rail is raised to the restart position after doffing and new bobbins are inserted, the machine restarts at a low speed for a certain period. The roving ends are automatically positioned and pressed against the bobbin where magic tapes catch the ends to begin winding.

6.3.2 Multimotor Drive System

The key to the success of any spinning mill operation is its ability to respond quickly to changing product requirements and production conditions. Minimal lot change times and the simple re-instatement of previously applied settings are required. Separate electrical drives for the drafting system, flyers, bobbins, and bobbin rail, all coordinated by the central machine control system, have replaced conventional machine components such as cone gearing, builder motion, differential gear unit, and mechanical change gears. Lot changes are performed at the operating panel of the roving frame. This facility offers a data archive able to manage up to 10 data batch records. The 4-axis multimotor drive system eliminates unnecessary power losses in mechanical transmission systems. There is also a substantial reduction in noise emission.

- Four-axis multimotor drive systems with microprocessor control enable easy setting of the roving frame for different materials and counts.
- The data for a particular lot can be stored and recalled at any time.
- The drafting system, the flyers, the bobbins, and the bobbin rail are driven by four independent motors integrated with an inverter.
- All the drives are coordinated via a central machine control system.
- The conventional machine elements such as cone drum, builder motion, differential gearing, and mechanical change gears are totally eliminated.

- The suspended flyer drive and bobbin drives are through toothed belts.
- After every doffing process, the roving is deposited automatically on the bobbins for initial lapping.

6.3.3 Automatic Winding Tension Compensating Device

The tension of the roving is detected by the optical sensor, and the feed rate of the cone belt is compensated automatically by a computer so that the roving can be wound with most suitable tension. It is even more effective when combined with the flyer speed control device with inverter, which gives the most efficient running pattern for different types of fibers.

6.3.3.1 Pneumostop Unit

This unit stops the machine whenever there is a roving break. The broken roving is carried along with the suction into the pneumafil and reaches the filter box. The pneumostop is fixed in the filter box so that whenever the roving interrupts the optical screen of the pneumostop, the machine is stopped. Productivity is increased, and top roller lapping is totally eliminated. Fly waste scattering is avoided as the machine is stopped immediately, and so efficiency is increased.

6.3.3.2 Photo Master or Back Creel Stop Motion Unit

This unit operates for sliver break at any point on the creel by stopping the machine immediately. Piecing of the sliver is easy as the machine stops before the sliver enters the draw frame.

6.3.3.3 Roving Stop Motion

This unit is used for sensing the roving break for suspended flyer speed frames. The sensors are aligned so that any build-up of roving yarns on the false twister block the beam path to stop the machine. This results in an immediate stoppage of the machine for a roving break.

This new generation machine applies intellectualized control technology and has high-level electromechanical integration. Driven separately by seven motors, process changeable gears are not required, and a simple mechanical structure provides convenient adjustment and maintenance. The special tension control model ensures stable-roving tension, effectively restrains yarn faults at machine start and stop, and guarantees the stability of roving quality. Applying patented power cut-off technology prevents yarn breakage or faults happening when unexpected disconnection occurs. Adopting a color LCD pro-face as the man–machine interface, all technical data can be set in real-time. Very simple and convenient to operate, this machine's display shows operational status and fault warnings.

6.3.3.3.1 Other Controls

Door safety switch: The main door head stock and the back doors are safeguarded with limit switches. As soon as the door is opened, the machine is switched off.

Pressure monitoring: The base pressure of the drafting arrangement and the air pressure for the bobbin shield control are individually monitored by diaphragm pressure switches. If the air pressure drops to the value adjusted on the switch, the machine is stopped. It should be possible to start the machine unless the top rollers are under pressure. This is ensured on the machine by the provision of the minimum pressure switch.

Over running safety device: If the bobbin rail reversing control switch fails to function during reverse of the bobbin rail, micro switches provide for stopping the machine. Even if the micro switch fails, the two limit switches stop the machine. This prevents the bobbin carriage from running beyond the intended reversing point and stops the machine. These two are overrun safety limit switches, which prevent damage to machinery parts and breaking of ends.

6.3.3.4 Roving Eye

This is an electronic eye used to indicate any roving breakage. It is based on the photoelectric principle using infrared light. In the head stock side of each roving spindle, there is a sensor whose intensity is affected when a roving break is indicated by an alarm. Different lights are used for indication.

6.3.4 Monitoring

In the creel zone, electronic eyes are suitably located to continuously monitor the movement of the sliver and take care of stopping the machine whenever breakage occurs. Suspended type flyers ensure an equal roving angle for uniform twist flow, and automatic roving tension controls precise bobbin build by a digital encoder.

This PLC continuously monitors the drives and machine elements, and adjust the speeds dynamically for optimum performance. The system also controls the flyer and bobbin speed. It displays:

Flyer speed instantaneous/average, delivery rate in m/m, twist in TPI/TPM, run time

Doff time, idle time, machine efficiency, total cumulative production in Kgs, production/shift

Machine data

6.4 Automatic Doffing

The integrated, fully automatic doffer RoWeMat has been designed to ensure easy automation within the context of an integrated overall concept. This means that all the automation components have been incorporated within the machine. The doffing operation and storage of both the full packages and the empty tubes are performed by the machine without the need for additional external components. The benefit of this arrangement lies in the very short doffing times achieved. The doffing process and the machine restart operation are performed fully automatically. The full roving bobbins are stored in the machine so that production can restart immediately. The operating area remains completely unaffected, and no operative is required for performing the doffing operation.

- Once the preset roving length has been achieved, the Zinser RoWeMat is automatically shut down.
- The bobbin rail holding the full packages moves downward, causing the rovings to be separated. To ensure safe handling of the roving, the bobbin can be provided with a defined roving tail overwind.

- The integrated rail segments are joined together, and the trolley trains carrying the empty tubes enter the flyer section to perform the doffing process.
- The bobbin rail moves upward in order to suspend the full packages in the train.
- The bobbin rail then moves downward again to extract the spindles from the bobbins.
- The trolley train then moves half a gauge length so that the empty tubes are located over the spindles.
- The bobbin rail moves upward to pick up the empty tubes.
- The bobbin rail then moves down once again to allow the trolley trains with the bobbins to leave the flyer section.

6.5 Summary

At the completion of the chapter, the user will understand the working of draw frame and speed frame machines. Detailed explanation of sensors, transducers, and measuring devices used in these machines were given. The importance of the monitoring various parameters used in the machines were explained.

References

1. http://www.rieter.com/en/rikipedia/articles/fibre-preparation/the-card/autoleveling-equipment/the-principle-of-short-term-autoleveling/autoleveling-in-the-infeed/.
2. http://www.fibre2fashion.com/industry-article/165/laser-based-autoleveller-draw-frame-concept-and-review?page=3.
3. http://www.freepatentsonline.com/5463556.html.
4. http://www.rieter.com/en/rikipedia/articles/spinning-preparation/the-drawframe/operating-principle/.
5. http://www.rieter.com/en/machines-systems/products/spinning-preparation/sb-d-22-draw-frame/.
6. http://www.rieter.com/cz/rikipedia/articles/spinning-preparation/the-drawframe/monitoring-and-autoleveling/lveling-drawframes-with-open-loop-control/.
7. http://textilelearner.blogspot.in/2011/07/draw-frame-actions-involved-in-draw_7896.html.
8. https://www.tradeindia.com/fp1421844/Silver-Watch-For-Carding-Draw-Frame-And-Comber-Machine.html.
9. https://www.uster.com/fileadmin/customer/Instruments/Yarn_Testing/USTER_Sliverguard/SP_Sliverguard.pdf.
10. http://www.rieter.com/cz/rikipedia/articles/spinning-preparation/the-drawframe/operating-devices/coiling/sliver-coiling/.
11. http://gpktt.weebly.com/uploads/3/7/5/5/37553617/draw_frame1.docx.
12. http://www.rieter.com/en/after-sales/technology-parts/draw-frame/.
13. http://www.rieter.com/en/rikipedia/articles/rotor-spinning/machine-and-transport-automation/transport-automation-in-the-rotor-spinning-mill/can-transport-between-the-drawframe-and-therotor-spinning-machine/.

14. http://www.rieter.com/en/rikipedia/articles/spinning-preparation/the-drawframe/logistics/.
15. http://www.ptj.com.pk/Web%202004/02-2004/rieter.html.
16. http://fs-server.uni-mb.si/si/inst/itkp/lttkt/izpiti-zs/tmp%20-%20otm/raztezalnik.pdf.
17. http://www.textiletoday.com.bd/recent-developments-in-draw-frame/.
18. https://www.truetzschler-spinning.de/en/products/combing/detailed-information/draw-frame-td-8/.
19. http://www.rieter.com/en/rikipedia/articles/technology-ofshort-staple-spinning/attenuation-draft/the-draft-of-the-drafting-arrangement/the-drafting-operation/.
20. http://www.directindustry.com/prod/rieter/product-172425-1763645.html.
21. http://www.directindustry.com/prod/rieter/product-172425-1763637.html.
22. http://www.rieter.com/en/machines-systems/products/spinning-preparation/e-26-servolap/.
23. https://www.lakshmimach.com/textile-machinery/range-of-products/ring-spinning-system/speed-frame-lf1400a/.
24. https://www.bibus.ro/fileadmin/editors/countries/birom/Yaskawa/documents/DRIVES_AND_CONTROLLERS_FOR_TEXTILE_MACHINERY.pdf.
25. https://www.d3engineering.com/technologies/motor-control.
26. http://www.rieter.com/en/machines-systems/products/spinning-preparation/f-16-f-36-roving-frame/.
27. http://www.rieter.com/en/rikipedia/articles/spinning-preparation/the-roving-frame/accessories/monitoring-devices/roving-tension-monitoring/.
28. https://www.lakshmimach.com/textile-machinery/range-of-products/ring-spinning-system/speed-frame-lfs1660v/.
29. https://textilestudycenter.blogspot.in/2015/09/speed-frame-and-its-function_2.html.
30. http://www.rieter.com/cz/rikipedia/articles/spinning-preparation/the-roving-frame/automation/doffing/manual-doffing/.

7

Instrumentation and Control Systems in Ring and Rotor Spinning

LEARNING OBJECTIVES

- To know the working of ring and rotor spinning
- To understand the working of instruments and sensors used in ring and rotor spinning machines
- To identify the control systems used in ring and rotor spinning
- To be familiar with the process of automation used in these machines

7.1 Control Systems in Ring Spinning

Ring spinning is the process of converting roving into yarn. It is a long, rectangular, double-sided frame in which the upper part is designed to hold the bobbins, the middle part for drafting, and the lower part gives the necessary amount of twist and winds yarn on the cops. The objectives of the Ring spinning process are:

- To attenuate the roving until the required fineness is achieved
- To add strength to the fiber strand by twisting it
- To wind up the resulting yarn in a form suitable for storage, transportation, and further processing

7.1.1 Existing Manual Operations

The requirements of the manual operation are:

- To creel the roving bobbins
- Placing optimum setting as per the count and material processed
- To change the required change wheels
- To change the traveler according to the count
- To change aprons, spacers, and other spinning accessories according to needs
- To change the building mechanism arrangement
- To piece the yarn at the time of end breaks
- To take care of all the maintenance activities

7.1.1.1 Need to Automate

Automation is the replacement of human activity by mechanization and electronic control; in economic terms, this appears as substitution of capital for wage costs. Automation is therefore a rational step when:

- A great deal of manual work is necessary
- The manual work is economically unfavorable
- There is a shortage of personnel
- The human operator represents a source of error that must be eliminated

7.1.2 The Possibilities for Automation

- Transporting roving bobbins to the ring frame. This would be a very useful step of installations enabling the usage of area economically available.
- Exchanging roving bobbins in the creel of the ring spinning machine. This would also be as useful step but raises difficult problems; the first installation for this purpose is also now becoming available.
- Threading rovings and piecing roving breaks
- Taking up and removing waste
- Piecing end breaks
- Stopping roving when an end breaks
- Doffing
- Cleaning
- Transporting cops to the winder
- Monitoring the machine
- Monitoring production
- Monitoring quality

7.1.2.1 Drive of Highest Operational Reliability-PLCV

Optimal and stepless control of the spindle speed by means of PLC variator control system (optionally inverter control system).

7.1.2.2 Electrical Controls

A PLC is used to control the complete sequence of the machine (m/c). A user-friendly, four-line display with keyboard is used as a man–m/c interface. The PLC control enables easy fault diagnoses through fault code numbers. MCB's are used instead of conventional fuses. The MCBs are more reliable and easily maintained. A separate motor protection circuit breaker and a power contractor are built into the panel for OHTC control. The speed control runs smoothly with minimum speed/length entries. The closed-loop control ensures consistent speed patterns for any variation, in mains frequency. The advantages of the PLC control are

- Ten sets of speed pattern program can be stored in memory. It is also possible to copy b/n program and edit one or two values in the new program, which saves time while changing the values.

- Speed pattern program can be copied from one m/c to another without a computer.
- It is also possible to network the m/c's to a central computer.
- Shift report is available for seven shifts along with a cumulative report.
- Doff report is available for four doffs.
- Proximity switches are used instead of limit switches in certain areas for more reliability and user friendliness.
- The direction of spindle rotation is controlled through contractors instead of a rotary switch for easy maintenance.
- Ring frame with smooth speed control device.
- Ring frame with inverter control system.

7.1.2.3 Other Automations

- High performance pneumatic guide arm with automatic, centralized pressure regulation and release assures top yarn quality.
- The programmable delayed drafting in the system accomplishes the end breaks to very minimum during starting of the spinning (Spg) frame.

7.1.3 Individual Spindle Monitoring

The individual spindle monitoring (ISM) sensor helps achieve higher yarn quality because slipper spindles are signaled and ring-spun yarns of insufficient twist can be systematically separated out. Accurate evaluation of the end-break frequency for each spindle brings problem spindles to light and leads to rapid elimination of errors and thus to improved yarn quality. Yarn breakages are signaled directly to the operators via the three-level guiding system (machine–section–spindle). Therefore, routine checks are a thing of the past. The operator is easily able to identify the side of the machine and the section where intervention is required. A light signal indicates, directly at the spindle, whether the problem is a yarn break or a slipper spindle. Data are archived automatically by SPIDER web and will allow long-term and successful analysis.

7.1.3.1 Three-Level Operator Guiding System

The ISM features a three-level operator guiding system. This system leads the operator v to the place of the event and will therefore help to reduce the number of routine controls. This will improve the efficiency on the ring spinning machine.

1. *Level: Machine*: Each machine side is equipped with two lamps on head and foot stock. These lamps are visible from a distance across the floor and will guide the operator straight to the right machine and right machine side. Their function is to provide permanent light on foot and head stock if the machine side needs attention.
2. *Level: Section*: Every section is equipped with an extra bright shining light emitting diode (LED), which is visible along the whole machine. The permanent lightning section LED indicates a yarn break in the corresponding section.

3. *Level: Spindle*: Finally, each spindle is equipped with its own LED. This indicates events on the particular spinning position. This allows the operator to aim directly at the spinning position. This permanent light on the spindle will indicate a yarn break.

A flashing spindle LED indicates a slipper spindle (a particular spindle is running under a defined limit). If the spindle LED is off, the spinning position is running properly.

7.1.4 Piecing Devices

Traveling piecing carriages are provided on rails fitted to the machine. The piecing carriage has to perform mechanically the same complicated operations as the operative performs manually:

- Watch for broken ends while patrolling the spindles.
- Stop at the right spinning position.
- Take up an exact location relative to the spindle.
- Search for the broken end.
- Stop the spindle.
- Bring the traveler into suitable position for threading up.
- Thread the yarn through the traveler.
- Release the spindle.
- Piece the yarn with fiber strand issuing from the front rollers.

During its patrolling movement along the ring spinning m/c, the FIL-A-MAT monitors each individual position for an end down. If a yarn is present, the process will be continued and the next position is checked. If a broken end is detected, the device stops in front of the spindle, swings out a frame carrying the operating elements, and centers it exactly on the spindle bearing. The spindle is braked by the special mechanical action. A further operating unit is lowered onto the ring rail and follows its movements during the subsequent operations. The broken end is blown from the cop upward into the trumpet-shaped opening of a suction tube; prior to this step, the broken end may be locked anywhere on the wound circumference of the cop.

A hook grasps the yarn the top of the tube and the thread guide, in the same way as the operative's hand in manual piecing. This hook lays the yarn on the ring, and the piecing arm joins the yarn to the fiber strand at the front rollers of the drafting arrangement. The superfluous yarn section is served and sucked away. The success of the operation is monitored by a photocell.

7.1.5 Automations with Auto Doffer

7.1.5.1 Speed Control Through Inverter System

The important parts of the machines are

- Main drive system with stepless control of inverter and maintenance free belt.
- The same electrical control of LR6 ring frame holds, good for LR6AX.

The cycle time is less than 3 min irrespective of the number of spindles. Sensing of the gripper rail position through the encoder and inverter driven main drive of the doffer leads to zero maintenance and optimum cycle time.

The positive gripping of the cops or empty tube is effected by the smooth individual gripper operating pneumatically without touching the yarn wound. The full bobbins and empty tubes are conveyed through an individual peg tray system.

Autodoffer operating cycle sequence:

- Emptying tubes on the peg tray
- Loading empty tubes on intermediate rail
- Unloading full cops from spindles (cycle starting)
- Unloading full cops on the peg tray
- Picking up empty tubes from intermediate rail
- Loading empty tubes on spindles
- Lowering gripper rail

The various automations found in this machine are:

- SERVOtrail
- ROBOdoff
- Cop transportation
- Link up

7.1.5.2 *SERVOtrail*

The servotrail logistics systems transport the manually or automatically doffed roving bobbins to the ring spinning machine. In the automatic version, a microprocessor controls the SERVOtrail system and ensures that the roving bobbins are brought to the right position at the proper time.

The SERVOtrail system is a quick connection for manual or automatic transports between the flyer and the ring spinning machine. It consists of an overhead system with rails, switches, trolleys, and a microprocessor control. It starts with the bobbin transfer station on the roving frame where the bobbins are loaded into the trolleys of the conveyor system. A selectable number of bobbins, that is, a full doff, a half doff, and so on, are assembled to a trolley. The trolley can then be moved to an intermediate storage area. A computer takes over the material management for ring spinning and therefore also controls storage.

Then, coded trolleys are directed to the appropriate ring spinning machines by a programmed or manual recall. A manual recall may be actuated via a contact bar extending the full length of the ring spinning machine. The same system conveys the empty bobbins back to the intermediate storage area where they pass through a cleaning station for the removal of roving remnants. The trolleys are moved via friction wheels located along the rails. These friction wheels are driven by small motors. SERVOtrail is also able to bridge differences in height. It may be adapted to any layout.

> *Drive and suction system*: Uniform suction and a constant level of vacuum are maintained by the integrated aspiration system, even on extremely long machines. The most important elements are a single suction tube and an automatically progressing, self-cleaning drum filter. The machine is driven by the Rieter variator located in the headstock. It allows continuous adoptions of the spindle speed during

start-up of spinning, production, and under winding before doffing. A certain amount of air from the suction system is used for motor cooling.

Quick doffing system: The fully automatic ROBOdoff cop removing system is a new, inhouse development of Rieter spinning systems. The entire doffing cycle takes only 2–2 1/2 min. The newly designed microprocessor system controls all functions of the doffing sequence. In addition, it stops the machine automatically when time for doffing is reached, starts the ROBOdoff, monitors doffing, and restarts the ring spinning machine after completion of the doff. The frequency delays caused by waiting for attendance are therefore eliminated.

The actual doffing operation still is according to the normal sequence:

- Cop seizing
- Withdrawing
- Placing the cops on the pegs of the conveyor belt
- Advancing the belt half a pitch
- Seizing the tubes
- Lifting the tubes above the spindles
- Placing the tubes on the spindles
- Returning the doffer beam into its rest position
- Restarting the machine
- Conveying the cops to the winding machine or into a container at the end of the machine

7.1.6 Ringdata

There are two types of ringdata:

- Production data collection
- Production and single data collection

7.1.6.1 Production Data Collection

The sensor is fitted at the delivery of each machine, from blowroom to simplex m/c. The production of each m/c is fed into central unit. We can find out the production, efficiencies, and down of these m/cs in each shift, previous shift, week, and so on.

7.1.6.2 Production and Single Data Collection

In addition to data collection, a further important extension configuration offers the spinning mill yarn break information. The traveling sensors check every spindle of every m/c side.

7.1.6.2.1 Configuration of Ring Data System
This includes the following:

1. Sensors
2. Machine station

3. Machine entry station

4. Concentrator

5. Central unit

6. Terminals

7. End break indication

Sensors: The sensors convert every rotational speed of the shaft into a electrical signal called a pulse. As shown in the Figure 7.2, there is a metal flap or stirrup on the rotating shaft. Every rotation is observed by the sensing point, and the signal is sent to the central unit memory.

Types of sensors:

1. Production sensor

2. Doffing sensor

3. Traveling sensor

Production sensor: This sensor determines the rotational speed of the front roller of spinning m/cs from blowroom to ring frame.

Doffing sensor: This sensor determines the doffing position of the ring rail. It is registered for the detection of the number and duration of doffing.

Traveling sensor: One traveling sensor per m/c side is provided along the ring rail. It detects the rotational movement of the ring traveler without contacting it and provides information about end breaks at each spindle. It also provides average piecing time and the machine rotational speed of the ring traveler and those spindles with a rotational speed that is too low.

Machine station: All m/c stations are connected to central unit over a common line and are called up periodically.

Concentrator: This is used for intermediate storage of signals from 16 spg m/cs.

Bus feed: The bus feed provides central current supply for the detection system.

Central unit: The computer controlled central unit is the heart of the data system and performs the following functions:

1. Periodic call up of the concentrators

2. Continuous processing and storage of the m/c signals

3. Control of dialogue with the user

Terminals: At the end of the shift, the printer automatically issues a number of freely selectable reports.

End break indication: A front lamp is installed at either side of the ring frame at the headstock and the pneumafil end of the m/c. As soon as the number of stopped spindles exceeds the selected value, the appropriate lamp lights up.

As described in the relevant Uster technical leaflet, the Uster ringdata essentially consists of the following parts.

1. *Production sensor*: Which determines the speed of the front roller of the drafting system and provides basic information regarding production, frequency, and duration of longer stoppages.

2. *Under wind sensor*: This registers the under wind position of the ring rail and detects the number and duration of doffing.

3. *Traveling sensor*: This monitors one side of the ring frame; detects the rotational movement of the ring traveler without contacting it; and provides information concerning end breaks/spindle, average piecing time, mean rotational speed of the ring traveler, and spindles rotating at very low speed.

4. *Machine station*: This is controlled by microprocessor, collects the machine and spinning position data, controls the movement of the traveling sensor, and includes optical end-break indication to save the operator time spent on finding spindles with an end-break. All machine stations are connected to a central unit over a common line for periodic call up of data.

5. *Machine entry station*: This monitors the important production data such as efficiencies, out-of-production times, production capacities, and so on, and facilitates reduction of machine stoppages.

6. *The concentrator*: In which a microprocessor takes over the tasks of combining, preprocessing, and intermediate storage of the signals from 16 machines.

The program is stored in read-only memories (RMOs) and cannot be lost.

Terminals: Which is a printer and issues a number of freely selectable reports at the end of the shift. The ringdata reports spinning position with high end-breaks, and machines with low production and low efficiency are marked for remedial action, which leads to increased production and improvements in quality.

7.1.7 Magnetic Spinning

In today's spinning technology, at least four types of spinning systems are commercially available. These are the traditional ring spinning, rotor spinning, air-jet spinning, and friction spinning. Among these types of spinning systems, ring spinning stands alone as the primary reference to high quality yarn suitable for any type of textile end product. Other more recent systems enjoy much higher production speed than the traditional ring spinning, but they are restricted to only narrow ranges of textile products by virtue of their technological limitations.

In this regard, three specific issues must be addressed to overcome this limitation: (1) the dependence of the yarn linear speed (or delivery speed) on the rotational speed of the traveler, (2) The continuous need to stabilize yarn tension during spinning and the dependence of this stability on the traveler speed, and (3) the impact of traveler speed on fiber behavior in the spinning triangle.

The traveler is a C-shaped, thin piece of metal that is used for a limited period of time, disposed, and replaced on frequent basis. Our approach is to replace the traveler with a magnetically suspended, lightweight annular disc that rotates in a carefully predefined magnetic field. This will result in a super high spinning rotation that is robust against all the traditional limitations of the current traveler system. The idea here is to create a non-touching environment of the rotating element of the spinning system.

7.1.7.1 Active Magnetic Levitation Principles

Magnetic levitation of a rotating disk typically incorporates four or more electromagnets to levitate a ferromagnetic disk without contact, where levitation is accomplished through computer control of the electromagnet coil currents. The controller uses position sensor outputs to apply stiffness and damping forces to the rotor (annular disk in our case) to achieve a desired dynamic response. In more sophisticated systems, the controller can automatically compensate for disk imbalance and selectively damp disk resonant modes.

More sophisticated controllers based on adaptive control techniques are also available for active vibration cancelation, automatic rotor balancing, and mechanical impedance synthesis. From simple electrically-biased systems to permanent-magnet multi-degree of freedom designs, Magnetic Moments can provide magnetic system technology for our application to guarantee reliable operation.

Besides the obvious benefits of eliminating friction, magnetic systems also allow some perhaps less obvious improvements in performance. Magnetic systems are generally open-loop unstable, which means that active electronic feedback is required for the systems to operate stably. However, the requirement of feedback control actually brings great flexibility into the dynamic response of the systems. By changing controller gains or strategies, the systems can be made to have virtually any desired closed-loop characteristics.

In the magnetic ring spinning system, a bias flux is generated from both permanent magnets across the air gap, supporting the weight of the rotating disk in the axial direction. In case the floating ring is displaced from its central position, the permanent magnets will create a destabilizing force that attracts the ring even further away from the center. The control system will read out this deviation from the center position, using two displacement sensors mounted radially to the floating ring, and generate a current signal to the power amplifiers.

The power amplifiers supply the electric coils with current to generate a corrective flux. This corrective flux subtracts and adds to the fluxes caused by the permanent magnets. By subtracting flux at the small gap side and adding flux at the large gap side, the total magnetic force will tend to bring the floating ring to its central position.

7.1.7.2 *Flux Density and Force from Circuit Theory*

This type of magnetic system is amenable to analysis via magnetic circuit theory. Assuming negligible leakage and fringing and neglecting the small reluctance of the iron parts of the flux path yields the following magnetic circuit (Figure 7.1):

A magnetic circuit for a horseshoe looks just like an electric circuit with batteries and resistors. However, instead of current, there is flux Ô, and instead of resistance, there is

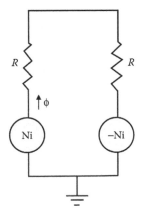

FIGURE 7.1
Magnetic circuit for a horseshoe.

reluctance R. Rather than voltage sources, there are magneto motive force (MMF) sources of strength Ni and –Ni. In terms of the geometry of the system, the reluctance is:

$$R = \frac{g}{\mu_o a}$$

where:
   g is the nominal air-gap between the rotor and the tip of each leg of the system
   a is the cross-sectional area of each leg
   μ_o is the permeability of free space (1.2566e-6 Tesla*M/Amp)

One can solve for the flux in the circuit:

$$\phi = \frac{Ni}{R} = \frac{\mu_o a Ni'}{g}$$

The flux density, B, in the air gap is the flux divided by the area of the air gap:

$$B = \frac{\mu_o Ni}{g}$$

We can now use Maxwell's Stress Tensor to figure out the force. For each horseshoe, the force in terms of flux in the gap is, from the Stress Tensor:

$$F = \frac{B^2 a}{\mu_o}$$

Substituting in the expression determined for B yields

$$F = \frac{\mu_o a N^2 i^2}{g^2}$$

In addition, the flux density obtainable in the gap should be limited by the flux that can get down the legs of the system. Say that the core material saturates at a force of B_{sat}. Then, the maximum force that can be produced by particular system geometry is

$$F_{max} = \frac{B_{sat}^2 a}{\mu_o}$$

For the purposes of system design, useful corollary results are the maximum current required in a horseshoe:

$$i_{max} = \frac{B_{sat} g}{N \mu_o}$$

And the inductance of a horseshoe:

$$L = \frac{2N^2 \mu_o a}{g}$$

which is a function of air gap g.

7.1.7.3 Typical Magnetic System Geometry and Control

The typical magnetic system is composed of four of horseshoe-shaped electromagnets. The four magnets are arranged evenly around a rotor that is to be levitated. Each of the electromagnets can only produce a force that attracts the rotor to it. The four electromagnets must act in concert to produce a force of arbitrary magnitude and direction on the rotor.

The coils may be arranged in the same plane of the rotor or in a plane normal to that plane. There are N turns of wire around each leg of the system. So that the system behaves functionally like a system composed of discrete horseshoes, sets of adjacent coils are connected in reverse series. Another innovative idea, which will be realized within the time frame of this proposal, is the way of placement of the permanent magnets and the electric coils. This placement helps us to minimize overall dimension of the suspended disk. It also helps in supporting the disk in the axial direction without adding additional setup to perform that function (Figures 7.2 and 7.3).

A bias flux is generated across the air gap, shown in paths R1 and R2 from both permanent magnets, supporting the weight of the rotating disk in the axial direction. If the disk is not centered, the permanent magnets will create a destabilizing force that pulls the wheel even further away from the center. The control system will detect this motion through position sensors at the disk's outer diameter and generate a corrective flux Y by sending current through the electric coils. In the air gap, this control flux Y subtracts and adds to the static fluxes R1 and R2 caused by the permanent magnets. By subtracting flux at the narrow gap side and adding flux at the wide gap side, the magnetic system produces a net restoring force to center the disk.

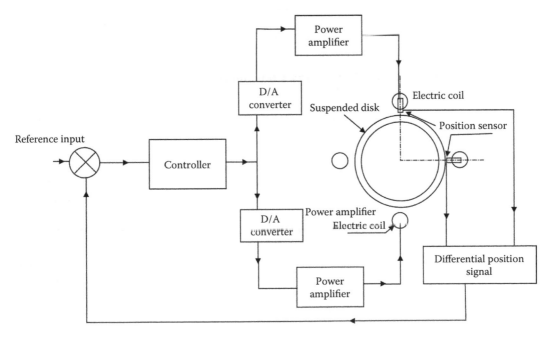

FIGURE 7.2
Main component of the control system.

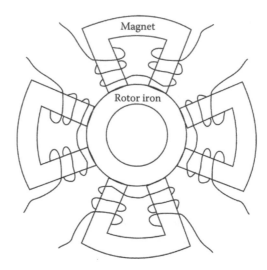

FIGURE 7.3
Magnetic system composed of four discrete Horseshoes.

The control system consists of the controller, power amplifiers, and a power supply. Additional circuitry is present for conditioning of the signals from the position sensors, and conversion of the digital outputs from the controller to the analog amplifiers.

Control systems for our system can be of the analog or digital type. Analog control systems have been used in control systems for over 30 years but are rapidly being displaced by digital control systems. Digital control systems take advantage of the tremendous developments in digital signal processing over the past decade and permit more types of control algorithms to be used, enabling more results to be achieved. Typical sub-systems in magnetically suspended disk control include position signal processing, digital signal processing, D/A conversion, power amplifiers, and power supply.

7.1.8 Monitoring and Control of Energy Consumption for Ring Frames in Textile Mills

The energy costs to produce quality goods have been constantly increasing. The conservation and efficient management of energy has become very important to keep the industry productive, profitable, and competitive. Hence, it was desired to design a microprocessor-based instrument system to monitor and control the actual energy consumption pattern in textile mills. The system could provide an excellent tool to mills for planning energy costs in their production and setting the process parameters in order to manage the consumption with a high degree of efficiency.

The power consumption of ring spinning frames represents the major part of total power requirements of the spinning mills. The energy consumption in spinning the yarns depends on yarn count, ring diameter, twist, spindle speed, process efficiency, machine type and its conditions, and so on. In short, the consumption of energy depends on machines as well as the desired quality of the product. There are different types of machines being used to spin the yarns for which energy consumption varies from machine to machine and their present conditions.

A computerised online information system for efficient energy management in mills. It is designed using microprocessor system, which provides features like keyboard entry, visual display of desired data, print-outs of summary reports, and so on. The system is a unique development to help mills to receive information in real time regarding energy consumption by the machine and corresponding production of yarn, which enables management to take corrective steps in time to control energy consumption and improve production.

7.1.8.1 Design of the System

The microprocessor-based circuit design is developed using an Intel microprocessor. The central processing card consists of circuits for data and address bus decoders, keyboard controller, EPROM, and RAM memory. The total software is stored in EPROM and 8 KB RAM memory is used for data accumulation and analysis. The real time clock battery backup facility on 8 KB RAM is contained in a memory clock card. The real time clock circuit keeps track of real time base analysis, periodic acquisition of energy and production-related data, and storing the same on an hourly or shift-wise basis. The sensors used are inductive proximity switches for front roller, spindle speed, and idle status. The transducer for energy meter, which provides pulses according to units of energy consumed, is also used. The sensors are interfaced with a CPU card. The port and timer counter are used for interfacing the sensor's output with the central processing unit. The power supply card is also designed to provide DC regulated power supply to various circuits of the product. The various connectors are used to interlink different signals for overall functions of the instrument. The dedicated software is developed to achieve various functions. The software is designed for various logical operations of the instrument, keyboard entry of parameters, digital display of desired result, analysis of data, and preparation of a summary report in a format suitable for mill management.

7.1.8.2 The Input/Output Data of the System

The input data for time, date, front roller speed, spindle speed, count, front roller diameter, Wharve diameter, ring diameter, ring frame number, and number of spindles are fed through the keyboard to estimate the energy consumption, production, and UKG. The microprocessor-based system is able to monitor actual energy consumption, yarn production, and UKG for the ring frame. The actual energy consumption measured by the system is compared with estimated energy value and deviations in the UKG detected. The monitoring data is analyzed and a print out of the summary report is available. The report output includes production in kg/hr, consumed energy/hr, accumulative production in kg., accumulative energy consumed in kWh, UKG hourly, estimated energy, UKG average, spindle speed, and TPI.

The various parameters that are required to predict the production of yarn in g/spindle/h are spindle speed, twist per inch, and count value of the yarn; whereas, to predict the energy consumption per hour, the parameters required are yarn count, ring diameter, twist factor, spindle speed, and so on. The energy consumption pattern using the tuned model for the ring frame machine under varied conditions were evaluated. The data for production of the machine was also estimated and the value of UKG obtained. The estimated energy through the tuned model is found in agreement with the monitored data under optimized machine conditions. If the machine performance is set to standard working conditions, UKG actually observed in optimized conditions will nearly agree with the UKG predicted through the software using the tuned model.

The sensor to measure production of yarn online was identified. The sensors to measure current, voltage, and phase online of the machine to find energy consumption and

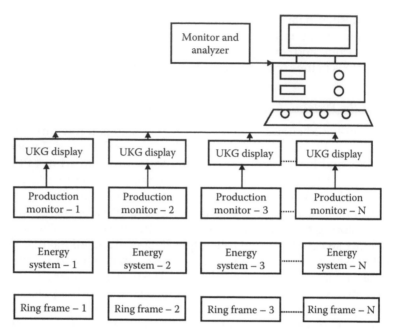

FIGURE 7.4
Prototype of analyzer.

production measurement were selected and fitted in the machine. The interface of the sensors with microcontroller circuit sense their output and convert it in digital form with the help developed hardware circuits. The actual production and energy consumption give the online value of UKG, that is, units per kg of production. The value of UKG is continuously monitored. This value is compared with the predicted value of UKG for that count of material and machine settings (Figure 7.4).

7.1.8.3 The Features of the Instrument System

The monitoring of production for UKG estimation: The microprocessor-based instrument monitors the production under running conditions of the machine. The actual value of the doff per shift in real time conditions is obtained. The units of energy consumption per kg of yarn production (UKG) is calculated and compared with standard (estimated values) of UKG. The deviations, if any, in UKG are detected and shown in a printed report.

The factor causing higher energy consumption: When power consumption or UKG monitored by the instrument deviates very much, the factors affecting higher energy consumption need to be attended by the mills. The factors that are echecked are: tape width, oversize motors, ring diameter, spindle speed, tin roller pulleys, spindle Wharve diameter, alignment of motor and bending resistance of spindle tape, doff time, and so on.

Contribution toward corrective action: The system gives hourly and shift-wise energy consumed, production, and UKG values in the printout. The deviations can be analyzed using the data of spindle speed, TPI, which is also given hourly in the printout.

The instrument system is designed using various transducers, interface, and microprocessor circuits, which provides features like keyboard entry, visual display of desired data, printout of summary report, and so on. During the field trial, it is found that the instrument system monitors the power consumption and UKG on a real-time basis, which is very useful for mills to keep constant supervision on production cost and optimize machine performance. It would be very useful if these instruments are fitted for more machines and data is transmitted to a central computer (Figure 7.4) so they may analyze the relative pattern of power consumption for all machines. The particular machine drawing more energy and giving higher UKG as compared to others can be identified. There are large numbers of ring frames in mills that can be monitored for their energy consumption so that energy cost with respect to production can be optimized. To achieve this, it is necessary that data generated by the instrument for each machine are transmitted to a central computer. The status of energy consumption in kWh and production in kg per hour for all machines in mills could be established on continuous basis. The system is a unique development to help mills to receive information in real time about energy consumed, corresponding production of yarn, and estimated standard values of UKG. The real-time monitored values are compared with standard values, that is, estimated values of a higher consumption of energy, if any are detected. This enables management to take corrective steps in time to control costs and improve production.

7.2 Controls in Rotor Spinning

The idea of producing yarn by the rotor spinning technique is far from new. Furthermore, the rotor spinning machine is the first final spinning machine to be almost fully automated.

7.2.1 Tasks of Rotor Spinning

The rotor spinning machine is unlike any other machine in the short staple spinning mill in the range of tasks it has to perform, namely, all the basic operations:

- Opening (attenuating) almost to individual fibers
- Cleaning
- Homogenizing through back doubling
- Combining, that is, forming coherent linear strands from individual fibers
- Ordering
- Improving evenness through back doubling, imparting strength by twisting

The total electronic control of the yarn winding conditions ensures the production of excellent "certified" packages to feed the loom with unparalleled working parameters and with a density up to 30% higher than what can be obtained on conventional open end machines. The high performance of the spinning unit, together with the intelligent, highly efficient and reliable end-piecing trolleys, translate into production capacities that far exceed all those previously possible. Better yarn quality is obtained with a "certified package." Total control of yarn winding conditions guarantees the production of packages with an even and variable density as wanted thanks to a system totally unhindered by any mechanical parts. Possibility of take-up at variable density, centralized adjustments of the counterweight and friction, speed twists, cross-winding angle, anti-patterning, and tension are all operations that can now be done by simply pressing a key on the computer on the spinning manager's desk. Yarns with different counts can be spun on each front, obviously with different speeds, twists, cross-winding angles, modulation, and tension. In addition, the intelligent trolleys go to the aid of the front requiring more interventions. Rotor spinning has proved in the course of history to be an extremely innovative process. This is true not just of the spinning technology itself, but also of the rotor spinning machine and the entire rotor spinning system. Developing complex systems to the point of market readiness takes a long time. In this sense, the Autocoro 288, which was launched in 1992, represents a crucial milestone in the development of rotor spinning. It was the first machine to be provided with a comprehensive control and information system with data acquisition and transfer between the machine, the individual spinning units, and the automation units. This intelligent system is the basis for the Autocoro 312. The Autocoro 312 is the continuation of the successful Autocoro series and has been developed with the aim of meeting increased market requirements regarding yarn and package quality, reliability, and flexibility as well as productivity and efficiency.

7.2.1.1 Automations in Rotor Spinning

The demand on quality and productivity in processing a wide variety of raw materials in rotor Spg has long driven the development toward automation of m/c processes. The sequence of innovations on BT903 are modern to match the yarn quality of automated m/cs.

- AMISPIN
- QTOP
- IQCLEAN

7.2.1.1.1 AMISPIN

AMISPIN is an electronically controlled device that carries out and controls the piecing process on otherwise manually operated m/cs. With AMISPIN, the critical point of synchronization of the yarn take–off and sliver feed and thus the piecing is fully controlled by electronics. The operator simply cleans the rotor and closes the spin box.

7.2.1.1.2 QTOP

The result is a consistently high piece quality and yarn appearance for all yarn counts. It maintains the yarn quality for even very fine yarn counts. The QTOP functions pneumatically and removes the damaged fibers from the opening section; therefore, the piecer body is formed only from new fibers. As a result, there is no need to increase the number of fibers in the cross-section to maintain piecer strength.

7.2.1.1.3 IQCLEAN

This yarn clearer uses a modern optical CCD camera technology, which guarantees extremely precise measurement. It directly measures the width of the yarn shadow projected onto the CCD linear sensors, which consists of the numerous light sensitive pixels. The number of pixels covered by the shadow equals the yarn is scanned several times, and the average value is used for sophisticated fault evaluation.

7.2.2 Compact SpinBox

The Compact SpinBox possesses the highest possible degree of technological flexibility. The BYPASS allows the adjustment of air balance according to the type of fibers processed. Rotors, take-off nozzles, and opening rollers can be replaced within seconds. Channel inserts are available for each rotor diameter. The maximum spinning speed is limited by fiber technological factors and not by rotor bearing technology. The TwinDisc bearing is capable of higher speeds than the fibers will allow.

Equipping the machine with the new Compact SpinBox is not sufficient. It is also essential to consider the frame in its entirety. With increasing spinning speeds, it is essential to strengthen the existing gearing for the yarn traverse. Additionally, old traverse bar guide brushes need to be replaced by guide rollers. On the piecer carriage, the following adaptations are to be carried out:

- Replacement of the rotor cleaning, due to the use of smaller rotor diameters
- Installation of a new feed roller drive with an air pipe for the feed roller cleaning
- Installation of two air pipes for the cleaning of the opening units while the piecer is traveling
- Replacement of the circuit board for rotor acceleration
- Re-centering of the piecer at the spinning positions

7.2.2.1 Adjustable BYPASS

For the best separation of fibers from the opening roller and their straightened conveyance through the fiber channel, the air velocity should be as high as possible. On the other hand, for efficient trash extraction the air velocity should be as low as possible. It can be problematic to satisfy these opposing requirements as long as the total volume of air has to enter through the trash extraction chute. The continuing demand for increasing air velocity in the fiber channel has, up to now, been a double-edged sword. The improved fiber separation from the saw-tooth profile and straightening went hand in hand with deterioration in trash extraction.

The Compact SpinBox solves this problem. It incorporates an adjustable air intake in the area after trash extraction, designated "BYPASS," which regulates the air sucked into the box. The volume of air entering the trash extraction chute is correspondingly reduced in accordance with the selected BYPASS setting. The setting of the BYPASS intake determines the amount of trash extracted. Without disadvantage to the transport of fibers into the spinning rotor, the amount of trash extracted can be decided with a high degree of accuracy. The extraction of good fibers decreases, resulting in better usage of cotton. The rotor groove remains clean for a longer period.

Yarn quality and spinning stability are thus improved by the BYPASS:

- "Merry-go-round" fibers are considerably reduced, that is, fewer yarn faults.
- "Dolphin jumps" are substantially reduced, hence there is less dirt in the yarn.
- Imperfections, clearer cuts, and ends-down are appreciably reduced.

The favorable effect on the end-breakage rate and, consequently, on operating efficiency of the machine, has been confirmed by numerous mill tests. When processing 100% synthetics, the BYPASS intake will be closed.

> *Turbulence-free airflow in trash extraction chute*: The geometry of the trash extraction chute has been aerodynamically optimized. The chute is as wide as the opening roller teeth profile. The absence of air turbulence at the rims has a positive effect on the trash and dirt extraction (self-cleaning effect).
>
> *Relief of traverse gearing load with "Ping-Pong"*: For relieving the load still further, SUESSEN has developed a new device, Ping-Pong, which will be installed at the end of the frame, and which is synchronized with the traverse gearing. For reducing the load on the gearing, it is in some cases adequate to replace the existing steel traverse rod with a carbon fiber bar, which substantially decreases the inertia forces.

7.2.3 Automatic Piecing Devices

The most conspicuous innovation is the Coromat, which combines the functions of piecing and package doffing. Starter spools are no longer required, thus satisfying a customer request which has been put to us frequently in the past. We have spent a long time planning this development and have thought it through carefully, as it makes sense only if two requirements are met:

1. Shortest possible cycle time both for piecing alone and for piecing following package doffing
2. Maximum possible piecing and functional reliability

The Coromat performs package doffing particularly quickly due to overlapping movement sequences. Package doffing, including piecing, takes only one second longer than a normal piecing cycle, whereas the systems on other automatic rotor spinning machines need 10–15 s longer for package doffing.

7.2.3.1 Piecing Principle

1. The yarn from the Coromat feed package is fed into the rotor. The piecing process is carried out.
2. The pieced yarn and piecing are sucked out through a suction tube. The yarn is cut after the piecing and laid at the same time on the tube holder.
3. The indentations on the tube holder catch the yarn, and the transfer tail is formed. The first piecing following package doffing is removed and the package contains one piecing less than before.

To increase piecing reliability following package doffing, a safety piecing is carried out with different piecing parameters. The piecing can be thicker and the piecing reliability is virtually 100% thanks to this measure. The piecing quality is immaterial in this case, as the first piecing is removed.

Several colors can now be processed on an Atocoro using this principle. The yarn on the feed package does not have to match the spun yarn and thus does not normally have to be replaced at a lot change. The Coromat checks the operating status of the spinning unit before and during piecing or package doffing using the Event Identification System. This intelligent process control system of the Coromat monitors the operating cycles of the Coromat by means of sensors and optimizes the sequences. All piecing parameters can be entered at the Informator or Coromat and selected according to the respective application. Lot data can be stored and retrieved using the up- and download functions. The maintenance and diagnostic functions can be called up directly at the Coromat control system. These include displaying the measured piecing parameters, rotor acceleration times, the evaluation of the piecing tester and piecing faults, and various Coromat test functions. The data can be stored externally in a notebook. No terminal is required, as these functions have been integrated. All events can be retrieved in several languages. The structure of the new machine control system is modular, with autonomous control units. Control system components are clearly allocated to the machine functions. The tube magazine and tube supply unit have their own control system with microprocessor. The same is true of the frequency inverter, machine control system, and Coromat. The package removal and bearing temperature monitoring functions are combined in the machine end frame along with the automation interface for package transport systems.

Maintenance and diagnosis are made easier by the arrangement of the components. The structured assembly facilitates swift troubleshooting. Three newly developed frequency inverters are used to drive the rotors, suction system, and tension draft of the yarn take-up shaft. These are distinguished by their greater efficiency and lower energy consumption as well as a sturdy construction and high level of reliability. Fault messages and diagnostic functions can be called up in plain text at the Informator.

7.2.4 Event Identification System: Electronically Controlled Yarn Transfer

- After each piecing or package doffing operation, the Coromat checks whether other spinning units in the continuation of its direction of travel are inactive. If not, the Coromat changes direction. Unnecessary waiting times of idle spindles are thereby avoided.

- If a sliver is missing, the Coromat switches the spinning unit immediately to red light status and leaves it again without waiting.

- The self-optimizing variable cycle times for the piecing process always ensure the fastest operating sequences.

The Event Identification System (EIS) also guarantees a high level of Coromat reliability. Piecing reliability is improved by the electronically controlled yarn transfer, especially at high take-up speeds and rotor speeds, or with a low winding tension (e.g., for dye packages). Here the yarn guide position is detected via a sensor. The Coromat control system ascertains the exact moment at which the lifter wire is lowered in order to transfer the yarn precisely to the yarn guide, thereby reducing the load on the yarn. The piecing parameters are adapted by the Coromat to the requirements, for example, depending on the package mass, and to the rotor acceleration. This facilitates constant and precise control of the timing sequences of the piecing process, thus ensuring a consistent piecing quality. The system is also immune to contamination and no longer uses reflectors, which can gather dust.

Efficiency gains: The efficiency gain can be achieved by the maximum production of weaving yarn. Efficiency is increased on the one hand by fast package doffing without double waiting times of the spinning units and by improved piecing reliability. Optimization of the direction of travel also increases the efficiency rating somewhat.

Fast response drives: In the Coromat, the fast response drives for the sliver feed and yarn take-up functions ensure precise operation during the piecing process. These new high-performance units are characterized by the fact that the actual speed values of the motors are almost instantly corrected to the setpoints issued by the control system. The excellent dynamic response of these drives enables precise control of the drafts and take-up speeds during the piecing operation. The result is optimum piecings in terms of both strength and uniformity. Naturally, the quality of the piecing is monitored, as is common, by a particularly sensitive piecing tester based on the Corolab measuring system. The piecings are thus indistinguishable from the parent yarn and are invisible in the woven or knitted fabric. Every time piecing or package doffing is performed, the rotor and navel are cleaned by mechanical and pneumatic means. A blowing device cleans the opening unit and doffing tube. Cleaning of the rotor using compressed air combined with mechanical cleaning by means of a scraper offers decisive advantages. It is only in this way that clinging residues such as trash particles or sticky substances in the rotor, navel, and doffing tube are removed effectively and automatically. Other effects caused by contaminated rotors and the resultant clearer cuts are safely avoided.

7.2.5 Automatic Suction Devices

The new suction system has a frequency-controlled fan with electronic vacuum adjustment (EVA) vacuum regulation. The spinning vacuum can be set steplessly as required by means of the frequency-controlled drive. Once set, the spinning vacuum remains constant, as EVA adjusts the fan speed when the spinning vacuum decreases. By contrast, the vacuum level on conventional rotor spinning machines without regulation drops uncontrolled as the chambers fill up. With a control system, the vacuum level can be reduced to the lowest possible level at which a raw material can be processed without limitations, resulting in energy saving. Furthermore, the spinning conditions remain constant, giving a higher level of spinning stability and better running behavior. Constant spinning conditions also produce a constant yarn quality.

The fact lack of need for a generator results in further energy savings. Should the mains voltage fail, the functions are safeguarded by the suction system motor, which continues to operate in this case as a generator. A short mains failure can be bridged in this way by utilizing the kinetic energy of the suction system. The new L- and U-shaped screens in the filter chambers of the end unit ensure an optimum through-flow of air.

7.2.5.1 New Suction System: Optimum Through-Flow

The separated waste is distributed on the surface of the screen in such a way that it is evenly covered and the air can flow through it virtually unobstructed. The capacity of the filter chambers has been doubled, making the cleaning intervals twice as long. The filling level of the waste collection chambers is monitored continuously, and the machine operator is notified when a chamber is full. The filter chambers can be cleaned on both sides independently of one another. The suction system can be connected optionally to a central waste removal system.

7.2.6 Foreign Fiber Detection System

The rotor machine onally with the familiar Corolab or the newly developed ABS foreign fiber detection system (Figure 7.5).

7.2.6.1 Optical Measuring Principle Using Infrared Light

The transmitter sends a beam of light through the measuring field to the receiver. Simultaneously, part of this light falls upon a reference receiver. The yarn in the measuring field throws a shadow onto the receiver, while the reference receiver always receives 100% of the light. The amounts of light transmitted with and without the yarn in each case are compared. From the difference in values, Corolab measures the yarn diameter with an accuracy of 0.01 mm. By means of this continuous comparison during yarn production, Corolab detects all irregularities in the yarn and thus ensures that the desired clearing degree is maintained. In order to obtain true quality data, it is necessary to have continuous information about the yarn being spun.

Even when there are no faults, that is, fault-free yarn is being produced, information must still be provided to this effect. This is achieved by the digital processing of the measured values in conjunction with the high scanning frequency of Corolab. The result of the measurement is directly proportional to the yarn diameter produced; Corolab shows the actual diameter for each millimeter of spun yarn.

Digital processing of the measured values makes it possible to obtain precise data on every meter of yarn produced. Faults that previously led to problems in subsequent processing such as thick places, thin places, moiré, and poor or incorrect sliver no longer exert a detrimental effect on quality.

7.2.6.2 Individual Corolab ABS System

In addition to the clearing of thick and thin places, an increasing number of spinning plants are demanding the elimination of foreign fibers during the spinning process. To meet this requirement, Schlafhorst has developed the absorption sensor ABS for the detection of colored foreign fibers on the Autocoro.

This new generation of foreign fiber clearers permits complex three-dimensional monitoring of the outer yarn body by means of Multi Focus detection due to the integration of

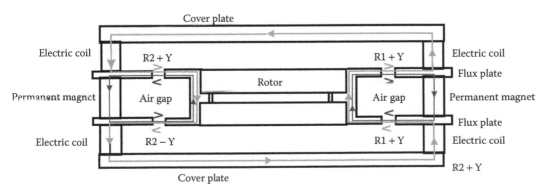

FIGURE 7.5
Optical and digital measuring principle.

the measuring head into the yarn withdrawal tube of the SpinBox. Corolab ABS permits the detection of foreign fibers in length, intensity, and frequency. The exclusion of intrusive factors such as dust or light substantially increases measuring accuracy compared with conventional systems.

ABS can be installed as a standalone system that is independent of the normal yarn clearing function on all Autocoro machines with the Informator. Foreign fiber detection is thus possible on the Autocoro without regard to the existing capacitive or optical systems of various manufacturers.

Inspection of the spun yarn has taken place up to now outside the SpinBox, thereby permanently exposing the measuring heads to external environmental influences. This had a significant influence on the accuracy of measurement. The ABS sensor is located inside the SpinBox. It is integrated into the yarn withdrawal tube directly behind the torque stop and is thus protected against dust and fiber fly generated in the machine environment.

7.2.6.3 The Measuring Principle

7.2.6.3.1 Measuring Head

The measuring head consists of an opaque ring of synthetic material with four LEDs as transmitters and four photodiodes as receivers arranged on the outside. The eight diodes are distributed equidistantly on the circumference in an annular arrangement. A die-cast casing provides additional protection against environmental influences for the ABS sensor.

The ABS sensor (absorption sensor) works according to the absorption principle. The transmitting diodes illuminate the inside of the withdrawal tube, and the amount of light present is measured by photodiodes. Yarn that is devoid of foreign fibers absorbs less light than yarn contaminated by these fibers. The quantity of light absorbed is digitalized and compared with the reference value for the amount of light absorbed by a clean yarn.

As the foreign fiber signal is amplified by ABS 20 times more than the clean-yarn signal, any deviation from the reference value indicates an impurity in the yarn. The annular arrangement of the diodes ensures that yarn is scanned all round, facilitating accurate determination of the foreign fiber length and of foreign fibers that are twisted in. ABS describes the length, color intensity, and frequency of foreign fibers by means of functions. It is thus even easier for operators to set the desired degree of clearing reliably and individually at the Informator.

7.2.7 Fault Recognition and Clearing

Corolab assesses faults in the same way as the human eye does. The user can thus set the yarn monitoring system to remove those faults that are considered objectionable. Each user can set the clearing level for his or her own specific purposes. On lot changing, Corolab determines the diameter mean value at each spindle over the first few yarn meters. This reference value is the reference diameter value for all further evaluations. It is retained and only recalculated when a new command is entered at the central control. There is no subsequent self-adjustment. The machine mean value results from the reference diameter values of all the individual spindles.

These values are compared and spindles with considerable reference diameter deviations are detected and stopped. The wrong sliver, for example, due to mixed up cans, can thus be immediately discovered at the start of the lot. Corolab 7 recognizes contaminations of the measuring head and takes them into account by means of an efficient compensation mechanism that is independent of the yarn count. Neither the measured results nor the desired degree of clearing are falsified because of contamination. The particular spindle stops only automatically when a tolerance limit is exceeded—a red light is displayed.

7.2.8 Moiré and Nep Detection

Detection of moiré within a yarn length of 2 m. Moiré is a periodic yarn fault that is caused by partial contamination of the rotor. The contamination in the rotor causes a thick place in the yarn, occurring periodically at a distance equal to the rotor circumference.

To detect moiré, Corolab needs data about the rotor diameter in mm and the lower limit of the diameter deviation. If thick places in the yarn occur several times consecutively at a distance corresponding to the rotor circumference and exceed a tolerable diameter deviation, Corolab detects these as moiré and stops the spindle. The piecer carriage receives the command to clean the affected rotor. The measuring principle of Corolab permits the detection of moiré within as little as 2 m of yarn. Other systems require up to 100 m to detect this fault.

7.2.8.1 Nep Detection

Neps are short, prominent thick places in the yarn that show up very markedly in the fabric, especially in the case of colored fabrics. This fault occurs very infrequently and at irregular intervals and is therefore extremely difficult to detect during the spinning process.

Corolab detects these short, thick places and stops the spindle when five or more neps with a diameter deviation in excess of a set value occur over a yarn length of 20 cm. Values of between 5% and 170% can be set. If frequent stoppages occur, for example, five neps stoppages within 1000 m of yarn, the associated alarm function blocks the spindle, indicating to the operator that manual intervention is required at the blocked SpinBox.

Due to the short reaction time of Corolab, the faulty piece of yarn is automatically removed in its entirety from the package during the piecing process. In other systems, up to 95 m of moiré yarn are left on the package on account of the much longer reaction time. This is all the more serious since each periodic thick place is preceded by a thin place, which reduces yarn strength.

7.2.9 "Sliver-Stop" Function

The "sliver-stop" function detects long thick and thin places caused by sliver faults, double slivers, and so on. In the event of count variations in the sliver, the diameter deviations of the resulting thick and thin places are less significant than the yarn faults registered in the quality matrix. That is why a flexible mean value is calculated over a set yarn length (e.g., 3 m) and is constantly compared with the reference diameter. If the difference between the reference diameter and flexible mean value exceeds the set value of the diameter deviation (e.g., 20%), the spindle is stopped. The detection of faults of this kind is assured only because Corolab operates without the use of a self-adjusting mean diameter value. With other systems that automatically adjust the mean value, these faults run gradually into the spun yarn without being detected by the clearer. The result is faulty yarn.

7.2.10 Monitoring Devices

7.2.10.1 Precise Monitoring of Yarn Counts

The sliver channel is designed to detect larger scale sliver faults arising due to double slivers or slivers that have been spliced off. The detection of thick and thin places in the sliver depends on the recognition of variations in the yarn diameter, which are averaged over yarn lengths ranging from 1 to 9 m and deviate from the reference diameter by between 2% and 30%. Owing to normal variations occurring in the yarn, too narrow a setting of the percentage deviations permitted would lead to unnecessary stoppages. For this reason, the sliver monitoring mechanism is oblivious to slight deviations in the yarn count. In systems in which no quality check is implemented at the final draw frame, for example, deviations of this kind can be caused by defects in the autoleveler or a lost sliver in the drafting zone of the draw frame.

Corolab therefore incorporates a separate function for detecting yarn count fluctuations, which forms average values over yarn lengths of between 10 and 1000 m and for which percentage yarn diameter deviations of between 1% and 10% from the reference diameter can be set. The yarn count channel is therefore distinguished from the sliver channel by checking of a greater yarn length and the possibility of setting smaller deviations in relation to the reference diameter. This function facilitates very precise monitoring of the yarn counts. If one of six slivers is missing in the last draw frame passage, for example, this will result in a reduction in the diameter of the spun yarn of 8.7% over a yarn length of more than 10 m. However, the value for thin places in the sliver is set in the example at −9% over 9 m, as a stricter setting of the deviation tolerance would have a negative effect on the production level due to the occurrence of accidental changes in the yarn diameter. The normal sliver monitoring algorithm is therefore unable to detect the lost sliver. For this reason, the permitted count deviation for the yarn is set in the yarn count channel at −4% over 50 m for thin places. With this setting, Corolab detects the missing sliver reliably, and efficiency ratings are not affected.

7.2.10.2 Spectrogram

With the aid of the spectrogram, periodic faults can be detected very easily. At the same time, it helps to judge the sliver and thus the previous working stages. The controllable range of wavelengths extends from 2 cm up to a maximum of 300 m. When exceeding a selectable limit, it is possible to issue a spontaneous report and/or to block the spindle. The asterisks depicted denote the mean values of the machine as a whole. The crosses indicate that the machine mean value and the spindle value are identical. A total of four different spectrograms can be displayed and printed out:

- Difference spectrogram
- Spindle spectrogram, last calculated spectrogram
- Spindle mean value spectrogram, mean value of a spindle
- Machine spectrogram, mean value of the machine

7.2.10.3 Histogram

Corolab measures diameter values absolutely; these values are evaluated by means of a frequency diagram. The frequency diagram reveals the percentage diameter distribution above and below the reference mean value. Deviations from the normal distribution of a fault-free yarn indicate thin or thick place zones occurring.

7.2.10.4 Variation-Length Curve

In addition to the spectrogram, nonperiodic deviations in the yarn can be detected by means of the variation-length curve (CVL) curve. Corolab analyzes CV values within the testing length range from 2 cm up to a maximum of 300 m. If yarn quality deteriorates due to randomly distributed accumulations of thick and thin places that are below the clearing limit, the variation length curve will be affected directly. The histogram and the length-variation curves are also continuously registered at each spinning unit.

7.2.10.5 Alarm Functions

If a predetermined number of stops within a specific yarn length produced is exceeded, Corolab blocks the respective spindle, thus preventing piecing. The red light gives a clear indication that operator intervention is required. The two diodes (red and green) on the measuring head indicate the reason for the alarm to the operator. If the yarn is not passing through the measuring head, or is not detected for other reasons, Corolab stops the spindle, thus excluding the production of unchecked yarn.

Corolab detects and reports on the following alarm functions when preset limit values are exceeded:

- Quality alert
- Nep alert
- Moiré alert
- Sliver alert (thick and thin)
- Yarn count alert (thick and thin)
- Mean value deviation
- CV alert
- Spectrogram alert
- Difference spectrogram alert
- Soiled measuring head

7.2.10.6 Online Hairiness Monitoring on OE Rotor Spinning Machines

Hairiness plays an important role in the textile industry. Hairiness variations in yarns can substantially affect the appearance and the hand of woven and knitted fabrics. With the introduction of compact spinning, the hairiness monitoring on the machine became more and more a must. Since the hairiness of compact yarns is very low, it is important that bobbins that deviate in hairiness can be recognized immediately. Otherwise the fabrics have to be downgraded.

7.2.10.7 Principles of Operation of the Hairiness Measuring Systems

The oldest hairiness monitoring system represents the counting of the number of protruding fibers at a distance of 3 mm from the yarn body. The result of this monitoring depends very much on the yarn speed. This monitoring is not reproducible and cannot be utilized for benchmarks such as the USTER® STATISTICS. Inter-laboratory variations (CVB) can be as high as 50%–200%.

Figure 7.6 represents the hairiness of yarns from the point of view of the receiver. The yarn body is dark, but all the loose and protruding fibers are bright and contribute to the hairiness measurement. The light intensity along the yarn is permanently measured by the receiver. Since the yarn body is dark, it does not contribute to the hairiness monitoring. It is possible to evaluate hairiness, calculate the absolute hairiness and the hairiness variation, and print out a diagram and a spectrogram of hairiness with this measuring principle.

Adjustments to lots are handled by quick, easy, and user-friendly operation directly at the control panel. Positioning of the robot is achieved with precise, contactless positioning of the robot using frequency-controlled traverse drive and laser control, thus reducing wear on the traverse drive. Efficient rotor cleaning with special module of the robot is prepared for efficient cleaning of the rotor and the nozzle at every piecing cycle.

7.2.10.8 UNIfeed®: The Universal Tube Supply System

- Tube magazine to hold 310 empty tubes for all customary tube formats.
- Tube feed by a fast conveyor belt system. The robot receives the empty tubes even before the next package change.

7.2.10.9 Automatic Can Transport

In the fully automatic SERVOcan transport system for CUBIcan® rectangular cans between draw frame and rotor spinning machine, empty cans are automatically replaced by full ones.

7.2.10.10 Package Removal

The package lift, the cost-effective system for ergonomically optimized manual package, consists of removal at the end of the machine. It is an interface for connection to external palletizing or transport systems for automatic removal of packages.

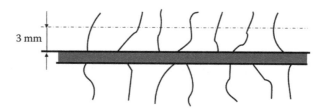

FIGURE 7.6
Hairiness in yarn.

7.3 Summary

At the completion of the chapter, the user will understand the control systems used in the ring and rotor spinning machine. The workings of different instruments and measuring devices were explained in detail.

References

1. http://www.rieter.com/en/rikipedia/articles/rotor-spinning/the-importance-of-rotor-spinning/the-principle-of-rotor-spinning/.
2. http://www.rieter.com/de/rikipedia/articles/rotor-spinning/machinery-and-process/structure-of-the-rotor-spinning-machine/.
3. http://nptel.ac.in/courses/116102038/23.
4. http://www.rieter.com/en/rikipedia/articles/ring-spinning/automation/monitoring/individual-spindle-monitoring-ism-by-rieter/.
5. http://textilecentre.blogspot.com/2017/02/ringdata-system-used-on-ring-frame.html.
6. https://textlnfo.files.wordpress.com/2011/10/f01-ae02-04e1.pdf.
7. http://www.autexrj.com/cms/zalaczone_pliki/1-06-3.pdf.
8. http://nptel.ac.in/courses/116102038/33.
9. https://pdfs.semanticscholar.org/61f7/ca14281f704a402a75574b1d3d147c547c77.pdf.
10. http://www.autexrj.com/cms/zalaczone_pliki/1-06-3.pdf.
11. http://www.yarnsandfibers.com/preferredsupplier/reports_fullstory.php?id=219.
12. http://www.indiantextilejournal.com/articles/FAdetails.asp?id=5644.
13. http://www.rieter.com/en/rikipedia/articles/rotor-spinning/machinery-and-process/operating-principle-of-the-rotor-spinning-machine/.
14. http://textilelearner.blogspot.in/2012/04/what-is-rotor-principle-of-rotor.html.
15. https://www.suessen.com/products/open-end-rotor-spinning/.
16. http://www.rieter.com/en/machines-systems/products/rotor-spinning/r-66-rotor-spinning-machine/.
17. https://www.textileweb.com/doc/compact-spinbox-sc-1-m-0001.
18. http://www.indiantextilejournal.com/articles/FAdetails.asp?id=5383.
19. http://nptel.ac.in/courses/116102029/37.
20. http://www.rieter.com/en/rikipedia/articles/rotor-spinning/machinery-and-process/the-spinning-box/rotor-cleaning/.
21. http://www.rieter.com/en/rikipedia/articles/rotor-spinning/technology/yarn-structure-and-physical-textile-characteristics/yarn-hairiness/.
22. http://nopr.niscair.res.in/bitstream/123456789/34375/1/IJFTR%2041%282%29%20129-137.pdf.

8

Control Systems in Cone Winding Machine

LEARNING OBJECTIVES

- To recognize the working principle of the cone winding machine
- To know how the control systems are used in the cone winding machine
- To understand the automation process used in the cone winding machine
- To identify the various sensors and transducers used in the process

8.1 Introduction

For a long time, winding was considered a simple, unimportant process because it merely transfers yarn from cop to cone. But today winding is considered important because it provides an opportunity to eliminate imperfections (thick and thin places, neppiness), but it may impart these faults if the process is not controlled properly. Thus, it can enhance the quality of the product.

Modern cone winding machines are designed to produce consistent quality of yarn at higher productivity levels. Cones with perfect winding-off qualities and long knot free yarn length are required for producing good quality fabrics. The yarn must be free from imperfections like slubs, neps, thick and thin places, and knots. The cone should be wound with uniform tension and homogeneous in density.

The machines are essentially equipped with a precision control system, which is electronically operated to achieve required quality with higher speeds. The features are auto tense yarn tension control, upper yarn sensor, direct drive with ATT, vacuum controlled suction facility with AVC system, informatory with touch screen and integration of clearer operation, caddy system, sensor control, and monitoring of the material flow.

8.2 Electronic Yarn Clearer

The electronic yarn cleaner (EYC) is firmly established as yarn fault detector in all ordinary and automatic cone winding machines. There are basically two principles: (1) capacitance type and (2) photoelectric type.

8.2.1 Capacitance Type

The block diagram shows the principle of the capacitance-type electronic yarn clearer. In this type, the yarn is passed through an air-spaced measuring condenser consisting of two parallel plates with air as media. The yarn fault limits according to yarn cross-section, length,

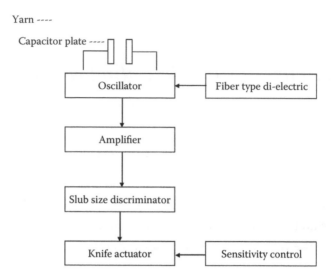

FIGURE 8.1
Capacitance type electronic yarn clearer.

and spinners' double are set by using suitable knobs on the instrument. As the yarn passes through the capacitor, according to the length and cross section of the yarn, the capacitance value is changed. This is related to the mass of the material between the plates. The condenser measures the material, and electric signals are fed to an amplifier that amplifies the signals. The amplified voltage is transferred to a discriminator that measures the preset fault limits. If the signals exceed the limits, the signal is fed forward and through a cutting impulse, and a knife is actuated to cut the yarn. A fiber type dielectric and sensitivity controls are used to get the required sensitivity during the operation (Figure 8.1).

8.2.2 Photo-Cell Type

The block diagram shows the principle of the photo-cell type electronic yarn clearer. The yarn is passed through a light source and is scanned by the light source. Opposite to the light source, a receiver is provided to receive the light rays emitted from the light beam (Figure 8.2).

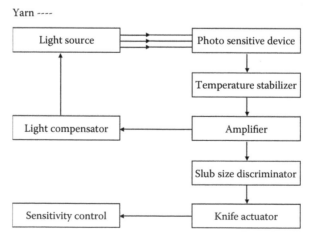

FIGURE 8.2
Photocell type electronic yarn clearer.

The yarn fault limits for thickness, length and spinners' double are set by using suitable knobs in the instrument. As the yarn passes through the light source, the yarn is scanned by the light beam. The scanned portion interrupts the light beam, which changes the signal from photo-cell. These signals are fed to an amplifier that amplifies the signals. The amplified voltage is transferred to a discriminator that measures the preset fault limits. If the signals exceed the limits, the signal is fed forward and through a cutting impulse, and a knife is actuated to cut the yarn. Uster Peyer, Uster Quantum clearer, and Loepfe clearer can be used as yarn clearers.

8.3 Electronic Anti-Patterning Device

The device is used to control both acceleration and deceleration of the drum. The duration of the anti-patterning cycle is adjusted in direct relation to the package diameter as the package builds up. This anti-patterning adjusted to all phases of the package winding cycle ensures optimum unwinding performance even in the case of critical and large package diameters, as effective anti-patterning can be achieved also for these. The intensity of anti-patterning can be set centrally at the Informator.

8.3.1 Drum Lap Guard

In the winding unit, nearer to the drum, an optical sensor is provided to monitor the traverse of yarn. If there is any deviation in the frequency of traversing, the unit is stopped immediately. Another sensor is provided to monitor drum lapping. With this electronic control device, the yarn traverse is monitored directly and thus the development of laps on the drum or package is detected quickly and the drum is stopped immediately. The drum lapping is also reduced by the yarn trap, the drum lap brush, and electronic clearer monitor.

8.4 Length and Diameter Measuring Device

Modern cone winding machines are fitted with an electronic length and diameter measuring device. The package diameter can be set centrally at the Informator. An additional quality package checks whether the diameter correlates to the length wound or the length to the diameter wound. Furthermore, the number of cuts per bobbin can be set so that the package quality is also monitored in terms of the number of yarn joints being wound onto it. Thus, bobbins of insufficient quality can be detected quickly and eliminated from the winding process.

8.5 Sensor-Monitored Winding Process

At a clearer cut, yarn break, or bobbin change, the upper yarn search is terminated as soon as the integrated optical upper yarn sensor has detected the sucked-in yarn. The cycle is then continued immediately due to the constant level of vacuum at the suction arm, swift, and reliable pickup.

8.6 The Informator: Central Operating and Control Unit

The Informator consists of the following: (a) large display with graphical user guidance and integrated input via a touch screen, (b) Clearer operation integrated into the Informator. It is possible to switch between machine and clearer operation simply by pressing the company logo on the touch screen. Thus, machine and clearer operation has been further centralized, standardized, and simplified. (c) PC card drive. A PC card can store a variety of information. Lot specifications and the complete Autoconer software can be read easily and centrally into the Informator via the PC card drive. (d) Output of reports via a thermal printer and (e) Standard specification incorporates an Ethernet interface for exchanging data between an external computer system and the Informator. A data flow in both directions is possible.

The Informator is distinguished by some special features. Graphical representation of the functions and contents selectable ensures simple machine operation. Production groups, which may be varied in size from one winding head to several machine sections, obtain the nominal data for the winding operation centrally. This applies, for example, to package diameter, winding speed, and splicing parameters. All nominal settings for the tensioner and yarn tension control system are also entered centrally at the Informator. This guarantees uniform settings at all winding heads. Another significant innovation is the ability to set the desired vacuum level centrally at the Informator.

The Informator registers and evaluates the actual data produced during winding. The Autoconer 338 can be divided, for data purposes, into as many as six production groups. Production data is logged and graphically displayed both for the production group as a whole and for each individual winding head.

Bobbins and packages are monitored during production by entering quality parameters, facilitating intervention by operators exactly where needed if quality margins are exceeded. This ensures that high package quality is maintained. Off-standard parameters permit limit values to be assessed for production, clearer cuts, tension breaks, and downtimes. Excess values are highlighted in the report. Detailed clearer data is also available to the Informator for better evaluation of the yarn quality and clearer cuts.

8.7 Automatic Tension Controlling Device

This device is used to maintain uniform yarn winding tension throughout the package formation. At the tensioner, the yarn passes in a proven manner between two tension discs driven against rubbing yarn. The tensioning system is the central adjustment of the yarn tension and electromagnetic application of force. The main component of the yarn tension control system is the yarn tension sensor. It is situated in the yarn path after the clear on each winding head to register the actual yarn tension at the package and provides continuous measurement of the yarn tension. These values are transmitted in a closed loop to the tensioner, where the pressure is increased or lowered according to the information received, that is, the yarn tension is not only measured directly but also regulated directly by the tensioner pressure and maintained at a constant level (Figure 8.3).

8.7.1 Direct Drive System—Auto Torque Transmission

The direct drive system of the drum—auto torque transmission (ATT)—in conjunction with the newly designed winding head control facility represents one of the most innovative

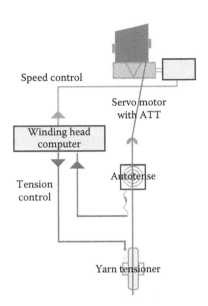

FIGURE 8.3
Autotense device.

developments with regard to the winding unit. The combination of winding head computer and direct drum drive (ATT) facilitates smooth, jerk free startup; slip-controlled acceleration; high winding speeds; and improved anti-patterning. The advantages of the optimized winding head control and the new direct drum drive are (a) tangled and loose layers are avoided on startup and acceleration of the package; (b) improved package build; (c) acceleration time is reduced to a minimum according to package diameter, feed material, and other winding parameters; (d) the winding speed achievable is, at a high level, commensurate with the yarn and bobbin quality; and (e) electronic anti-patterning is improved.

Every winding head is equipped with an adjustable servo-motor for the drum drive. On the Autoconer, the drum is mounted directly on the shaft of the drive motor. This ensures that the drum is under the complete control of the winding head control system so that the drive torque is transmitted reliably to the package. The acceleration of drum and package is optimally coordinated by the new type of slip control. The maximum permitted speed difference is entered, and a constant comparison is made between the drum and the package speed. The winding head control system then regulates the acceleration time automatically. The slip that occurs is utilized for the purpose of anti-patterning during acceleration.

The possibility of controlling the tensioner pressure facilitates precise yarn laying even in the acceleration phase, as precise yarn guidance is achieved due to the uniform yarn tension throughout winding. Loose layers due to insufficient yarn tension are avoided. These could cause yarn tangles or end breaks in subsequent processing.

The advantage of this system is that it allows direct integration into automatic electrical control systems. The main component of the newly designed yarn tension control on the Autoconer is the yarn tension sensor. It is situated in the yarn path after the clearer on each winding head to register the actual yarn tension at the package and provides continuous, direct measurement of the yarn tension. This value can be displayed for each winding head using the winding head tester.

The measured values are transmitted in a closed control loop to the tensioner, where the pressure is increased or lowered according to the information received, that is, the yarn tension is not only measured directly but is also regulated directly by the tensioner pressure and maintained at a constant level. It should be emphasized that the Autotense yarn tension control not only prevents an increase in the yarn tension toward the end of the bobbin but also compensates for the lower yarn tension during acceleration by increasing the applied pressure. Autotense provides reliable compensation of the yarn tension variations occurring in the unwinding cycle from top to bottom of the bobbin. The ability to set the desired yarn tension and all tensioner centrally at the Informator provides the basis for achieving optimum build uniformity from package to package.

8.8 Yarn Clearer

With the Autoconer, one can target specifically and eliminate only those faults that actually have an adverse effect on downstream processing and the end product. Selective setting of clearer curves allows integration of the clearer operation into the Informator, upper yarn sensor technology and removal of only a precisely controlled length of faulty yarn from the package, and consistent practical utilization of clearer information in the winding head control system. These measures prevent unnecessary clearer cuts that would otherwise result in production losses and yarn waste (Figure 8.4).

The quality of the splice is recorded separately by the clearer, and the spliced joints are assigned to the appropriate clearing categories. At a clearer cut, the yarn length to be removed from the package is determined by the length of the yarn fault signaled by the clearer. The winding head computer ascertains from this information the yarn length that has to be unwound from the package. The upper yarn sensor is used for the first time on the Autoconer winding head The end of the yarn is detected precisely, thereby ensuring

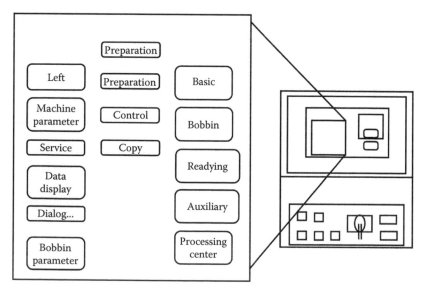

FIGURE 8.4
Informator.

that the complete fault length is unwound from the package without creating any unnecessary waste. A unique control system not only enables the Autoconer to detect and cut out short faults in interaction with the clearer, but also to effectively remove from the package longer and periodically recurring faults.

8.9 Variable Material Flow Systems

The modular design of the basic Autoconer machine means that different material low systems for bobbins and packages can be integrated into the machine. The basic functions of the other assemblies such as winding heads, energy unit, and suction system remain the same. Bobbins are readied for the winding process by the winding operatives and placed into the pockets of the circular magazine, guaranteeing a reliable supply of feed material to the winding head.

The Autoconer can be fitted with two different conveyor systems for the removal of spinning tubes. The machine comes as standard with a full-length tube transport system with a central sorting table at the energy unit. To make the handling of empty tubes easier in multi-lot processing, the Autoconer offers an alternative, as an option, in which the empty tubes are deposited in a tube box at each section. It is also possible to combine these two conveyor systems on one machine.

The decentralized material flow concept is a typical feature of the Autoconer D/V machine. Decentralization means that the units are arranged adjacent to the main conveying path, ensuring a smooth flow of material.

Bobbin readying station: Gaiting-up ends and backwinds are freed, and the yarn end is placed on the tube as a tip wind.

Tube inspector: This separates bobbins with yarn residues that can still be wound, those with unwindable residues, and empty tubes, A high cycle rate (up to 60 cycles/min) guarantees high throughput rates.

Tube stripper: Unwindable residues are stripped off here. Spring-loaded stripping units strip residues effectively in a multiple cutting and stripping motion.

Piece bobbin readying station: This station involves specific and concentrated readying of bobbins that are difficult to ready, such as piece bobbins and deformed bobbins. The suction arm moves vertically across the entire bobbin, can be positioned in particular at the tip of the bobbin, and operates more closely there than the stationary suction arm at the bobbin readying station.

Bobbin removal station: The manual readying track on the type D machine can optionally be complemented by the addition of the bobbin removal station, which removes the piece bobbins passing through manual readying automatically and places them in a container. Further, it assists in emptying of the machine during automatic lot changing.

Manual readying track: Bobbins that cannot be readied mechanically or tubes with residual windings that cannot be removed by the tube stripper are stored ready for manual cleaning by winding operatives. Tubes with residual windings are prepared here for manual stripping on machines with no tube stripper.

8.10 Winding Head Control

The winding head control unit receives from the Informator the required parameters for the winding process and controls the operations such as drive of the yarn guide drum, braking and stopping, reversing, yarn joining, bobbin change, smooth startup (acceleration yarn-protecting, non-slip acceleration of the package after a splicing cycle by the adjustable servo-motor of the yarn guide drum ATT), infinitely variable winding speed (adjustable between 300 and 2000 m/min depending on type of yarn, bobbin builds), electronically controlled anti-patterning system, drum lap detector, length measurement and package diameter computation Autotense yarn tension control, and yarn monitoring with sensors.

8.11 Full Cone Monitors

In cone winding, maintaining uniform and specified cone weight is a major problem in almost all the mills. Various methods are followed to maintain the cone weight.

The weight varies between cones occur mainly due to:

- Variations in cop weight
- Yarn breakages between the cops

This problem has become acute after the introduction of the electronic yarn clearer. This results in low cone winding productivity, weight complaints from the customers (especially warp counts), and increased cost of production. Hence, full cone monitor has become essential to manage the problems.

8.12 Automatic Package Doffer

The package doffer replaces a full package with an empty tube and deposits the package either in a trough at the back of the machine or on a package conveyor belt. The automatic package doffer removes every package that has attained its predetermined length or diameter. It operates in request mode, that is, the doffer remains in the parking position until it receives a doffing signal from a winding head. The doffer then moves by the shortest route directly to the winding head involved, thus minimizing waiting time and increasing machine efficiency. The doffer provides each full package with a tip tail. Once the doffer has taken an empty tube from the tube magazine and placed it into the package cradle, the transfer tail is wound and the beginning of the yarn is fixed in place by over winding. The conveyor belt is concave and carries the packages without them coming into contact with one another or with the machine.

A variable speed controlled belt allows the packages to be conveyed to the suction end swiftly and gently without jerks and without toppling. If more than one doffer is used, the number of winding heads allocated to the patrol range of each doffer can be selected freely. This enhances efficiency considerably, particularly in multi-lot winding involving different yarn types and counts, as flexible, lot-related operation is possible. Inputs and outputs can be affected directly at the doffer via a plain text display or at the Informator.

A magazine located above the winding heads holds the empty tubes in readiness for the package doffer. The Autoconer is equipped for the automatic feeding of package tubes. Because the number of empty tubes in the package tube magazines and also the number of full packages on the conveyor belt are monitored, a high level of functional reliability in doffing is ensured.

8.13 Cleaning and Dust Removal Systems

The cleaning and dust removal system of the Autoconer is comprised of bobbin dust removal, traveling cleaner, and a multi-jet blowing device. The bobbin dust removal device is a continuously operating cleaning system that sucks away fiber fly and dust arising at the winding head. Dust particles blown downward from the upper part of the winding head by the traveling cleaner are also sucked away. Studies have shown that around 80% of the dust and fly generated in the winding process arises in the area of the bobbin and the yarn balloon.

Schlafhorst has developed a bobbin dust removal system that sucks in a large part of these dust particles and removes them from the winding head. Installed at each winding head is a dust removal nozzle, which opens into a suction channel for each section. The air sucked in flows from there through a cylindrical filter screen, which traps the fly and dust. The dust collected in the filter is removed automatically and carried to the yarn/dust chamber of the suction system at variable intervals. The filtered exhaust air is blown into the base area of the machine section.

An advantage is that the exhaust air can alternatively be directed into the ducts of the air conditioning system. Thus, the fine dust that passes through the filter can be removed from the working area. The traveling cleaner patrols along the super structure of the machine over the doffers and winding heads. The surface of the winding machine is cleaned by means of blowing hoses. A suction hose is provided for cleaning the floor area. The traveling cleaner then disposes of the dust either into the yarn/dust chamber of the machine suction plant or into external disposal systems.

Traveling cleaner can be switched to intermittent operation at the Informator. The multi-jet device cleans specific dust-sensitive points on the Autoconer 338 winding unit with a blast of compressed air at each splicing cycle. The air blast is directed into the tensioner zone and toward the clearer measuring head and the waxing unit.

8.14 Automation Variants

The bobbins arrive at the Autoconer, type D machine in large containers and, following singling in the flat-circular conveyor, are donned onto the Caddies and conveyed to individual processing units in the machine. The empty tubes are returned to the flat circular conveyor, where they are removed and placed in crates or large containers standing ready.

A direct link with the ring spinning machine is created by the Autoconer 338, type V machine, the interface being formed by the continuous transfer station (CTS) facility, which transfers the bobbins onto the Caddies and the empty tubes onto the carriers of the ring spinning machine respectively. With its high speed (up to 50 cycles per min), the CTS is geared in particular to longer ring spinning machines and short bobbin unwinding times.

The bobbins and tubes are interchanged between the carrier systems of the ring spinning and winding machine, while the carriers themselves remain in their separate material flow systems. The high level of flexibility, transparency, and speed of the transfer process offer convincing proof of the performance of this concept. The CTS transfer facility is easy to maintain, extremely reliable, and resistant to wear. The rotating function principle of the CTS is a characteristic feature of the continuous transfer process. Linked systems can also be realized with a steamer (optional) integrated between the ring spinning and winding machines.

8.15 Caddy Identification Systems

Certain performance features of the Autoconer 338 are dependent on the integration of freely programmable data media into the base of the Caddy. A variety of quality and material flow data is read in and out of these in read/write stations. Lot data can also be stored. The Caddy identification system is characterized by extensive flexibility of information and ease of operation.

8.16 Spindle Identification

In cases where the Autoconer 338, type V machine is linked directly to the ring spinning machine, online quality monitoring of the ring spinning plant is possible with the aid of the spindle identification system. The number of the ring spindle and the batch are stored in the Caddy chip so that the quality data obtained by the clearer can be attributed to the individual ring spindle. Spindle identification provides continuous monitoring of the quality trend, as not only clearer cut data but also yarn quality data are evaluated. It also and makes it possible to detect accurately ring spindles that are producing faulty yarn and permits targeted, preventive quality monitoring and control, as operatives are able to intervene at an early stage.

8.17 Package Quality Control

Yarn tension and package cradle contact pressure are the two process parameters having the greatest influence on the properties of a staple fiber yarn package. The integration of these parameters into a feedback control loop is a prerequisite for high and uniform package quality. Propack optimizes the winding process by improved package quality. The features include exactly determined package contact pressure that is the same at all winding units to centralized data input; no differences in package density owing to the elimination of pattern zones; higher flexibility for sales yarn spinners, as there are no more diameter limitations due to the elimination of pattern zones; and prompt availability of the package doffer, as Propack exerts pressure on the package cradle right at the moment when winding on the package starts. The doffer is ready in less time for dealing with the next winding head, which reduces package acceleration time due to higher contact pressure at the start of winding.

All parameter data are input at the Informator. The contact pressure of the package cradle and the additional contact pressure at package startup are input as numerical values.

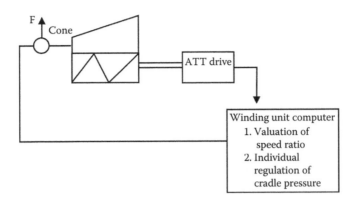

FIGURE 8.5
Propack system.

This eliminates the need for time-consuming mechanical adjustment of each individual package cradle. All setting parameters are stored in the Informator. This ability to retrieve and adjust settings centrally and quickly reduces the risk of incorrect settings being entered and ensures reproducible settings at different machines, even over fairly long periods of time.

The winding head computer records the speeds of package and drum continuously. Shortly before the arrival at a pattern zone, that is, a zone with a critical speed ratio between the drum and the package, a specific amount of extra slippage of the package relative to the drum is produced by reducing the cradle pressure (Figure 8.5).

The package thus skips the critical speed ratio, and winding is done below the critical patterning speed. In this way, the winding ratio is changed abruptly immediately before the pattern zone is reached, and maintained at a constant level that generates no patterns, until the pattern zone is left. The package is "lifted" over the critical pattern zone. The process is coordinated to the specific conditions of the different types of drums used. The drum speed resulting from the set winding speed remains constant throughout the winding process.

8.18 Variopack System

By the intelligent combination of the two independent automatic control loop-systems, Autotense and Propack are integrated into the Variopack system, which guarantees outstanding package quality. This high-performance module of the new series represents a new standard for the efficient processing of elastic materials in the textile production chain.

8.19 Summary

On the completion of the chapter, the user will understand the process involved in the cone winding machine. The various control systems and automation process are discussed in detail. The sensors and transducers used in the machine are explained clearly to help users understand the working of the cone winding machine.

References

1. https://textilelearner.blogspot.in/2011/08/yarn-clearer-in-winding-types-of-yarn_9929.html.
2. http://nptel.ac.in/co urses/116102005/13.
3. http://www.indiantextilejournal.com/articles/FAdetails.asp?id=5192.
4. https://www.slideshare.net/fahadrabby/yarn-clearer.
5. https://www.scribd.com/doc/23665185/Uster-Quantum-Clearer.
6. http://saurer.com/fileadmin/Schlafhorst/pdf/Spulen/Autoconer_6/Autoconer_6_Prospekt_150219_en.pdf.
7. http://www.leclairmeert.be/files/3114/5770/4696/AC6.2_Bro_Aktualisierung_EN_1215_Web.pdf.
8. http://www.textileworld.com/textile-world/new-products/2015/03/the-new-autoconer-6-from-schlafhorst-the-best-original-ever/.
9. http://www.indiantextilemagazine.in/technology/new-autoconer-6-off-to-a-good-start/.
10. http://schlafhorst.saurer.com/fileadmin/Schlafhorst/pdf/Spulen/ACX5_Brochure_Saurer_en.pdf.
11. http://www.indiantextilejournal.com/articles/FAdetails.asp?id=395.
12. http://www.somewhereinblog.net/blog/arafa/30126228.
13. http://textilelearner.blogspot.in/2014/07/an-overview-of-winding-machine.html.
14. https://www.scribd.com/document/326494710/Study-on-Cone-Winding.
15. http://pmkvyofficial.org/App_Documents/QPs/Cone-Winding-Operator.pdf.
16. http://textilelearner.blogspot.in/2014/07/an-overview-of-winding-machine.html.
17. http://www.rieter.com/en/rikipedia/articles/fibre-preparation/the-blowroom/summary-of-the-process/basic-operations-in-the-blowroom/dust-removal/.
18. K. Srinivasan, P. Jeevapriyadharshini, A. Karpagam, and S. Pavithra. Quality improvement of yarn by automatic waste removal in autoconer. https://www.ijareeie.com/upload/february/22_Quality%20Improvement%20of%20Yarn.pdf.
19. http://www.rieter.com/cn/rikipedia/articles/alternative-spinning-systems/the-various-spinning-methods/air-jet-spinning/winding/.
20. https://www.saviotechnologies.com/savio/en/Products/Automatic-Winders/Documents/EcoPulsarS_en.pdf.
21. https://www.saviotechnologies.com/savio/cn/Products/Automatic-Winders/Documents/PolarI_dls_en_cn_032012.pdf.
22. https://www.scribd.com/doc/81691688/AC-338-Type-RM.
23. http://www.equitystory.com/Download/Companies/saurer/Annual%20Reports/saurer_ar_e_99.pdf.
24. http://www.crcnetbase.com/doi/abs/10.1201/b18910-5.
25. https://www.slideshare.net/hammamasyed/auto-cone-winding-machine.
26. https://books.google.co.in/books?id=zIxqBAAAQBAJ&pg=PA248&lpg=PA248&dq=Propack+system+in+cone+winding+machine&source=bl&ots=Xorv0lvSi-&sig=hZRUuaLZ6bD-3laWYByfoilNQ4c&hl=en&sa=X&ved=0ahUKEwj0kdiHvNnVAhUJuo8KHW8cCIMQ6AEIKjAB#v=onepage&q=Propack%20system%20in%20cone%20winding%20machine&f=false.
27. http://www.ptj.com.pk/Web%202004/01-2004/general_artical.html.

9

Instrumentation and Control Systems in the Warping and Sizing Machine

---

LEARNING OBJECTIVES

- To identify the control system concepts used in the warping and sizing machine
- To describe the working of various sensors, transducers, and measuring devices used in the warping and cone winding machine
- To understand the importance of automation in the warping and sizing machine

---

9.1 Control Systems in Warping

9.1.1 Introduction

Warping is the process of winding warp threads on to a beam with the help of a warping machine. In this process, the supply packages in the form of cones or cheeses are mounted on a creel, and the number of supply packages in the creel determines the number of ends in the beam. The warp yarns are fed in the form of a sheet wound on a beam in which all ends are parallel to each other. The problems related to warp in weaving are mainly due to the improper warping that contains slack warp, tight warp, cross ends, uneven density, insufficient/excess ends, and varying yarn length per beam. These defects lead to huge waste in sizing in the form of lappers and creel waste as well as defective cloth production in weaving. Moreover, they cause frequent stoppage of the downstream machines, adversely affecting their efficiency.

Today's warping is no more just assembly of threads. Good warping alone can result in good sizing. The requirements are uniform tension between and within yarns, continuity of all yarns, correct width of the beams, uniform density of the beams, accurate staring position of sections, use of right cone for a given style, and reproducibility of beams.

Warping can be classified into direct beam warping and the sectional warping machine. In direct beam warping, the warp threads are wound onto the beam directly. This beam is then taken for sizing. Sectional warping machines are where different colors of warp threads are used to produce a fabric with multicolored stripes or checks. The warp threads are wound as sections on a beam, and the threads from this beam are wound onto the warper's beam. The role of control systems in this process is significant and directly influences the quality like the appearance of the fabric that is finally woven in a loom.

Computer technology is transforming our working world. Its entry into mechanical engineering and machine functions marks the "new machine generation." The computer in the machine makes new concepts possible, with enhanced security from monitoring of functions and higher quality thanks to all-automatic control, maintenance of constant process data, and much simplified attendance.

Programmable logic controller (PLC) is a microprocessor-based device used to perform every function precisely. It works based on user-friendly software that can be updated for future developments. Warp specifications can be stored for different parameters that are memorized by the PLC and can also be obtained even during power failure. This PLC base control can also produce the data in a printout form. A digital numerical control (DNC) ensures that all beams are of the same length. It calculates the number of meters by continuously measuring the circumference of the beam. It operates at 99.9% accuracy and helps to reduce the amortization period by saving yarn. The DNC control system also permits faults to be analyzed reliably, and therefore guarantees the highest level of efficiency during direct beaming. A touch screen and a clear open layout also enable the machine to be operated quickly and easily, and prevents operator errors. The computer assisted process monitoring system checks that the required processing sequences are following one after the other. A network connection enables data to be transferred by clicking the mouse, which is simple, quick, and paperless. The advantages of all this include shorter change-over times, longer production run times, and reduced operator errors, which all contribute to the production of top-quality beams.

In the direct beaming machine with the dust extraction system, textile dust, which tends to be created at the very high beaming speeds, is condensed in the air stream from the creel and removed. This protects the warping operatives from the adverse effects of dust, and the improved beaming conditions allow high-speed beaming that is free from fiber and fly.

The direct beamer is of rugged construction and powered by a three-phase drive and frequency converter. This allows high beaming speeds with a beam width up to 2200 mm. The electronic control systems not only ensure speed is held constant, but also provides high torque in the lower speed range, leading to a short run up phase. A blow-off device across the entire comb width keeps it free from fluff by means of intermittent blasts from air jets. A straightening rod raises the warp sheet at the expanding comb and directs it onto the guide roll. All adjustment-monitoring controls usually are displayed on the operating terminal.

An optional built-in printer produces a beam protocol after every beam change as well as after lot or job changes. This is accompanied by a comprehensive job protocol. Fly is continuously extracted from the winding zone so that neither fluff balls nor fiber dust can be wound onto the warp. Often the beaming speed has to be reduced because of the dust contamination and therefore the performance of the machine and the materials are not fully utilized.

Because the speed of the air is fastest close to the warp sheet, the direct beaming machine and the operator's stand become polluted with fly and textile dust. If the air stream is not extracted then it distributes it unwanted "freight" throughout the entire room to the disadvantage of the production machines, the room, and the people running them.

9.1.2 Automatic Feed Control System

Cylindrical winding will only be assured if the material build-up and feed per drum revolution are matched to the cone angle. The Supertronic section warper measures the take up behavior of the warp material shortly after the first section is started and then adapts the feed automatically. The measuring system employs a high precision electronic micrometer. The so-called measuring phase proceeds automatically as follows.

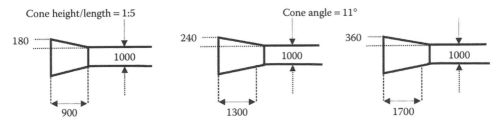

FIGURE 9.1
Electronic micrometer for precise measurement of the yarn uptake staple fibers, texture filaments, and coarse filaments.

After a small amount of material has been wound on, the machine stops at a defined drum position. The measuring system relays the material height to the computer. The machine is then restarted and continues warping until the second measuring stop, which again takes place after only a few millimeters have been built up. The measuring operation is repeated, and the computer works out the correct feed from the two measurements. This is now taken over by the machine and is retained until the end of the section. Because this feed corresponds exactly to the actual yarn build-up behavior, a cylindrically wound warp package is assured. With subsequent sections, the first revolutions are run with the feed adjusted as originally, and the changeover performed on the first section is repeated automatically at full speed.

The long-standing Benninger principle of "no revolution without defined feed" has been retained here. From the very start, accurate winding conditions are provided on the first section owing to the predetermined starting feed—even with the finest yarns, such as for screen printing or ultra-fine filter cloths, where one is dealing with 8 dtex monofils and build-up during subsequent sections, the precisely uniform package build is maintained over the entire warp width (Figure 9.1).

9.1.3 Automatic Tension Controller

The productivity of a beaming installation is determined to a great extent by the creel concepts. The controlled yarn tensioner permits high production speeds at low yarn tension. Yarns being processed may vary widely in their properties, and this places considerable demands on any direct beaming installation and seriously affects the yarn tensioner. The fiber "shifting" tendency, formation of snarls, creation of dust ballooning hang-up, and end-to-end catch-ups are effectively handled by the BEN-Stop.

The BEN-Stop model V-Creel is computer controlled and permits beaming speeds up to 1200 m/min. During the start and stop of the machine, uniform yarn tension is maintained across the entire warp sheet to avoid overrunning. The tensioner controlled yarn tensioner is open during beaming so that there is no twist accumulation, so snarling cannot occur.

The yarn tension, while warping continuously, changes during the course of the process. The creel extends to a certain length in front of the warping machine depending on the number of supply packages used. The tension should be constant regardless of the amount of yarn present in the supply package. The required yarn tension in the warp section is precisely set on the electronic tension sensor meter. The set value of creel tension is continuously monitored by pressure load cells mounted on the warp table. When there is a loss in tension, it gives a signal to the creel PLC, which in turn regulates the yarn tension very precisely so as to achieve perfect warping in all sections. This is the most important feature required for preparing the highest quality sectional warp beams for onward sizing operation.

9.1.3.1 Optostop Tensioner

For V-creel and parallel creel, the optostop tensioner is used. This is a universal stop tensioner for staple fibers between 5 and 170 tex. This is suitable for high direct beaming speeds with low thread tension. This device can be highly sensitive to end breaks and stops the machine without slack ends. It is an integrated optoelectronic thread detector that works on the optical principle.

The new surveillance device tensoscan automatically controls the thread tension of all the individual threads in a yarn sheet during the warping process. An electronically-driven measuring carriage travels at a constant speed over the complete width of the yarn sheet, and an integrated thread probe measures the tension of each individual thread.

The scan device has a measuring speed of 2 m/min and minimum thread separation from 0.7 mm.

9.1.3.2 Reliable Fault Detection

It will give a warning by clearly visible warning display in case of incorrect tension. On detection of a fault, the measuring carriage is immediately stopped. The detection of broken threads can be a problem whenever high running speeds on textile machines are encountered. The further development of the well-known digital thread counter camscan for warping, beaming, sizing, draw warping, and similar machinery prompted us to bring onto the market an improved system, which in addition to the present control of missing threads in the yarn sheet or the incorrect number of threads, also has the possibility to detect and display the actual position of the missing thread in the warp sheet. Now there is also the possibility to establish the independent positional frequency of thread breaks and with its help, for example, a faulty package that is always causing the defect can be identified at once and then be exchanged. If the preset fault count for one package is exceeded, then the machine can be locked and then, for example, the package exchanged. Camscan supports avoidance of costly losses caused by the late or never detected faults in the warp sheet.

AMSCAN II is a modular camera system in working widths from 21″ up to 85″. The cameras are mounted above the lamp illuminated warp sheet. In this way, every thread is monitored. The resulting light impulses are read from the camera and passed to the computer for evaluation. Basically, a target/actual comparison between the pre-set and actual number of threads is carried out. If the camscan detects a variance between these two values, the machine will be stopped. As an additional safety measure against false stops, the number of surveys before stoppage of the machine is adjustable. The decisive development in the camscan is the possibility to exactly determine the position of the missing thread. The fault position will be displayed on the LCD screen. In addition to the LCD display, a fault signal is given by the signal lamps that are connected to the control unit. All necessary settings can be input with a hand terminal. These input values as well as operational parameters (computer signals) are shown on the LCD screen.

9.1.4 Automatic Warp Divider

Splitronic, the cleaning device, is controlled from the central PC. The cleaning sequence, together with warp auto clear, shed opening, and shed change with synchronized drum motion, are programmed and monitored.

The functionalities of the automatic warp driver are

- Electronically controlled feeder that ensures that all sections in the conical area are the same height, this ensuring that every end of the beam is the same length.
- The lateral traverse, which is computed electronically. A feed roller is laid against the drum surface and its position registered. The section is then run-on, pushing the roller back in accordance with the build-up. An electronic transmitter device divides the distance traveled into timed impulses, which are fed to the computer that calculates the rate of traverse and displays it digitally.
- Motorized leasing operation in a programmed order.

9.1.5 Automatic Warp Stop Motion

The yarn inspector Warpstop Types 3010 (mono), 3011 (major/minor), and 3012 (major/minor/length selector) are used to detect yarn faults during the warping process. By using the latest light wave conductor technique, a high operational standard of the surveillance system is guaranteed. Thanks to the digital sensitivity setting with a stepped calibration possibility from 0,1%, even the smallest faults are detected. The yarn inspector is comprised of the control unit Warpstop series 3000, with integrated operator panel, an inspection head bed, and supporting frame. The version 3012 requires, in addition, an impulse sensor to monitor the yarn speed. The inspection head works using the latest light wave conductor technology for the evaluation of the data transfer to the control unit. The optical head guarantees a high linearity of the light beam and also provides a stable sensitivity level over the full working width of the unit.

Since no electronic parts are contained inside the inspection head, the unit is not sensitive to any electromagnetic influences. Should it be necessary to exchange the transmitter or receiver of the inspection head, no adjustment is needed since all electronic parts are inside the control unit. An improved guidance of the yarn through the light beam is ensured by means of the rounded section overrun profile on the inspection head, as well as a reduction in the soiling of the inspection head caused by slub and yarn residue. The signal received in the control unit is compared to the pre-set threshold stopping signal. The warping machine will then be stopped. The control unit is provided with an integrated operator panel with LCD display and keyboard that can be placed in any position to suit your requirements. Depending on the model of the unit, the following information is displayed on the LCD screen: yarn noise level value, size of the last stop signal, preset stopping and counting thresholds, length information, and total number of faults indicated. All operational parameters can be entered via an easy-to-use menu control. In addition, it is possible to provide the control unit with a printer interface.

Safe and fast detection of thread breaks in the warp sheet helps to reduce losses. The newly developed laser light barrier system laserstop sets a standard in reliability and safety in the control of warp sheets on elastomeric equipment. This newly developed system uses the most up-to-date laser technology for the light barriers and an evaluation incorporating the most modern digital signal processing in the control unit. The special characteristics of the system are:

- Fast and safe detection of broken threads
- Compact construction of the laser light channels
- Visible, safe red-light laser (laser class 1)
- Vibration insensitive receiver

The Control unit consists of digital signal evaluation and computer supported, automatic system. The laser light barriers operate with a visible red-light laser (670 nm). These 6 diode lasers are used because of their long operating life and low mechanical sensitivity characteristics. The high homogeneity of the light beam guarantees a constant sensitivity over the full working width. In the receiver, a newly developed measuring system is used, which gives excellent results in terms of vibration independence and sensitivity. The laser light barriers are mounted parallel to the yarn sheet. When a broken thread comes out of the yarn sheet, it will break the laser beam. The resulting impulse will be digitally processed and the elastomeric machine will be stopped immediately.

The laserstop control unit contains all the necessary components to run the surveillance device and allows the connection of up to two laser light barriers 480 EL. At each of its large size display fields, the current operational condition of the connected laser light channels is displayed in a color coded form so that the system status is easily seen even from a distance. All settings to the control unit are carried out with the help of a hand terminal. This hand terminal operates via an LCD display and a keyboard. All settings are carried out with the help of an easy to understand operating guide on the LCD screen. The special ELAS-Software for the surveillance system is contained in a new type of memory chip so that when updated, the new software can be loaded via the hand terminal socket. In this manner, the system is prepared for any future expansion.

Flawless beams are a necessity for a high production standard of textile fabrics. Only the accurate scanning of the yarn sheet to detect broken or run-out ends during the beaming or warping process will provide complete success. As a result of the current high warping speeds, the human eye is unable to detect faults in the yarn sheet. The end break detector laserstop type consists of the detecting head bed, which is a metal blower housing mounted on one or two floor stands depending on the width of the yarn sheet and the spacing between the threads the laser light barrier type 480, the control unit laserstop, and the integrated air blower.

The integrated air blower consists of a row of individual blowers. These blowers provide a high air output together with a low noise level and a low power consumption. The air stream is directed onto the yarn sheet with the help of deflector plates. In case of a broken or run-out end, the loose end is blown out of the yarn sheet and moved through the laser light beam. The resulting impulse will be digitally processed and the machine will be stopped immediately. The height of the detecting head bed is adjustable according to the height of the yarn sheet. Where a lot of dust caused by fluff is a problem, two laser light barriers can be mounted onto the detecting head bed as one operational DUO channel. In this automatic warp stop motion, a stopping the two lasers light barriers which are mounted in parallel to the yarn sheet within an adjustable time. The start delay period can be adjusted up to 10 seconds to avoid false stops caused by loose threads entering the laser beam when unit 4080.

9.1.6 Precision Length Measuring Unit

A presser roller presses on to the circumference of the beam and senses the built-in diameter of the beam. This type of presser is used in case of grey yarns. When the set value is reached, the machine is automatically stopped and the beam can be removed from the warper. When dyed yarns are used, the presser roller may abrade the color on the yarn, so a laser-sensing system is used for sensing the diameter of the warp beam. The computer controlled wind-up of the beam guarantees the identity of all sectional beams of the warping set. During the warping process, the diameter of the beam is measured by a laser system.

By the computer, the current warping values are compared with the given values of the master beam and can be adjusted by alternating of the revolutions. The integrated relax control compensates the stretch differences inside of the cone depending on the unwound thread length by changing of the speed.

The warping process is based on the superior thread length principle, so the end stretching is obtained by the length difference of the elastic yarn between the creel and the warper. Thus, the disadvantages of the thread tension principle, such as tension varying of the yarn by ageing, climate, or surrounding influences, are avoided.

The sectional beams are driven hydraulically. The expansion reed allows an easy thread-in of the ends. This reed rises and falls, automatically conforming the winding diameter of the warped yarn. The yarn is safely relaxed to the end stretching due to the relatively low pre-stretching.

The threads are prestretched up to 200%. The pre-stretching is obtained by the speed difference between the prestretching unit and the creel. The speed is infinitely variable by a drive; an adjustment by change wheels is not necessary. To stop the warping unit in case of yarn breakage, the prestretching unit can be equipped with a yarn inspection device type camscan or type laserstop. A camera (camscan) scans the ends or laser-light barriers (laserstop) indicate the wavy breaking yarn.

9.1.7 Automatic Braking System

In the event of a yarn break, the machine has to stop instantly so that the broken end does not get itself buried in the subsequent layers of yarn that is wound during the halting of the machine. This may cause missing ends in the warp sheet while sizing and weaving. The warping machine consists of a braking system that must function during a thread break while warping. To achieve this control, a powerful, air-cooled disc brake on each side of the drum is provided. The hydropneumatic circuit present in this arrangement ensures soft but extremely effective instantaneous braking during a warp break. In addition, a servo-assisted hydropneumatic system helps to regulate brake pressure automatically for uniform yarn winding onto the beam that can be mounted on the loom for weaving a fabric.

In an efficient braking system, all controls and function regulation are provided by a 16-bit microcomputer system. Electronic control is provided for warp length, piece length, constant warping and beaming speeds, each stoppage location of drum at the end of each section, and automatic determination and input of the exact feed rate for the beam particulars at the commencement of warping.

9.1.8 Beam Pressing Device

Pressure rolls made from light material with antistatic coating are mounted on frictionless, sealed ball bearings. Quick acting electromagnetic brake ensures synchronized braking with beam and guide roll. Auto kickback motion eliminates any friction between the pressure roll and yarn. The pressure roll automatically moves out of contact with the beam as the machine stops. After the beams come to a standstill, the pressure roll automatically is reversed back on the beam.

Roller pressure can be stepless regulated to suit the required beam density. An electronic measuring device is mounted inside the roller assembly. This system gives more accurate results when compared to measuring on a guide roller assembly. A battery backup with memory backs up the digital length counter. In case of power failure, the reading on the length counter is automatically stored in memory and the same reading reappears on the monitor on resumption of the warping operation. This prevents any possible error in length measurement. The same electronic device also indicates the machine running speed on the digital display board.

9.2 Measurement and Control Systems Used in Sizing

Modern sizing machines from Suker –Muller, Benniger, Zell, and Toyota have sensor controlled systems that monitor the process and operating condition and maintain at a set level, so the machine can run without stoppage and uniform size will be applied. The things to be monitored and controlled are speed; measure of a tension in the winding zone warp beam creel; measure of a stretch at the wet zone, dry zone, and winding zone; temperature of the sizing solution and drying cylinders; squeezing pressure at low speed and at normal speed and moisture content of the sized yarn.

All the above values can be easily set and controlled automatically when the actual value deviates from the set value. In the event that manual correction is required, the machine comes to a stop (Figure 9.2).

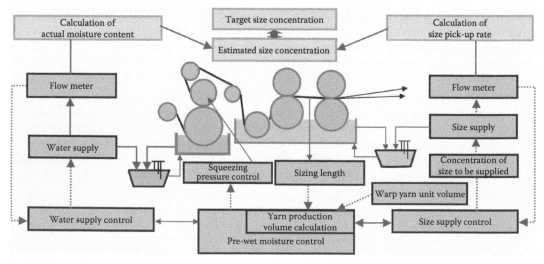

FIGURE 9.2
Sizing machine control elements.

9.2.1 Pre-Wet Sizing

Pre-wet sizing involves passing the yarn through hot water and squeezing out excess water before applying size. Running the yarn through a preliminary hot water box removes cotton wax and foreign matter, which allows the size to penetrate the yarn more smoothly and enables a thinner, more uniform size application. This makes it possible to reduce size consumption, and also provides the added benefit of reducing the amount of size removed in the finishing process that would normally enter the industrial waste stream, resulting in a cleaner environment (Figure 9.3).

This system controls the size concentration in the size box to ensure that a consistent amount of size is applied to the yarn. The system constantly checks the moisture content of the yarn following the pre-wet process and the amount of size pick-up to the yarn, and compares these values to the estimated size concentration in the size box. Based on the results, the pre-wet squeezing pressure is adjusted according to the type of product being loomed, and the size concentration in the size box is maintained at a consistent level.

9.2.1.1 Compact Roller Arrangement

The roller configuration is designed to minimize the distance between the point where pre-wet squeezing takes place and the point where the yarn is immersed in the size. This roller configuration makes the moisture content following squeezing less likely to be affected by factors in the external environment. It also offers the additional advantage of making it easier for the operator to deal with machine stops, such as repairing warp breaks.

9.2.1.2 Eliminates Pre-Drying

Super-high-pressure squeezing and precise control of squeezing pressure in the pre-wet section ensure consistent, stable moisture content in the yarn following squeezing. This eliminates the need to pass the yarn through a pre-drying stage and provides sufficient pre-wetting effect resulting in excellent yarn handling characteristics and greater energy efficiency.

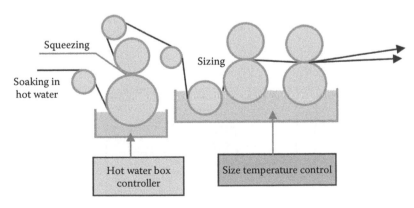

FIGURE 9.3
Pre-wet sizing.

9.2.2 Temperature Control

Temperature determines the viscosity of the size paste, which in turn determines percentage of size pickup, size penetration, and end breakages during weaving.

The thermostat tube is inserted into the size paste through a hole on the side of the size box. It is connected to the throttling valve by means of a capillary tube. The thermostat is equipped with a temperature setting unit that can be adjusted by means of a key at the end of the thermostat tube. The throttling valve is situated in the steam supply line. A bypass arrangement is also provided to supply the steam directly to the size box. In such cases, the valves X and Y should remain closed and the valve Z should be opened. When temperature of the size mixture in the size box reaches the predetermined level, the fluid in the thermostat tube expands along the capillary tube and closes the throttling valve, which has a double set ring, thus cutting off the steam supply.

When the temperature falls below the preset level due to the cutting off of the steam supply, the fluid in the capillary tube contracts and the spring at the top of the housing causes the valve to open, allowing the steam to pass from the supply line back to the size box.

9.2.3 Size Level Control

The electrically operated controlling device uses two electrodes located in the size mixture (Figure 9.4).

A, A': Electrodes (A' at a slightly lower level)	D: Relay
B: Diaphragm value	E: Size box
C: Electropneumatic relay	G: Strainer

The electrode conductivity of the size paste itself provides the basis for the application of the level control system. A slight differential between the electrodes prevents the system of cyclic opening and closing of the size-flow value due to turbulence or foaming in the size box. When the size level falls below the lower electrode, the circuit is opened, the electropneumatic relay is deenergized, and the air-operated valve controlling the flow of the size paste is open. When the size paste level rises and reaches the upper electrode, the electric circuit is closed and the relay closes the control valve.

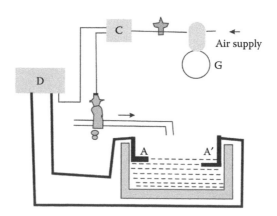

FIGURE 9.4
Size level control.

The system maintains the size level within + or −5 mm of the desired level. The disadvantage of this system is the deposition of the size material on the electrode, thus affecting the working of the control device. This problem can be overcome by coating the electrode with Teflon material. In the place of the air-operated control valve, a solenoid-operated control valve can also be used.

9.2.4 Automatic Tension Control on Single End Sizing

The headstock is designed to operate at high speed of 140–150 m/min. The winding tension is maintained high at 6200 N by 45 kw DC motor. Each beam has a powder brake impulse from the beam tension controller. The powder brake not only deals with emergency stops but also controls the creel tension by progressively reducing the braking torque as the beams are reduced in unwinding diameter. The draw roll in the head stock is driven through a powder clutch, which is imposed from a second tension controller. Both controllers have independent settings and are interconnected and separately linked to the Tacho generator. The creel powder brakes are also capable of providing taper-tension or a tension gradient down the weaver's beam (Figure 9.5).

9.2.5 Stretch Control

BTRA (Bombay Textile Research Association) has developed a digital stretch meter specially designed for continuous monitoring of stretch level on sizing machines (the display unit along with the sensing units). The unit consists of two special transducers and the display unit. One of the transducers is mounted on the guide roller preceding the sowbox. The other transducer is placed on one of the rollers of the drag roller assembly at the headstock. The pulses generated by the transducers are processed by an electronic circuit and the result is digitally displayed. The gadget is useful as it gives the operating stretch instantaneously and therefore is facilitating corrective action wherever necessary.

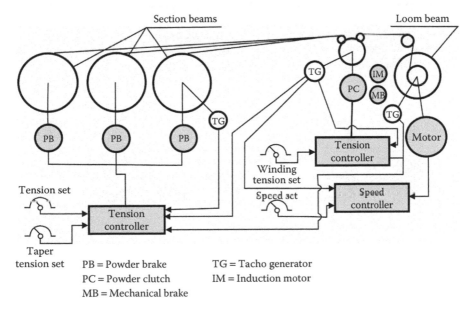

FIGURE 9.5
Single end sizing.

One display unit can cater to three sizing machines, while a pair of transducers is required separately for each machine. The salient features of this stretch meter are its fully solid-state circuitry, high accuracy, and easily readable bright red display. The construction of the unit is robust, which helps it to withstand vibrations and hot, humid conditions.

9.2.6 Size Application Measurement Control

This control continuously measures the moisture content of the sized warp directly after the sizing trough and before the drier without contact and free of inertia by means of microwave absorption. With constant size liquor concentration, a change in moisture content means an immediate change in the degree of sizing (Figure 9.6).

The microwave measuring heads are mounted on a stainless-steel measuring frame. Since steam clouds and high temperature occur behind the sizing trough, the measuring heads are protected by a temperature-controlled warm air cushion, a warm air unit producing hot air, which heats the entire measuring frame and washes around each measuring head. In this way, measuring faults due to steam clouds are avoided, the measuring heads are protected against extremely high temperature, and condensation on the measuring frame and consequent droplets on the fabric are prevented.

AS measurements can be processed visually with PLC for controlling squeeze roller pressure in the sizing trough. In order to meet the special sizing requirements (e.g., switching over from fast speed to creep speed in the case of warp end breakages), special control algorithms have been developed, guaranteeing a uniform degree of sizing at any time and thus maximum weaving efficiency.

Size application is affected by numerous parameters. If we start from a known size concentration, these parameters are, for example, speed, squeeze pressure, squeeze roller hardness, and liquor temperature. Therefore, parameters can be identified online with the size application meter AS.

By means of squeeze pressure control, which linearly changes squeeze pressure relative to warp speed, the sizer must specify a squeeze pressure per product for fast and

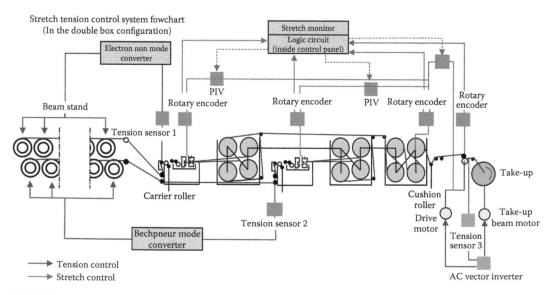

FIGURE 9.6
Size application measurement control.

creep speed. Achieving the correct squeeze pressure setting is a very laborious process since the degree of sizing has had to date to be determined by desizing. The results of measurements with the AS in practice have revealed that a significantly higher degree of sizing is achieved by very many seizers at creep speed than at fast speed.

With the aid of the AS, pressure setting is extremely simple. The degree of sizing is seen immediately, and exactly the same value can be set as fast speed by changing creep speed pressure.

9.2.7 Computer Slasher Control

The preparation foreman or sizing room supervisor types the code of the style to be run into the microprocessor, which locates it in the memory and supplies all the technical data necessary for the sizing operation the computer. This data includes all the settings necessary to control each sizing parameter, including temperatures, pressures, tensions, speeds, stretch ratio, and all details of size constituents with mixing instructions. The size mixing information, including a complete listing of ingredients and quantities, is displayed on the screen adjacent to the mixing cisterns. The computer verifies the valve status on the drain and transfer valves, and if not closed, instructs the operator to adjust them via the machine apparatus screen. The computer then admits the correct volume of water specified in the size formula, that is, the volume which with calculated condensate will give the correct final volume, and starts up the agitator drive motor. The operator's screen then sequentially displays each mixing constituent and a specified quantity to transfer to the weighing hopper and will only instruct the transfer to the cistern when verified as correct. If an incorrect weighing is loaded to the hopper or any constituent omitted, the error is displayed on the machine operator's screen. When the last constituent has been transferred, the steam control valve is automatically opened and the temperature is held at the specified cooking temperature for the specified time. When the cycle is complete, the computer checks that the valve on the receiving vessel at the machine is closed, and if the vessel contains size, checks that there is sufficient available capacity to accept the mixing. If all indications are positive, the transfer valve is opened and the transfer pump started by the computer. The storage kettle agitator is started, the steam control valve is opened, and the mixing is controlled to the specified holding temperature. When the transfer operation is complete, the valves are automatically repositioned, and the size is pumped to the machine size-boxes as indicated by the box level-control instrumentation.

The computer calculates the total volume of size required for the set from the data, supplied by the memory. If more than one mixing is required, which will be the normal situation, the formulation is repeated on the size preparation screen. If a part mixing is required to complete, an adjusted formula is displayed showing the reduced quantities. The discarding of unused mixing residues is eliminated, along with the associated effluent treatment costs associated with sizing effluent. During the preparation of the mixing, the machine operator receives instructions on his or her screen regarding the warper's beams to be loaded to the creel. The set is loaded, pulled over, and established in the wraith and an empty loom beam correctly positioned in the headstock.

The computer then fixes the following variables as prescribed in the data bank:

1. Creel brake setting to achieve the necessary warp sheet tension
2. Level of size in the boxes
3. Temperature of size in the boxes

4. Squeeze roll loading, running, and crawl

5. Size box speed differential for required stretch

6. Temperature of the drying cylinders, by group or individual cylinders

7. Drying cylinder/head-end speed differential for creel tension

8. Winding-on tension

9. Moisture content for automatic speed control

10. Beam press-roll loading

In addition to these major control functions, the yardage on the weaver's beam can be monitored, the machine changed to crawl speed and then stopped, and the operator informed by screen display and light signals. The residual length on the section beam can be more accurately determined than is possible by comparing length sized with the warper's beam length. The length entering the size box is accurately measured and corrected for creel/box stretch, which has been previously established for the style and included in the style operating data. The operator is alerted when the creel is almost exhausted. The operating data required by management such as average running speed, number and duration of stops, and steam flow measurements for the set; as each full weaver's beam is doffed, an operating condition report is produced by the printer. This goes to the weaving machine with the entered warp showing the condition experienced throughout the sizing of the warp, thus providing an opportunity for correlation of weaving performance. All the information normally made available to management, as presented in the manual system described in previous section, can be provided by computer slasher control. Such information is accurate and reliable.

The success of size application depends on proper selection and utilization of control systems in the machine. The non-contact type sensor monitors the weft pick up and size concentration. The size add-on is naturally controlled.

The sensor immediately after wet splitting, automatically regulates squeeze roller pressure and also checks its hardness and based on that it uniformly squeezes across its width. The uniform pressure of the squeeze roll is monitored by piezo-electric sensors. Online yarn hairiness sensors are used at unsized yarn and after sizing (at leasing section) to check that encapsulation is maximum at certain add-ons only.

The sensors after dryers check the residual moisture in yarn and correct the cylinder temperatures. If moisture is excessive, speed is reduced. The various parts of the machine are driven by servo motors. Their speeds are monitored and controlled by programmed microprocessors. Therefore, control of stretch below 1% is possible for cotton yarns and about 1.5% for viscose P/C and P/V yarns. If stretch is not property controlled, the irregularity in yarn will increase. In fact, sizing increases the basic irregularity present in yarn. The final moisture content at the leasing section is measured to control drying according to need.

9.2.8 Auto Moisture Controller

This instrument is especially designed for the Stentor, Sizing, and Sucker-Muller machines. For cotton type processing in Stentor, an accurate moisture controlling system is required to keep perfect drying of the processed cotton fabric and to avoid under-drying or over-drying of the fabric. In the Stentor machine, temperature of the system and speed of the machine are two variable factors. If temperature is more, the speed of the machine has to be increased and if temperature is less, the speed of the machine has to be decreased. The auto moisture controller has been developed to solve this problem. This controller is beneficial in terms of fabric quality along with fuel saving. It contains imported type mechanical structure below the fabric and above the fabric that continuously senses the moisture in the middle and at both the ends of fabric. The main function of this system is to automatically control the speed of the fabric according to the variation in temperature, thereby perfectly drying of the fabric.

9.2.9 Evaluation of Sized Yarn

Though assessment of yarn quality at various presizing processes is popular and well established, the evaluation of sized yarn and its correction with weaving performance is hardly practiced. This may be because of complexity of weaving process. The experiences at weaving are still the only method of assessment. An instrument such as the K–Z weigle abrader can be used to assess weavability, elongation with Instron, moisture content, film strength, as well as other properties of sized yarn.

9.2.10 Automatic Marking System (Diagram)

In the latest Hibbert sizing machine, an electrical marker has been incorporated in the measuring motion—the driving shaft is fitted with a special type of cam that actuates a marker solenoid snap-switch through a roll and forked bracket. The cam is set so that one of its high portions comes into contact with the forked bracket when the shaft is released and given a half-turn.

Contact of the high part of the cam with the forked lever engages the solenoid switch, and a mark is made on the yarn. On some machines, the solenoid and marking mechanism are situated between the tin roller and the big splitting rod, or just beyond the point where the yarn leaves the size box.

The cam that operates the solenoid switch is designed with one high portion for the full width, with the other high portion for the half-width marking. Normally, the cam is set so that the follower rides over both high points, operating the marking solenoid each time. The bracket that carries a follower can be set so that the follower is only raised by the high point occupying the full width of the cam. With this arrangement, the marking solenoid operates once for every full revolution of the cam, instead of twice, so the distance between marks is doubled. Above the solenoid switch is an isolator that can be used to cut out the solenoid for any required length of time.

9.3 Summary

The chapter describes the working principles of the warping and sizing machine. The control system concepts and the importance of measuring devices are explained in detail.

References

1. https://www.omicsgroup.org/journals/warping-parameters-influence-on-warp-yarns-properties-part-1-warping%20speed-and-warp-yarn-tension-2165-8064.1000132.php?aid=16450.
2. http://www.artisam.org/descargas/pdf/TERMINOLOGIA%20DE%20FIBRAS%20INGLES.pdf.
3. https://www.yumpu.com/en/document/view/23770058/ben-v-creel-karl-mayer-textilmaschinenfabrik-gmbh/5.
4. http://www.google.com.pg/patents/US2219155.
5. https://books.google.co.in/books?id=337NBQAAQBAJ&pg=PA76&lpg=PA76&dq=Sizing+Machine+Control+Elements&source=bl&ots=lsCKTSbrvq&sig=fbve9FdjJ2xbzBQ9Jm1Iwi7hwmY&hl=en&sa=X&ved=0ahUKEwiNuPLK0djVAhUMMY8KHSzGBAIQ6AEIRTAH#v=onepage&q=Sizing%20Machine%20Control%20Elements&f=false.
6. http://lumberquality.com/wp-content/uploads/2014/12/Performance-Excellence-EM8731.pdf.
7. https://www.textileweb.com/doc/tensoscan-automatic-warp-tension-monitoring-d-0001.
8. http://www.semitronik.com/shrinkage_elongation_auto_controller.html.
9. http://www.google.com.pg/patents/DE2519557A1?cl=en.
10. Ivana Gudlin Schwarz, Stana Kovacevic, *Krste Dimitrovski, University of Zagreb, Faculty of Textile Technology, Prilaz baruna Filipovica 28a, 10000 Zagreb, Croatia, e-mail: ivana.schwarz@ttf.hr, *University of Ljubljana, Faculty of Natural Sciences and Engineering, Snezniska 5, 1000 Ljubljana, Slovenia *Comparative Analysis of the Standard and Pre-wet Sizing Process*, 2013.
11. http://www.amritlakshmi.com/Brouchure/Multi-%20Cylinder%20Sizing%20Machine.pdf.
12. http://journals.sagepub.com/doi/pdf/10.1177/0040517507080261.
13. https://en.wikipedia.org/wiki/Software_sizing.
14. http://www.amritlakshmi.com/Brouchure/Multi-%20Cylinder%20Sizing%20Machine.pdf.
15. http://nptel.ac.in/courses/116102005/23.
16. https://www.google.ch/patents/US5149981.
17. http://www.tandfonline.com/doi/abs/10.1080/19447027.1926.10600003?journalCode=jtit20.
18. https://repository.lib.ncsu.edu/handle/1840.16/5780.
19. http://nptel.ac.in/courses/116102005/16.

10

Control Systems in Weaving

LEARNING OBJECTIVES

- To understand the importance of control systems in the weaving machine
- To identify the measuring devices used in the weaving machine and to understand its working principle
- To comprehend the concept of drive and its importance in the weaving machine

---

10.1 Introduction

The interlacement of warp thread and weft thread to form a cloth is called weaving. Rising wage levels make it essential for the weaving industry constantly to improve the production of weaving machines. This higher output must not, however, lead to any deterioration in fabric quality brought about by mechanical causes. Therefore high-speed weaving machines must be equipped with precise positive monitoring devices with quick response. The machines must moreover be easily maintained in order to keep down maintenance and training times for operatives and to eliminate faulty operation. The above requirements can only be achieved in practice by the use of electronic controls and monitoring devices in place of electromechanical or purely mechanical components (Figure 10.1).

10.2 Electronic Shedding

The treadle motion consists of crank gear for up to six weaving shafts positively controlled tappet motion for up to ten heald shafts, specifically for heavier fabrics and large weaving widths electronically and positively controlled dobby for up to 16 heald shafts with shedding by means of servo motors (E-shedding) (Figure 10.2).

Electronic shedding opens up completely new possibilities for producing ultra-fine, dense, and delicate fabrics of a quality never before achieved. As each shaft is driven by a servo motor, the movements of the shafts can be individually controlled. The crossing time and duration of standstill can be programmed easily at the terminal. With electronic shedding, shed separation can be improved and warp rollover prevented (Figure 10.3).

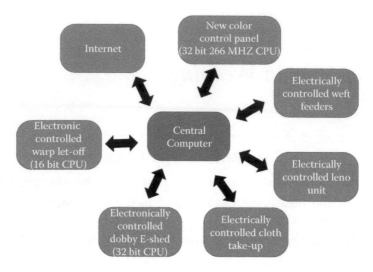

FIGURE 10.1
Application of microprocessor in weaving.

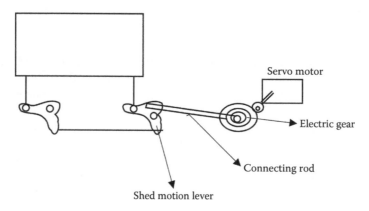

FIGURE 10.2
Electronic shedding devices.

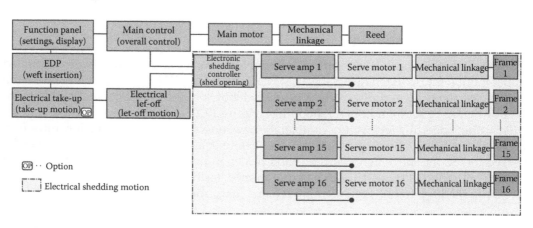

FIGURE 10.3
Block diagram of electronic shedding program.

10.3 Electronic Jacquard

Bonas has developed a very simple individual electronic control mechanism for electronic shedding. It is based on a simple pulley mechanism that multiplies the movement of the healds and, by the use of simple electronic magnetic contactors, selects whether a warp end has to be lifted or not. The use of durable, lightweight plastic components means that the system is self-lubricating, inexpensive, and capable of running at exceptionally high speeds—1000 selections per minute is well within range.

It is now possible to have up to 36 shedding points through a unique, patented six-bar linkage. This offers a wide selection of positions from parallel in oblique shedding. The rapid and shed changing a link pivot permits weavers to modify the shed easily.

10.4 Automatic Pick Controller

The computer automatically sets all conditions for air weft by insertion (pressure and timing) according to the weaving patterns and speed set by the automatic initial condition setting (ICS) system. Even experts find it hard to decide which guide system is best for flexible rapiers in particular applications. Using guide teeth means accepting reed marks and filament damage, and working without guide elements leads to more yarn breaks in the selvedge area and overruns caused by asymmetrical rapier entry in the weaving shed. This seriously restricts the flexibility, productivity, and reliable quality required by versatile, internationally active weavers producing a wide range of styles. The technically perfect solution for guiding mechanical elements is the contact-free form of the aerostatic bearing. The straightness of the rigid rapier offers ideal conditions for this "air cushion guide." A guide plate that directs air to the rod replaces the existing guide rollers. The plate also has a temperature monitor that, for the first time, provides automatic self-control on mechanical filling insertion systems. This opens up a new dimension in process reliability for users and provides a decisive technological lead for positive rigid rapiers.

10.5 Controls in Weft Insertion System

With the Flexible Insertion System (FIS), widely differing weft yarns can be woven using optimum air pressure. FIS is equipped with two independent systems for high and low pressure compressed air for inserting heavy yarns and light yarns and its respectively controlled and monitored weft insertion positively. The Automatic Brake System (ABS), Automatic Pick Control (APC), and Automatic Time Control (ATC) all help to ensure optimum weft insertion and braking. Simultaneously, weft strain is greatly reduced by synchronization and control of the components involved in weft insertion. The system is reliable even when weaving yarns of low and/or varying strength (Figures 10.4 and 10.5).

It consists of an automatic main nozzle pressure regulator, controlled via the weft arrival time in the weft detector (time controller). This is the technological heart of the air system; it reduces the energy consumption and weft waste, reduces the operator's workload, and increases the reliability in production. Two compressed air tanks with different pressures

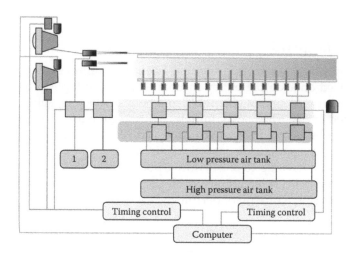

FIGURE 10.4
Flexible insertion system.

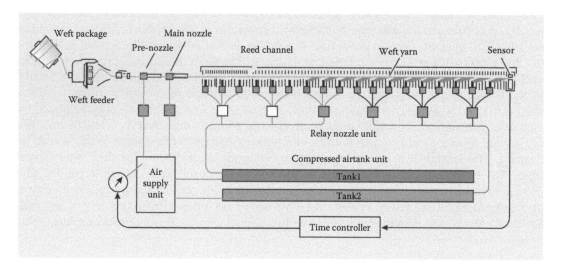

FIGURE 10.5
Compressed air system.

enable air to be distributed in two separate systems. Different air pressures can thus be applied on the insertion and output sides. Intelligent compressed air distribution to the valves, which supply three relay nozzles per valve, results in low air—and thus also energy—consumption.

10.6 Let-Off Electronic Control Warp Beam

- Terry weaving machines consists of automatic warp let-off motions of the ground and pile warps, by which the two warps are let-off by drive motors. Control of these motors is performed by signals from proximity switches fitted to levers of the suspended rollers; the switching cycle of the proximity switch is so small that on any slight change in the preselected warp tension, the warp let-off motor is switched on immediately for a short time.

- Filling insertion speed reaches 1300 mts/min corresponding to 550 PPM. A variable speed drive pulley enables weaving speed to be easily adjusted in increments of 8 PPM.

- Microprocessor-controlled filling tension system that allows filling tension to be varied during the filling insertion cycle, thus eliminating excessive or insufficient tension (peaks and valleys) that causes filling stops and/or fabric defects.

- Digital control of warp tension with input of setting data through main keypad. Programmable automatic compensation of warp tension at a warp or filling stop, to avoid start marks. Immediate reaction of the let-off motion to programmed changes in pick density by the electronic take-up motion, with even tension in forward or reverse modes and when manually advancing the warp sheet. Optional double electronic let-off motion for twin beams. Preset and actual warp tension displayed on console display screen. Capability of weaving different pick densities in the same pattern in combination and synchronization with color or weave changes.

10.7 Electronic Take-Up Motion

Weft density can be changed at will, even while operation is in progress, making border weaving possible for high-grade handkerchiefs, for example. All settings can be designated from the new function panel. Settings for up to 100 different patterns can be designated, and change-wheel changes are unnecessary, simplifying the weaving start-up operation.

10.7.1 Electronic Weft Detectors

The direct monitoring or sensing of the weft thread is still performed mechanically by one, or on wide weaving machines with two, control weft detectors. On electronically controlled weaving machines, the machine is no longer stopped by awkward mechanical means, but when weft is absent, the detector fork produces an electronic signal through a charge in the magnetic field. This is evaluated by the electronic control, and the machine is stopped in the desired position. The electronic control can also switch on the automatic reverse and stop the machine in the open-shed position when a weft thread breaks. Compared with the mechanical weft detectors, the reaction is appreciably quicker, and adjustment and maintenance work is largely unnecessary.

10.7.2 Weft Sensors

Conventional optical-electronic weft sensors operate on the principle of reserved reflection. They sense the weft pirn without contact but require a reflective coating on the pirn. The latest weft sensor, however, does not require any treatment of the pirn.

A light beam is emitted by a transmitter senses the weft pirn at a flat angle and axial direction. If there are still coils of yarn on the pirn, the receiver is in the shadow of the coils and only receives a small proportion of the light beam. With a clear surface, the pirn reflects sufficient light to actuate the pirn change.

10.8 Warp Stop Motion

The warp sheet is fed in the weaving process at a low speed and with a large number of the warp ends. Hence, checking the continuity of the individual warp ends requires an essentially different approach from that adopted in the weft checking.

The main problems of the warp stop motion design are:

1. The high number of warp ends to be checked; in cotton weaving the average warp set is 20 ends/cm, and in silk and filament weaving it is 40 ends/cm and more.
2. The large free length of the warp ends to be checked; the warp length in the weaving plane ranges from 1.2 to 1.6 m.
3. The many possible breakage points, scattered at considerable distances, that is, on the lease rods, in the drop wires of the warp stop motion, in the drop wires of the stop motion, in the healds, in the reed, and at the beat-up line. Moreover, the warp end may already be disrupted on the warp beam.

10.8.1 Classifications

The warp stop motions are classified in three categories:

1. The drop wire stop motions
2. The harness stop motions
3. The contactless electro-optical stop motions

With the drop wire and harness stop motions, the stop signal may be transmitted mechanically or electrically. The contactless stop motions are usually based on an electro-optical checking principle and an electrical stop signal transmission.

10.8.2 The Warp Stop Motions for Large Width Weaving Machines

On the large width weaving machines, the warp stop motions frame must be adequately rigid or supported in the middle to avoid deflection of the warp stop motion rods. To make the breakage location easier, the warp stop motion rods are divided in two zones, each provided with an individual breakage indication system.

The through rod helps to find the dropped wire more rapidly. The rod is swung out in the two indicated directions, and while moving freely under the drop wires pinned on the tension warp ends, it pushes the dropped wire outside the row.

10.8.2.1 The Harness Warp Stop Motions

If the warp end breaks at the front near the reed, it may get entangled with the other way ends. If it is not slackened at the rear of the weaving plane, the respective drop wire does not drop down and the weaving machine is not stopped. In this instance, it is convenient to use harness warp stop motion.

The mental shaft rods and carrying the healds are connected to one terminal of the electro circuit. The lower part of the heald shaft frame is provided with plate, which is connected to the other terminal of the electric circuit. The current is not fed in the circuit unless the heald shaft is at the lower position; then the spring-loaded contacts come to rest on the contact plate, which is mounted on the machine frame and insulated from it. Electrically connected contact are on one side, rods, and on the other plate. During normal operation, the healds are lifted by the motion of the warp tension when the heald is at the lower position and the electric circuit is disconnected. When a warp end breaks, the heald drops, and when current is switched on at the lower heald shaft position, the electric circuit is completed, releasing a stop signal, and the weaving machine is stopped.

10.8.2.2 The Electro-Optical Warp Stop Motions

One of the systems is based on the use of pulse counter that, mounted on a cursor, travels regularly across the warp sheet. If a warp end from the preset number is missing, the weaving machine is stopped. This type of warp stop motion is still in the phase of experimentation. It may only be placed where the warp ends are uniformly distributed, that is, close to the fabric fell or on the back rest roller or, possibly, on the lease rods. It may be expected that, in the future, the warp sheet will be checked by this method in two places, on a special single-pass leasing comb at the rear of the weaving machine and close in front of the beat-up line.

10.8.2.3 The Hayashi Optic-Electronic Warp Stop Motion

The light beams are emitted in the shed by a lamp fixed at the reed and a lamp mounted on the machine frame. The two light sources are operated by a magnetic switch within one part of the machine revolution as long as the shed is fully open. On the opposite side of the warp sheet, photoelectric transistors are fitted on the reed and the machine frame respectively.

In the warp stop motion, when a warp end breaks, or when spliced warp ends pass, the luminous flux is disrupted and the weaving machine is stopped. This type of warp stop motion is suitable for the weaving widths up to 4 m. It is a contactless checking of the shed cleanliness. The stop motion has outstanding reliability in silk and filament weaving. It appears that owing to the textile dust and fly connected with the staple yarns of natural fibers as well as to the water spray connected with the water jet weaving, it will be less reliable in the above two sectors.

10.8.2.4 The Warp and Reed Protector Motions

If the shuttle or gripper projectile, flying through the warp, were retarded or otherwise failed to arrive in the reversing box in time, it would be caught and braked by the closing shed. During the subsequent movement of the reed toward the beat -up position, the warp ends would break; the reed, but also the shuttle or gripper projectile and possibly the temples, would be damaged. The damage caused by it is extremely laborious, time consuming, and costly to repair.

To eliminate the risk, the weaving machines are provided with warp and reed protector motion, which stops the machines in time.

The design of the warp protector motion must meet strict requirements:

1. The stop motion must be absolutely reliable because any failure in operation would result in a heavy breakdown of the weaving machine.
2. The stop motion must interlock the machine start as long as the trouble is not removed.

3. While withstanding high stresses, the stop motion must have small dimensions and allow weight.

The warp protector motions, used at present, can be divided into two groups:

1. The stop motions with mechanical sensors, designed for a mechanical signal transmission on the weaving machines equipped with a mechanical clutch and brake system, or for an electrical signal transmission on the weaving machines equipped with an electromagnetic clutch and brake system
2. The contactless stop motions operating on either a photoelectric, capacitance, or induction principle

10.8.2.5 The Contactless Stop Motions

The photoelectric stop motion during the weft carrier, moving in the direction of the screen, when it attains the exit position, the light beam emitted by lamp and directed toward a phototransistor. A disadvantage of this system is that the stop motion fails to operate if the lamp is burned or soiled. According to modified design, the stop motion operates through two checking positions:

1. The screening of the light beam by the moving weft carrier between lamp and photoelectric sensor is merely registered by the stop motion and the weaving machine is not stopped.
2. When the weft carrier attains the exit position, the photoelectric sensor must again be exposed to the light; otherwise the weaving machine would be stopped. Both checks are interconnected so that when a short pick occurs or when the lamp filament is burned, the weaving machine is stopped.

A capacitance warp protector motion operates on the principle of the change of the sensor capacitance caused by a metallic weft carrier attaining the exit position. A block diagram of this stop motion is shown in the Figure 10.5, which includes a radio-frequency oscillator, stop motion sensor, signal amplifier, and signal translator. The double arrow indicates the output signal. The induction stop motion is arranged similarly.

10.9 Electronic Controls and Monitoring Devices on Shuttle Weaving Machines

With conventional mechanical starting and stopping devices, it was more or less up to the skill of the weaver to avoid cloth faults such as shuttle marks. In the push-button control system, electromagnetic clutch is used.

Drive is usually by a normal 3-phase alternating current motor through v-belts to a flywheel. This flywheel transfers the power through an electromagnetic clutch to the machine. In this way, it is ensured that the weaving machine reaches full speed at one half revolution of the crank shaft at the maximum. This prevents the troublesome starting-up marks. On the other hand, the electromagnetic brake stops the machine rapidly and precisely at any desired position of the sly. It is also actuated even if the electricity supply is interrupted.

The electromagnetic clutch and brake are electronically controlled, as well as various functions of the weaving machines such as

1. Switching on and off
2. Slow forward and reverse running
3. Single pick, controlled by push buttons

For precise control and monitoring, crank angle indicators are necessary. They give information to the electronic control of the machine position or give the signal at the time at which a process is to be completed. The indicators are adjustable within certain limits. In order to reduce weaver movement to the minimum, the push buttons are located at both ends of the machines and on wider machines additionally in the center. Stop buttons at the back of the weaving machine provide for stopping from there also. Three-section warning lamps indicate the causes of stoppages so that they may be remedied without a lengthy search.

In comparison with mechanical control, push-button control brings marked benefits. It permits simple, positive operation, shorter training times, shorter manual work for repairing weft thread breaks, and thus the allocation of a greater number of machines per weaver. Faulty operation that occurs most frequently with extra-wide weaving machines is eliminated. This means a reduction in cloth faults originating with the weaver (e.g., shuttle marks). The above benefits contribute, along with the increased output, toward reducing weaving costs and thus toward counteracting increased wages.

10.9.1 Shuttle Flight Monitoring

The heart of the whole control and monitoring of the shuttle loom is the electronic shuttle flight monitor. In the mechanical shuttle flight monitoring devices used earlier, for example, stop rod and loose reed, the shuttle is only controlled on entering the shuttle box. If the shuttle does not reach the shuttle box at the right time, the weaving machine is stopped by the stop rod or the loose reed. Control is relatively late and requires extensive mechanical work. The electronic shuttle flight monitor detects shuttle movement directly when it enters the shed. In this way, more time can be made available for shuttle flight, for example, in a machine on which the shuttle requires $280°$ of the crankshaft rotation from picking to completed checking in the shuttle box. With electronic shuttle flight control, an additional $30°$ are available for shuttle flight. This results in lower shuttle speed for the same running speed, or as already explained and practiced, shuttle speed is maintained and running speed is increased accordingly.

10.9.1.1 The Principle of Operation of Electronic Shuttle Flight Monitoring

The shuttle with an attached permanent magnet passes over the left side sensor on entering the shed from the left and on leaving the shed passes over the right-hand sensor. These sensors each pass a signal to the electronic control, which compares the actual time (2, 3) of the shuttle with the set time (I) of the shuttle with the set time (I) of the impulse transmitter attached to the main shaft. When the time does not agree, that is, if the shuttle arrives at the right hand probe too early or too late (III), the electromagnetic brake is actuated through the electronic control and the weaving machine is stopped so that the sley comes to a standstill at a positive distance from the fell on the cloth. Only a few milliseconds are required for this process.

The electronic control circuit is designed so that the weaving machine cannot be restarted directly but can only be first operated by the reserving button. This completely avoids cloth damages, for example, shuttle marks. The electronic shuttle monitor thus reacts appreciably more quickly than the mechanical systems. The latter require a lot of experience for optimum picking adjustment because of the complex lever arrangement. Every weaving technologist is fully aware of the consequences of excessive or inadequate picking forces.

The electronic shuttle flight monitor operates without any parts being subject to wear and is practically maintenance-free. A further advantage is that shuttle flight can be reproducibly adjusted at any time relatively easily and precisely. The weaving machine makers have developed further suitable adjustment and control devices, which will be briefly described in the following.

With the incorporated control device of the conversion kit Ruti C 1006 through sensor built into the shuttle race, the speed of the shuttle is measured and digitally displayed. This provides a precise value for exact picking adjustment. Measurement in the desired direction of shuttle flight can be undertaken and by the digital display compared with the set value. The device can be fitted stationary on the weaving machine or if required is available as a portable unit for checking a group of weaving machines.

The Saurer Monitor MS 300 electronic setting and control unit was developed for the precise "trimming" of the versa speed 300 high-speed shuttle weaving machines. The device consists largely of the transmitter, display, and sensing sections. The transmitter, a crank angle indicator, is linked mechanically to the central circuit of the machine and linked by a synchrony-tie to the display section. The display section has an up to-electronic device for the measurement and optical display of the sensor impulse and a connection for the sensor section and flash stroboscope. The sensor section consists of an up to-electronic measuring head for the precise determination of the end position of the shuttle. The electronic setting and control unit operates on the following principle.

On each crankshaft revolution, an up to-electronic impulse is given by the transmitter section in the position of the crank cycle scale shown on the visual display. If the crank cycle position of the measured impulses coincides with the position of the crank impulse, a coincidence lamp lights up on the display. The monitoring impulses are measured with continuous rotation of the scale of the display section and crank cycle zone recorded in which the coincidence lamp lights up. All results are recorded on a special record sheet.

For measuring the shuttle end position, a small piece of reflective tape is attached to the picker. When the shuttle is driven into its end position, the picker passes a light cell. The reflective tape transmits a signal through the light cell to the setting and control unit, which serves as an aid for picking adjustment. The parameter setting of the weaving machine is carried out by the picking force of the force of the right and picker being adjusted in such a way that the desired shuttle end position is achieved at the left side the crank positions of the signals front he probes fitted in the shuttle race are recorded. The picking force of the left-hand picker is then also adjusted so that in the shuttle flight from left to right similar impulses are produced by the sensors. Thus symmetrical picking forces or shuttle end positions are obtained.

After the above procedure, all other adjustments are made according to prescribed methods or experience values and are recorded on the test sheet. Once these measurements have been made, the electronic and many mechanical settings may be checked at any time and reset reproducibly in accordance with instructions. Optimum adjustment of shuttle weaving machines is possible using the apparatus described above.

10.9.2 Electronic Monitoring Devices on Rapier Weaving Machines

Rapier weaving machines are controlled almost exclusively by electromechanical means through punch buttons. If electronic control is used, as applied on one model, it works on the same principle as on shuttle weaving machines, so there is no need to describe it here. Electronics are used on rapier machines primarily for weft monitoring. Exceptions are those weaving machines operating on the Gabler system (loop insertion). On these machines, weft monitoring by piezoelectric weft detectors is not advisable as the thread draw-off speed in the second half of the weft insertion operation in which monitoring is needed is zero.

For rapier weaving machines operating on the Dewas system (tip transfer), the piezo-electric monitoring system is preferred. The weft thread without any additional tensioning is passed at low angle over the sensing unit. The start and duration of monitoring are determined by the control system (trigger). With the Loepfe system, this consists of a stationary coil and a switching quadrant rotating with the weaving machine crankshaft. A signal is thus produced by induction. Eltex uses a light cell for control, which operates in conjunction with a marker on the crankshaft as an electronic switch. On electronic weft detectors for rapier weaving machines, the required minimum speeds for the positive functioning of the device on the Dewas weft insertion system is dependent upon the following criteria:

1. The warp-round angle of the weft thread
2. The thread material
3. The yarn count
4. The potentiometer setting

It was demonstrated that a certain "pressure" must be exerted on the thread guide eye or the warp-round pin for the weft detector to be able to register any thread movement. The following advantages of electronic compared with mechanical weft yarn detectors may be quoted:

- Monitoring of weft yarns of all types over the full working width
- No movable sensing parts
- Simple mounting, adjustment, and maintenance
- Monitoring of single, double, and multiple weft insertion
- Low thread tension and no weft thread breaks in the center of the cloth

10.9.3 Real Time Monitoring and Planning for the Weave Room

Real time machine monitoring and production reporting is still one of the primary functions of a CIM system for a textile plant. WeaveMaster is a loom monitoring system. Through a graphical user interface (GUI), WeaveMaster users are constantly informed about the actual situation in the weave room, resulting in faster reactions to problems and an increased efficiency. Powerful analysis tools allow quick identification of poor performing machines and bottlenecks in planning, resulting in an optimal usage of production capacities. With today's small order quantities and short delivery times, scheduling has become a critical function for the textile mill. WeaveMaster offers the planning department an interactive tool allowing them to optimize loom loading based on real time information. Using the World Wide Web, the salesman can give immediate and correct information about deliveries.

10.9.3.1 Color Mill

The most important real-time analysis tool in WeaveMaster is the Color Mill. On this color-coded layout of the mill, the machines are pictured in a number of colors, each color indicating a certain machine status or alarm condition. From a data selection window, the user selects the type of information to be displayed: efficiencies, speeds, stop rates, and so on. For each data item, exception limits can be defined and problem machines are automatically flagged. User definable "filter sets" allow the user to display only those machines that correspond with a certain condition, for example all machines with an efficiency less than 85%, all machines waiting for an intervention, all machines weaving a specific style, and so on. A mouse click on a specific machine opens a window with a detailed report showing all required information for the selected machine. As a unique feature within WeaveMaster, these detail reports may contain any user defined mixture of text and graphics.

10.9.3.2 Cockpit View

The Cockpit View is the ultimate analysis tool for the plant manager and the industrial engineer. Four graphs show the vital information needed to make quick decisions at the glance of just one screen.

10.9.3.3 Film Report

The film report is the result of an automatic machine activity sampling carried out by WeaveMaster. Based on user definable time intervals, the film report makes a complete machine diagnostic and shows whether efficiency losses were due to one long stop or were caused by several short stops.

10.10 Electronic Yarn Tension Control

An additional tension meter allows you to quickly measure weft tensions accurately. The portable and precise sensor is connected to the microprocessor of the machine. The tension values are displayed as a graph, so the operator obtains the information needed to fine-tune and control settings very fast. Fine-tuning and troubleshooting are easy thanks to the immediate feedback. The results are repeatable and reproducible on other machines.

Using the tension meter will have a direct impact on productivity and quality. Each prewinder can be equipped with a new type of programmable filling tensioner (optional). This PFT is microprocessor-controlled and ensures optimum yarn tension during the complete insertion cycle. Reducing the basic tension is an important advantage when picking up weak yarns, while adding tension is an advantage at transfer of the yarns and avoids the formation of loops. The tension control enables you to weave strong or weak yarns at even higher speeds. It also drastically reduces the amount of filling stops, and enables you to set an individual waste length per channel and reduce the waste length for some channels.

10.10.1 The Production Sensor

The sensor determines the speed of a shaft that is rotating in proportion to the production. From the revolving impulses, the following information is available via the weaving machine:

- Running/stop
- Efficiency in %
- Number, frequency, and average duration of the short stops
- Out of production times
- Production speed (revolutions per minute)
- Production amount (in the number of picks or meters)

The production sensor has been arranged for the special ambient conditions encountered in textile mills.

- The signal determination is undertaken by means of an inductive sensor and is therefore not influenced by dust or dirt.
- The built-in light emitting diode signals the revolving impulses and provides for a simple function control directly at the weaving machine.

10.10.2 The Central Unit

The central unit is the heart of the Loom Data computer-controlled data system. It fulfills primarily the following functions:

- Periodic calling up of the concentrators and machine stations
- Continuous preparation and memorizing of the machine signals
- Control of the dialogue with the user via the monitor and printer
- Output of reports
- Output of pre-ordered data via the special online interface for further processing and central computer system
- Control of the program timings via a built-in timer
- Determination of the allocation and control

The central unit is arranged according to the special requirements of a process data system applied to weaving:

- The central unit is fully electronic and requires no special climatic conditions. It contains parts that are sensitive neither to dust nor to wear such as ventilators, disc drives, etc.
- The program cannot become lost as it is contained in fixed value memories (EPROM's).

10.10.3 The Three-Stop Connection

The stop signals for warp and weft yarn breaks are accepted by the three-stop connection box from the weaving machine electrical circuit. The three-stop connection box is only suitable for those types of machines where electrical stop signals are available.

- Provides the galvanic separation necessary for safety reasons between the machine electrical circuit and the data system
- Makes possible a flexible adaptation to the type and voltage of the electrical stop signals
- Makes possible a means of arranging connections in a simple manner of the production and stop signals via three-core cable with the concentrators

Special connection conditions include:

- With modern weaving machines having electrically available production signals (one impulse per pick entry), the fitting of a production sensor is not necessary. Disabling the production signals is undertaken in this case by an extra impulse entry circuit which is built in at the three-stop connection box.
- At weaving machines with electrical warp and mechanical weft break detectors, a three-stop determination can be achieved if the signal for "manual stop" instead of the signal for "weft break" is evaluated.
- With certain older types of weaving machines, it is sometimes necessary, from case to case, to include an extra relay circuit for separating the stop signals for "warp yarn break," "weft yarn break," and "other stop."

10.10.3.1 Standard Reports

The most important production data are summarized in so-called standard reports. These reports primarily serve the purpose of issuing the data collected for every machine according to certain selection characteristics. The evaluation of the data is considerably facilitated by the standardized and clear arrangement of the structure of the report.

The main characteristics under which the standard reports can be called up are:

- *Machine area*: The machine area is a higher-order characteristic that can be utilized if desired. This remits the arrangement of the reports according to the weaving sections or the issue of separate reports for preparation processes or looms.
- Machine.
- *Style*: With each of these reports, either a determined value of the characteristic is selected or all values can be set out in an ascending sequence.
- Weaver.
- Fixer.
- Group.

Report variants for the preparation of purpose-oriented standard reports are:

- All individual machines
- Only sum lines per characteristic

- All machines with (machines outside tolerances)
- All machines that are out of production
- All machines that were out of production
- Machines classified according to style
- Only sum lines per style and characteristic
- Selected information for economical fabric production

The Loom Data system presents the information collected in the form of clearly arranged reports.

- Concentration on key data, which can be utilized for forecasts, simplifies the evaluation and avoids the confusion of a mass of figures.
- For simple operation, the call up and structure of all reports is standardized.
- All types of reports can be called up independently at any time on the printer and the video terminal.
- Freely selected reports can be printed out automatically at the end of the shift.

List of reports available:

1. Standard reports
    a. Machine
    b. Style
    c. Weaver section
    d. Fixer
    e. Group
2. Special reports according to machine areas out of production
    a. Warps
    b. Style standards
    c. Warp store
    d. Control data
3. Service reports
    a. Service concentrator
    b. Service weaver
    c. Service online/offline
    d. ROM/RAM test
    e. Power failure

Reports:

- The machine report provides immediate information concerning the operating behavior of individual machines.
- Special report variants serve for the selection of machines with deviating operating behavior. The all machines out of production report, for example, covers all those machines that have been stopped for some time.

- The styles summary shift report permits an immediate comparison between the different styles.
- Styles with partly the same identification within the style identification can be selected accordingly and summarized.
- In the standard variants, the weaver section report provides the responsible people with a review of the operating conditions at a specific workplace.
- With the summary variant, all weaving areas can be rapidly compared with each other.
- The fixer report facilitates the selection of the machines according to fixers areas. With the variant machine, the person responsible can concentrate his or her attention on machines with unsatisfactory operating behavior.
- The group report is of particular value for mill trials. Examples include comparison of sizing recipes, yarn batches, and machine settings. Selected machines can be summarized as an experimental group and evaluated independently of the usual characteristics.
- The warp report serves for the allocation of warps. The warps are detected in all the stages of planning and preparation; for example, warps that are in the preparatory process are also shown next to the warps actually running in the looms as well as warps that have been woven out. The flexible structure of the report provides different variants for the adaptation of the report to the needs of the moment. In the warp report, selection of specific values or classification can be made according to:
  - Warp number
  - Machine
  - Style, etc.
  - Warps that are allocated to particular machines
  - Warps that are not yet allocated
  - Warps for which a succeeding warp has not been planned
- The report from diagram provides a graphic representation of the warp running time during a selectable period of time. The time slop shows the warps allocated to a machine area, classified according to machines.
- The report op-diagram provides a survey of the time slop of the out-of-production conditions. The graphical representation of the time slop improves the legibility and, for example, assists in recognition of overlapping effects. In the following example, those machines in a fixer's area are represented that show out-of-production times.

10.10.4 Connection Capacity of the Central Unit

The connection capacity of the central unit is up to approximately 1000 weaving machines, in each according to the number of warps, styles, and weavers. For all known types with three-stop detection, technical considerations are necessary from case to case. Based on the requirements, different type of data output instrument can be connected like one printer terminal and one to six monitors. The discrete stations or online communication consists of overall computer systems.

10.10.4.1 Data Assurance with a Mains Breakdown

Timer circuit and memory for the machine and production data contains a break-down proof feed for bridging voltage disruptions of up to five days.

10.10.4.2 Function Control

The function control consists of built-in-testing program and it can be changed according to the user/machine requirements.

10.10.4.3 Distributed Control System

The main control computer is a 32-bit CPU for quick data processing. A distributed control system connected by optical fiber cable can follow to the high-speed machine operation.

- Destination addresses are assigned to all transmitted data. Each system component will accept only data that bears the address of that component. Data that bears another address will be passed on. With this advanced distributed system, the data entered via the new function panel is transmitted quickly and precisely to each component.

10.10.4.4 Memory Card

Data entered in the new function panel can be instantly stored in the memory card and transferred to another loom, preventing common human errors in transferring data. Memory card writing can be executed via the machine's function panel, or remotely with the Handy Function Panel.

10.10.4.5 Handy Function Panel

The Handy function panel, in conjunction with a memory card, allows you to set patterns and machine conditions off-site, away from the loom.

10.10.5 Configuration of Loom Data System

In the standard arrangement, data are collected from each weaving machine by means of one single universal sensor. The sensor deducts, without contact, the number of revolutions of any particular weaving machine, for example, a drive shaft that turns proportionately to the pick entry frequency and produces per pick entry and electrical impulse. From this information, the installation calculates the data such as running/stop condition, the number of picks, the speed, efficiency, the out of production time and frequency, and down time of the short stops. The illuminated diode built into each sensor provides a simple method of carrying out the function control directly at the machine (Figure 10.6).

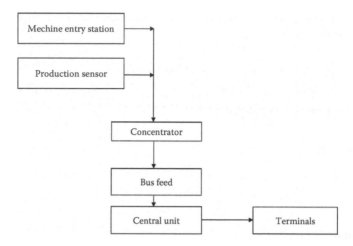

FIGURE 10.6
Design and function of loom data system.

10.10.6 Software for Loom Data

In order to be able to carry out purpose-oriented trials, selection possibilities are available according to the following groupings.

1. Machine wise
2. Style wise
3. Weaver wise
4. Over looker wise
5. Any freely chosen group

All these reports can be called up both for the running shift and also for the previous seven shifts. A number of other call-up possibilities cover further functions such as:

1. Out of production time
2. Warp stops
3. Quality data
4. Control data (shift change times, service reports)

In addition to the various production reports, it is also possible to get warp planning reports regarding warp beam change prediction and status of the warps stored. All these reports are automatically printed out at the end of the shift or can be printed at any time during the shift. It is also possible to provide security by means of passwords to protect certain sensitive data from being changed by unauthorized persons.

It is possible to store some important reports in a summarized form in a long-term memory. The system is capable of storing data for a maximum period of one month. In spite of the multiplicity of call-up possibilities, the installation is simple because the calling up of reports and the entry of the data is undertaken together with the calculator. This means that the calculator always guides the user what he or she should enter next.

This menu-based system institutes a question and offers a list of entry possibilities, thereby helping the user to choose the correct entry. Also by means of a service report, it is possible to locate any disturbances to the system or the sensor cable being disturbed or short circuited.

Functions of Loom Data System.

1. Increased production performance
   a. Higher efficiency
   b. Better running characteristics of machine
   c. Rapid detection of weak spots
2. Reduction of out-of-production time
   a. Positive recognition of the most important causes of stoppages
   b. Simple control of the effectiveness of organizational measures
3. More effective quality assurance
   a. Fewer second quality products
   b. Rapid attainment of optimum operating condition
4. More effective deployment of personnel
   a. Reliable decision basis for all levels of supervision
   b. Defined assignment of service personnel
5. More rational warp management
   a. Supervision of all warps at a glance
   b. Due dates, continuously calculated from the progress of production
6. More reliable cost control
   a. Comprehensive production data for all styles and machines
   b. Reliable actual values
7. Higher employee motivation
   a. Objective comparison possibilities between workplaces and shifts
   b. Practicability of fair remuneration
8. Cost-saving acquisition of raw data
   a. Error free data
   b. Direct transmission of the data

10.10.7 Data Processing in the Air-Jet Weaving Machine

The principle of operation reflects recent findings in ergonomics research. A browser-ready terminal with a color touch screen is the man-machine interface. Via simple menu prompting, the expertise stored in the machine provides the basic settings for the weaving process. Malfunctions are rectified by menu-assisted problem analysis. Service via the Internet offers additional online support. The weaving mill control and monitoring system can be integrated in the network. Machine and style data can be compiled on a PC outside the weaving room and transferred to the machine either online via a network or off-line via a memory card.

10.10.7.1 Intelligent Pattern Data Programming

"SmartWeave" offers fabric designers intelligent support in the preparation of weave designs and pick repeats. The prepared pattern data are combined with the setting parameters of the style and transferred to the G6500 via data carrier or intranet. Once the data has been transferred, the machine can be set up quickly. Weaves, pick sequences, and setting parameters can of course also be programmed and/or modified directly at the terminal.

10.10.7.2 Network-Ready Touch-Screen Terminal

The control interface is a user-friendly, Internet-ready touch-screen terminal. The logical structuring, with self-explanatory pictograms, guides the operator to the desired function simply and with a minimum of keying. To the Internet capability, remote interrogation of the machine parameters is possible. This allows potential improvements to be rapidly identified and put into effect.

10.11 Sumo Drive System

The SUper MOtor is a main drive motor with exceptional characteristics that has been recently developed by Picanol. The Sumo drives the weaving machine directly, without belt transmission or clutch and brake. The speed of the motor is variable and is electronically set and controlled. Automatic pick finding and slow motion movements are also executed by that same motor.

- Machine-speed setting is done accurately and completely electronically via the keyboard of the microprocessor. This reduces the setting time to zero.
- Speed setting is easy to copy to other machines either with the electronic SetCard or with the production computer with bidirectional communication.
- Automatic pickfinding goes faster, which significantly reduces the downtimes for repairing filling and warp breakage.
- The machines can always work at the optimum weaving speed in function of the quality of the yarn, the number of frames, and the fabric construction.
- Sumo has a wide speed range, which is useful for starting up new styles. Conventional drive systems can cover max.100 RPM, after which the pulley must be changed.

10.11.1 Programmable Filling Tensioner

The programmable filling tensioner (PFT) makes it possible to control and adapt the peak tension of the filling yarn. This results in fewer machine stops and higher fabric quality. The PFT also makes it more feasible to work with weaker filling yarns.

10.11.2 Automatic Pick Repair

Automatic pick repair (PRA) uses air blasts to remove faulty picks on air jet machines and then restarts the machine automatically, thus dramatically cutting down the number of machine stops.

10.11.3 Quick Style Change

With the Quick Style Change (QSC) system, a style change can be carried out in less than 30 minutes by a single person. The system drastically reduces machine downtimes, as well as raising productivity and reducing the number of operators required. Furthermore, it confers great flexibility, making it economically attractive to weave a variety of styles, even in small quantities.

The Picanol machines are designed for QSC, and customers are informed about it in depth. The success of the QSC system is demonstrated every day by the more than 4000 Picanol weaving machines on which it is installed.

10.12 Electronic Selvedge Motions

The unique electronic selvedge motions (ELSY) full leno selvedge motions are electrically driven by individual stepper motors. They are mounted in front of the harnesses so that all the harnesses remain available for the fabric pattern. The selvedge crossing and pattern are programmed on the microprocessor independently of the shed crossing, even while the machine is in operation, so the result of a resetting can be checked immediately. The easiest position for re-threading can be set by a simple push of a button. When the machine starts, the selvedge system automatically returns to its original position. At a style change, the ELSY can be removed and repositioned very easily.

Improved control of the weft insertion leading to fewer filling stops is afforded by the positive opening of the right-hand traction gripper with the new ERGO II system. This allows individual setting of the moment of opening according to each type of weft inserted. Both the length of the weft tail and the degree of opening of the right-hand gripper during the release of the weft can be adjusted.

Both settings are done at the display with a few keystrokes. This enables the immediate observation of the result. The feature also makes slow weaving of elastic yarns possible by delaying the opening of the gripper in that case until the ELSY selvedges have closed onto the weft, preventing it springing back into the shed. This perfect control of the waste length in all circumstances reduces the cost of weaving by ensuring top quality with minimal wastage of weft material.

10.13 Weave Master Reporting Report and Formula Generator

All data is stored in a relational database and is available for history reporting. By means of a report and formula generator, a textile manager without any programming or SQL knowledge can define his or her own calculations and reports. For every report item selected from the database, upper and lower warning and alarm levels can be defined resulting in color coded exceptions in the report.

Once a report layout has been defined, the user can select it for a variety of selection keys such as by machine type, by weaver, or by style, and for any period such as shift, day, week, month, or even year.

10.13.1 Integrated Graphics

Data to be presented in graphic format is selected by a single mouse click. The user has a set of graphic options available, such as bar charts, pie charts, line charts, and pareto charts. These graphical reports are useful for a quick identification of poor performing machines or styles. Further analysis allows the identification of the most important stop reasons within a group of machines or styles.

10.13.2 Automatic Printing and Data Export

All reports available in the WeaveMaster system can be printed on request, defined as automatic print reports to be printed at predefined time intervals or as an export file. The latter allows easy and transparent data transfer between the monitoring system and any other information system within the company.

10.14 WeaveMaster Production Scheduling

10.14.1 Planning Warps and Pieces: The Graphical Plan Board

With WeaveMaster, planners conduct their demanding job by means of an electronic plan board. Integrated with the style database and the monitoring system, the plan board software automatically calculates the exact time for every order and updates it based on real time information such as actual speed, efficiency, and stop level. The WeaveMaster scheduling software supports multiple planning levels. Some textile mills only require single warp planning; other companies, such as terry towel and upholstery weavers, require the scheduling and follow up of multiple warps as well as single pieces on every loom. By means of simple drag and drop functions, the planner can allocate pieces to warps, reschedule warps and pieces, assign them to another machine, and so on. The system calculates the consequences on the spot. Production orders can be entered manually in the system or can be downloaded from a host computer.

10.14.2 Printing of Warp Tickets and Piece Labels

From the plan board, the user has access to the style specifications database with all style-related technical details. Based on the loom loading and the information available in the style database, WeaveMaster prints the beam ticket and piece labels as well as instructions for warp preparation, drawing in, and so on.

10.14.3 Warp Out Prediction and Yarn Requirements Calculation

Based on the warp out prediction, WeaveMaster knows exactly when each warp has to be ready. This information allows the system to calculate backwards in order to produce a production schedule for the warp preparation department and to produce a yarn requirements list.

Microprocessor-controlled looms equipped with the VDI interface are connected by means of the DU7P interface board. Automatic stops are transmitted through the microprocessor's VDI-interface, and weavers enter manual declarations through the keyboard and display of the loom. As such, the weaver uses the same user interface for operating the loom as for communicating with the monitoring system. Unlike with other systems, no extra keypad is required. Through full bidirectional communication, the DU7P has access to all information and can activate any function within the machine's microprocessor.

10.14.4 Looms with Ethernet Interface

The latest generation of microprocessor-based looms is often equipped with an Ethernet interface for host communication. These looms are connected to the WeaveMaster system by means of a standard Ethernet network (UTP 5 cable) without the need for any additional hardware interface.

10.15 Summary

The chapter deals with the control system concepts used in the weaving machine. The necessary measuring device used in the machinery is explained with its working principle. The interface of the weaving machine with monitoring systems is also discussed in detail.

References

1. http://textilecentre.blogspot.com/2014/02/electronic-control-systems-in-weaving.html.
2. http://textilelearner.blogspot.in/2013/09/different-types-of-shedding-mechanism_4.html.
3. http://www.google.ch/patents/US6988516.
4. https://www.google.ch/patents/US7740030.
5. https://www.google.com/patents/WO2016134973A1?cl=en.
6. http://www.fibre2fashion.com/industry article/3194/weaving-weft-insertion-rapier?page=2.
7. https://www.scribd.com/document/60518531/Copy-of-35174000-Computer-Application-in-Textiles-3-Vig-Dhanabalan-Pavitra.
8. http://textilelearner.blogspot.in/2011/06/modern-weft-stop-motion-device-of-loom_7523.html.
9. http://nptel.ac.in/courses/116102005/33.
10. http://oecsensors.com/magnetic-proximity-sensors-for-weft-break-detection/.
11. http://yongjin-machinery.com/?gclid=Cj0KEQjwt8rMBRDOqoKWjJfd_LABEiQA2F2biLroQRzAbQ39yMgxLuteB8iMDgrXyMWvAAlTrdCzI04aAiel8P8HAQ.
12. https://www.lindauerdornier.com/global/mediathek/brochures/weaving-machine/dornier-rapier-type-p1-e.pdf.
13. http://www.bmsvision.com/sites/default/files/downloads/WeaveMasterEasy_BRCH_EN_A00694_0.pdf.
14. http://textilecore.com/production-control-and-analysis-in-the-weaving-rooms/.
15. https://www.google.si/patents/US6321576.
16. https://www.google.co.in/patents/US3093330.
17. http://www.textileworld.com/textile-world/new-products/2000/11/picanol-introduces-omniplus-with-sumo-drive-motor/.
18. S. Maity, K. Singha, M. Singha. Recent Developments in Rapier Weaving Machines in Textiles, India.
19. http://www2.gsu.edu/~wwwotc/WEAVE/Instructions%20for%20WEAVE.pdf.
20. http://strategicplanning.uthscsa.edu/User-Instructions-for-Accesing-WEAVE-and-Entering-Accomplishments.pdf.

11

Controls in Knitting

LEARNING OBJECTIVES

- To understand the control system concepts used in the knitting machine
- To know the automation process used in the knitting machine
- To identify the measuring devices used in the knitting machine

11.1 Designing and Patterning

Perhaps the most revolutionary event since Rev. William Lee's invention of a stocking frame in 1589 is the application of the computer to knitting technology. The computer is basically a simple tool that can be a powerful aid for industrial engineering, research. or management problems. It can perform various elementary arithmetical and logical operations. The main feature of a computer is its tremendous speed. The speed is measured in nanoseconds, which is thousandth of a millionth of a second. This increasable speed allows the computer to perform many millions of operations in a second. But a computer is not only a high speed calculating machine—it can also perform logical AND or OR functions (Figure 11.1).

The problem to be dealt with is broken into a number of the simplest logical steps. This breaking up of the problem into steps and setting the computer to work out the solution is called programming. If the problem is not properly defined and programmed, the computer will not be able to offer any solution to the problem. A special branch has been developed by computer manufacturers that is known as textile graphics, which is a computer-aided technique for developing a textile design or textile machine.

The use of electronics is not restricted only to replace the mechanical needle selecting device, but it is used in pattern preparation also. The well-known example is that of a television screen and a light pen. By using the television screen and light pen, the computer can generate random patterns on the television tubes in full color combination. The designers then select the appropriate and placing combinations, which can then be transferred onto a tape. This tape then helps to give signals to the electronic knitting machine for stitch and color combinations, which are amplified and conveyed to the needle selecting mechanism.

The designer can play with these devices. He or she can draw into a computer a design and program it in such a way that the computer can enlarge, rotate, and manipulate the patterns or can superimpose one pattern on top of the other until the designer is satisfied. There are many limitations in pattern designing, mainly restrictions imposed by the fabric used and knitting machine. If these constraints are defined, they can also be fed into the computer, which can then automatically adjust the pattern, taking into consideration the constraints.

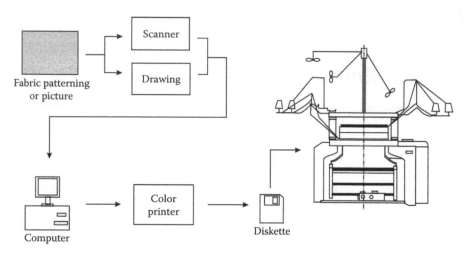

FIGURE 11.1
Designing and patterning.

The point paper designs are required to transfer into a pattern of jacks on the various control wales or drums of the knitting machine. It is a boring task requiring meticulous attention to detail. If the rules of transformations are well defined and specified to the computer, it will calculate the offsets for the jack arrangement for the various wales.

11.2 Electronic Jacquard

Similar to jacquards used in weaving or weft knitting, jacquards are fabricated for warp knitted fabrics. Particularly for Raschel lace warp knit structures, with the advent of electronics, a new generation of electronically controlled multibar Raschel machines has achieved a peak position in jacquard lace manufacturing. Karl Mayer has come out with a jacquardronic series of machines.

The main parts are pattern computer, binary patterning mechanism, electronically controlled jacquard device, and mobile loading device.

11.2.1 Pattern Computer

The pattern details are fed from a diskette into the memory cards of the storage system. High storage capacity and easy access to stored data are advantages of this device. The stored details are fed into the selector units of the totalizing mechanism.

11.2.2 Binary Mechanism

A selector mechanism has six crank units, and it coordinates these units with each pattern bar. The crank drives operate on the binary system and carry out the following needle jumps.

1st crank unit = 1 needle

2nd crank unit = 2 needles

3rd crank unit = 4 needles

4th crank unit = 8 needles

5th and 6th crank unit = 16 needles each

This can give a maximum displacement path of 47 needles.

Electronically controlled jacquard: The pattern details are transmitted from the storage and control unit to the electronically controlled jacquard device. An electromagnet is assigned to each jacquard guide needle. The transmission of the displacement path from the selector units to the patterning guide bars takes place via rocker arm units. The pattern information is transmitted to the storage unit of the microprocessor by means of a data carrier via a mobile loading unit.

Mobile loading device: The data display unit and diskette mechanism transmit the pattern information from the cassette into the pattern computer and also serves as a communication system between the pattern computer and the operating personnel.

The piezo element consists of a carrier plate with ceramic elements on either side. The individual guides are fixed to the guide holders. Sixteen or 32 piezo elements are cast together and attached to the jacquard guide base. The individual jacquard guides are moved sideways by one needle space. When an electric current is applied, the piezo crystals contract or expand and thus bend the jacquard guides out of their normal position and displace them sideways.

11.3 Control Systems

Control systems play a major role in improving the quality of the fabric and reducing the labor force in the knitting industry. Normally, the following areas can be controlled automatically.

11.3.1 Feeding Zone

- Top feeding
- Bottom feeding

11.3.2 Feeding Cone Indicator

Knitting zone
- Needle detection
- Cam setting

Winding zone
- Winding speed
- Winding length

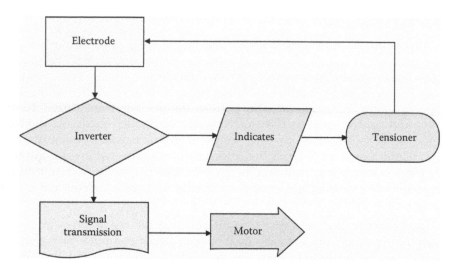

FIGURE 11.2
Feed package controls.

Auxiliary motion

- Oil level control
- Door stop motion

Feeding

- Regulates the correct amount of tension to the feeding yarn
- Controls the quality of fabric by reducing the fault

In the feeding section, there are two stop motion devices. If there is yarn on the tensioner, the circuit will be opened. If there is no yarn, that is if the yarn breaks, suddenly the circuit will be closed and the signal is passed to the inverter. Then the machine will be stopped (Figure 11.2).

11.3.3 Feeding Package Indicator

When the yarn in the cone is in the exhaust stage, the sensor senses the cone diameter and passes a signal to the controlling unit. Then the controlling unit will pass the signal to the lamp to indicate the exhaust stage of the cone (Figure 11.3).

11.3.4 Knitting Zone

This zone is the heart of the knitting machine in which the needle and cam play a major role. In case of needle breakage, a bulge is formed on the yarn because of continuous feeding of yarn. Now the needle detector is displaced by means of a bulge in the fabric; due to the bulge formation, the needle detector gives the signal to the LVDT, and the machine will be stopped (Figure 11.4).

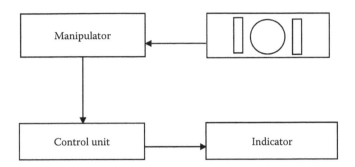

FIGURE 11.3
Feed package indicator.

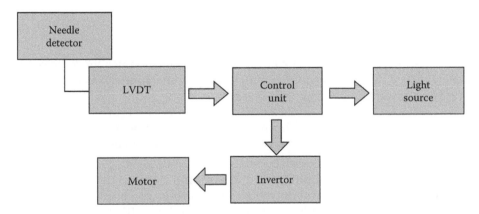

FIGURE 11.4
Knitting zone.

11.3.5 Winding Length

Measuring the correct amount of delivery wound on the take up roller is done by an electromagnetic device. The rigid beam consists of fixed magnetic coil and the revolving take up roller also having the magnetic devices. The magnetic flex created on each revolution of the take up roller. Then the magnetic flux signal is passed on to the controlling devices. This counts the number of revolutions, and if required, when the length is reached the machine is automatically stopped.

11.3.6 Oil Level Controller

In the oil level controller, the oil tank contains two electrodes: one for upper level oil control and one for lower level oil control. When the oil level reaches the minimum level, the signal will pass to the controlling unit and the oil is filled by the oil distributor to the tank. When the oil reaches the maximum level, the upper electrode will give the signal to the controller, and it will control the distribution of oil (Figure 11.5).

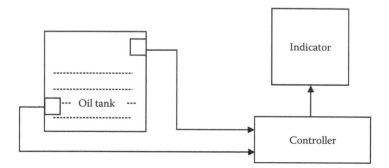

FIGURE 11.5
Level indicator.

11.4 Individual Needle Selection

The control system with piezoelectric bending transducer modules for individual needle selection is tailored for use in jacquard circular knitting machines. This unit combines the following functions in a highly compact design:

- Various pattern repeats, which are stored permanently on the controller, are available immediately after a pattern change, with no time-consuming loading from diskette.

- Functions for continued knitting to pattern after a power failure increase the productivity of machines and prevent production rejects. The piezoelectric ending transducer modules permit low-loss operation on the knitting machine, resulting in low heat generation and a long life.

11.5 Knitting Machine-Needle Detector

The needle sensor monitors the needle hooks on single, fine rib, jacquard, and interlock circular knitting machines without contact. It stops the machines immediately if a needle hook or butt has broken. Depending on the type of knitting and machine, the needle sensor consists of up to four optical heads and a control unit. The optical head projects a narrow spot of light onto the needle hooks running past, which reflect part of the light. These light pulses are fed to the control unit via fiber-optic cable. The control unit automatically adapts to the needle sequence corresponding to the machine speed and gauge of the needles.

If a pulse is not received due to needle breakage, the machine control is commanded to stop. Simultaneously with the stop signal, a needle counter begins to work. It indicates the number of needles running past until the actual machine shutdown. This makes the localization and replacement of defective needles much easier. Needles missing produces separating rows or creates patterns that are masked out. As additional security against false stops, the number of fault repetitions before stoppage of the machine is adjustable. The piezo modules are systems made from integrated piezo-ceramic bending transducers and matching actuation electronics used as selection elements for material patterns, for example, for circular knitting machines. The piezoelectric bending transducers are at the very heart of the piezo modules and are responsible for controlling the anchors and thus also the needles. The piezo modules are specially designed for use with the control system.

11.6 Knit Master System

The knit master control unit, which becomes a permanent portion of the knitting machine, serves three major functions:

- It counts all machine revolutions and will stop the knitting machine after a predetermined number of machine revolutions.
- It serves as a power source and control for the knit master electronic yarn meter.
- It serves as a power source for the knit master compu-Tach, the fast and highly accurate electronic RPM meter, which gives RPMs to 0.1 RPM.

Furthermore, the control unit can serve as a feed source for data collection systems of various types. The knit master electronic yarn meter measures and gives an instant digital read-out of the yarn consumed per machine revolution, in contrast to most other meters that are merely speedometers. As the knit master electronic yarn meter is plugged into the knit master control unit of a running knitting machine and the yarn is engaged in the wheel slot of the meter head, it will immediately start to accumulate a reading. This first reading may be inaccurate, due to the point of entry; the second reading will be absolutely accurate. It will alternately and automatically display a reading and accumulate a new reading every other revolution of the machine, without resetting the yarn meter.

11.6.1 Machines with Surface-Driven Packages

Surface-driven machines with a grooved drum traverse or a traverse geared directly to the drum can be controlled directly in yards or meters. Of all the machines available, those machines that have a common shaft for all or half of the spindles per frame are usually the least expensive to control. Conversely, machines of this type with individually driven spindles (where the various spindles run at different speeds) are the most expensive to control. On common shaft machines of this type, a multi-toothed sprocket is mounted on the shaft with a sensor scanning the teeth as the shaft rotates.

11.6.2 The Functions of the Knit Master System

The knit master system performs a number of different functions:

- It is used to "even-up" all feeds of a machine and to set each feed of a machine to a specified reading.
- It serves as a continuous yield (cost) control checking system for all yarns used and for all fabrics made, in contrast to the normally used laboratory or spot check methods; therefore, it affords total control.

- It is used to determine yarn percentages in any fabric without stopping the machine and without having to unravel and weigh yarns or having to weigh cones prior to and after sample roll knitting. The time saving is enormous. Different amounts of different color, size, or type of yarns in fabrics can be simply calculated by the use of fabric weight formulations.

- It is used to calculate settings for machines of different sizes to produce identical fabric in a different width.

- It is used to calculate production capacity per fabric style, per machine for industrial engineering and production planning purposes.

- It is used to write the specifications of knitted circular fabrics for later reference (such as in fabric development). These specs can easily be controlled and repeated on other machines, at a later date, or in another plant location.

11.6.3 All Solid-State Doff Counter for Knitting

The knit master Greige Yield System includes a control unit. This functions as a doff counter for ensuring that rolls doffed contain the same number of cylinder revolutions, and supplies power and revolution signals to the Model 104 Yarn Length Meter. The Model 403 has no mechanical revolution counters to load up and jam with lint, and no moving parts to wear out. It has a built-in ni-cad battery to retain count information should power fail.

11.7 Simodrive Sensor Measuring Systems

Simodrive sensors are rotary measuring systems (encoders) to sense the position, angle, and speed and velocity of machines. The encoders can be used in conjunction with

- Simatic control systems, plc
- Sinumerik CNC control systems
- Simodrive drive systems

A differentiation is made between incremental and absolute measuring techniques. Incremental encoders generate a defined number of steps (increments) per revolution, which are processed in the control system. Absolute value encoders provide, directly after power-on, the absolute position value without the machine moving. Simodrive sensor measuring systems are available in various versions and technologies, optimized for the particular application.

Absolute value encoders provide, directly after the control system is powered up, the accurate position value without the machine moving. The absolute position is determined by opto-electronically scanning several code tracks. Single-turn encoders sense the absolute position within a revolution, while multi-turn encoders code the number of revolutions. The position information is transferred to the control system, either via the serial SSI (synchronous serial interface) or via Profibus-Dp. The encoders with Profibus-Dp can also be parameterized online via the bus. The following values can be set:

- Resolution and traversing range
- Velocity signal
- Preset value
- Direction of rotation
- Limit switch
- Teach in

The connection as well as address setting is realized in a cover, which can be removed. This guarantees interruption-free bus operation, even when an encoder is being replaced. The encoders are certified and support Class 1 and Class 2 profiles, standardized by the PNO, as well as additional application-specific special functions.

11.8 Monitoring Yarn Input Tension for Quality Control in Circular Knitting

The prevention and quick detection of defects play an important role in modern knitting technology, since productivity and quality are closely related to it, and an undetected and uncorrected defect may result in machine down time and possibly fabric rejection. Effective monitoring of the knitting process enabling the rapid detection of a possible cause of a defect or a defect itself is therefore desired.

There are two major causes of defects in knitted fabrics: the yarn and the knitting elements used, which can be related to horizontal and vertical defects. While for the first category, the solution is related to careful yarn selection and good yarn housekeeping, for the second category the solution is related to the detection of a cause of a defect or a defect itself, as quickly as possible, in order to avoid the rejection of large quantities of knitted fabric. Currently, this detection is usually made with the aid of optical and capacitive sensors, which detect changes in the normal pattern generated by a specified knitting element such as a needle or a sinker.

In fact, the monitoring devices used until now for the detection of defects can be separated into two categories according to the approach used: knitted fabric monitoring devices and knitting elements monitoring devices. The first type is based in the detection of fabric holes. For this purpose, an optical sensor looks for a change in the intensity of the light coming through the fabric being produced, and when this change occurs, the machine stops. The second type monitors the knitting elements and evaluates the changes in the expected pattern that results from a non-defective knitting element. In the case of significant differences, the knitting machine stops and the problem is corrected. These systems are very specific, and they do not give any information concerning the knitting process. This inspired the proposal of an alternative approach to the detection of causes of defects and the defects themselves by analyzing and interpreting variations in the yarn input tension, as these are a reflection of the knitting process. When an abnormality occurs during the knitting process, this may result in a visible defect in the knitted fabric and should be normally reflected through variations in the yarn input tension.

11.8.1 Measuring System

Its main features are a 3.75″ diameter cylinder, 14 gauge (168 needles), single feeder with a positive feeding system (B) and adjustable machine speed. The positive feeding system is based on a rubber roller, and the yarn speed can be adjusted through a worm screw (Figure 11.6).

The measuring system consists of a Rotschild yarn tension meter (D) for measuring the yarn input tension with a measuring range from 0 to 10 cN, and an optical sensor (C) for triggering and allowing the separation of the waveforms acquired in individual cylinder revolutions. Both sensors are attached to an acquisition board from National Instruments, model LabPC+. The process of acquiring the yarn input tension is controlled by an application—KnitLab—developed with National Instruments LabVIEW 4.0. This application is organized in six modules: calibration, configuration, acquisition, visualization, storing, and reading.

The acquisition module is considered the most important for this purpose and will be briefly described in the following lines. The program core is a 2D array containing measurements of the yarn input tension for several revolutions by the order they were acquired. The process of acquisition can involve four steps: the calibration of the measuring system, the knitting machine speed adjustment, the configuration, and the acquisition itself. The calibration and configuration processes can be skipped, depending on the experimental conditions. The speed adjustment is always required and is performed on a step-by-step basis. Given the indication of the desired speed, the knitting machine starts and an acquisition is made, triggered by the optical sensor, for at least 10 revolutions. The average speed and its coefficient of variation are calculated and displayed. The user can alter the speed until the correct value is obtained. The revolutions are determined through

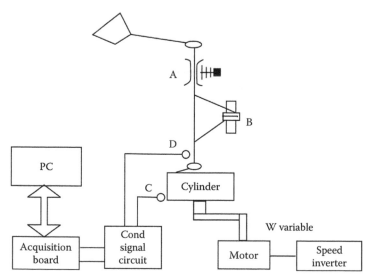

FIGURE 11.6
Monitoring yarn input tension.

the high to low transitions from the optical sensor. This process is also used to organize the acquired yarn input tension waveform in the 2D array. When the speed adjustment is accomplished, the measuring system is able to acquire the yarn input tension. The application waits for a signal from the optical sensor indicating the starting point of a cylinder revolution, thus triggering the acquisition. Then, as the tension is acquired and stored, the waveform is displayed on the screen. By means of a manual command or a predefined number of revolutions, the acquisition process will terminate. After this stage, the data is analyzed and organized in the 2D array, in which each column represents one cylinder revolution, being then ready for further analysis such as data manipulation, visualization, and so on. Along with the resulting matrix, a set of information is extracted from the acquired waveform: cylinder average speed and its coefficient of variation, time elapsed between needles, and so on. In the visualization module, the user is able to use a set of tools such as average waveform, digital filtering, spectral analysis, and differentiation, among others. The remaining modules allow saving of the experiments and loading of the waveforms for further analysis. The files generated by KnitLab are ASCII files, allowing its use on other applications like spreadsheets and statistical packages.

11.8.2 Representation of the Waveform by a Measured Value

The representation of the whole waveform, even though possible, does not seem to be the best solution for a quick observation of the knitting process behavior. It would be interesting to use a single measured value to summarize the behavior of the knitting process. Due to the excellent similarity between the revolutions acquired for the same knitting conditions, a measure based on a direct comparison between an expected pattern and the actual waveform acquired should be possible to obtain. Based on this assumption, SQD measurements were proposed and studied. This measure is based on the sum of quadratic differences between the last waveform (for a revolution) acquired and the expected standard waveform, taken from the average yarn input tension measurement of one hundred revolutions of the knitting machine under normal conditions.

11.8.3 Defect Identification by Stitch Formation Differences

The information contained in the resulting waveform acquired from the yarn input tension allows an analysis of the yarn input tension behavior during stitch formation. There is a correspondence between the variation of yarn input tension and the needle position in the knitting cycle. The yarn input tension increases until it reaches a maximum at knock over (just before due to roving back), and then it decreases to a minimum as the stitch is cleared. Some variability is expected as there are several factors that are not under control, such as yarn friction, individual knitting elements variability, and so on. The observation of the whole waveform for normal and abnormal knitting conditions shows that the normal pattern is not present when a defect appears. This behavior suggests the use of a frame comprehending one stitch formation for defect identification. In order to accomplish this objective, some parameters were extracted from a typical waveform. They are as follows: average tension, maximum and minimum tension, time necessary to cross the average tension twice, time necessary to reach maximum and minimum tension, slopes, and position in the knitting cycle.

11.9 Warp Knitting Machine Control

A safe and fast detection of thread breaks in the warp sheet will help reduce losses. The newly developed laser light barrier system laserstop sets a standard in reliability and safety in the control of warp sheets on warp knitting and Raschel machines. By using the most up-to-date laser technology for the light barriers and an evaluation incorporating the most modern digital signal processing in the control unit, a system has been devised that can be used for a multitude of tasks. Since the basic function of the system can be selected, the surveillance device can be used either at the warp yarn sheet position or close to the needle area. The special characteristics of the system are:

- Fast and safe detection of broken threads from 12 dtex in the warp sheet or at the needle position
- Compact construction of the laser light channels
- Visible, safe red-light laser (laser class 1)
- Vibration insensitive receiver
- Control unit with digital signal evaluation and computer supported, automatic system control function
- Software update via hand terminal socket

The laser light barriers operate with a visible red-light laser (660 nm). These diode lasers are used because of their long operating life and low mechanical sensitivity characteristics. The highly homogeneity of the light beam guarantees a constant sensitivity over the full working width (Figure 11.7).

In the receiver, a newly developed measuring system is used, which gives excellent results in terms of vibration independence and sensitivity. The laser light barriers are mounted at the needle area and, when required, parallel to the warp sheet. In addition

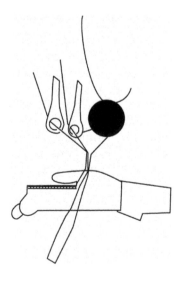

FIGURE 11.7
Warp knitting principle.

to the laser light barriers for the warp sheet control, sometimes it is necessary to install a blower device to assist in guiding the thread movement of a broken end through the light barrier. If a broken thread should come out of the yarn sheet, it will break the light beam. The resulting impulse will be digitally processed by the control unit, and the production machine will be stopped without delay. An optical fault display is provided for each of the separate channels so that the operating personnel can see immediately in which guide bar the thread break has occurred. The laserstop control unit contains all the necessary components to run the surveillance device and allows the connection of up to four laser light barriers. At each of its four large sizes display fields, the current operational condition of the connected laser light channels is displayed in a color-coded form so that the system status is easily seen even from a distance. All settings to the control unit are carried out with the help of a hand terminal. This hand terminal operates via an LCD display and a keyboard. All settings are carried out with the help of an easy-to-understand operating guide on the LCD screen. The software for the surveillance system is contained in a new type of memory chip so that in case of an eventual update, the new software can be loaded via the hand terminal socket. In this manner, the system is prepared for any future expansion in a most optimum way.

The ever-increasing technical development of textile machinery, together with the call for better quality standards at economic cost, has led to a demand for an optimal fabric checking system to provide a constant and efficient control of the production.

11.9.1 Scanner Head

The scanner head model 590 travels at a speed of 1–2 m/s at a distance of approximately 6–7 cm above the knitted fabric. In this manner, the fabric can be scanned directly after it has been formed by the needle bar. The scanner head travels over the selvage spreaders in order to ensure control of the full fabric width.

11.9.2 Hand Terminal

Operational data such as numbers of stops and the reason for them (machine stops/scanner stops/selvage faults) can be called up. This provides a basis for use in a production data processor. The processing of the measured data is undertaken by means of a modern microprocessor. The use of the latest algorithm technology has made possible a previously unknown level of fault detection, interference suppression, ease of use, and operational safety. Even the very small faults made on four-bar machines can be quickly and reliably identified. When changing the fabric, the device will retain its sensitivity automatically. Furthermore, any service needed following the mounting of the unit is kept to a minimum. FIL-STOP segments (optional) for the selvage control can be linked directly to the control unit.

11.10 Summary

At the onset of the chapter, the user will understand the working concept of the knitting machine and the control process used in the knitting machine. The computer interface of the knitting machine is explained in detail.

References

1. https://www.newagepublishers.com/samplechapter/001326.pdf.
2. http://textilelearner.blogspot.in/2014/03/yarn-feeding-mechanism-in-knitting.html.
3. http://www.google.com.pg/patents/US3539782.
4. https://www.researchgate.net/publication/293004155_Needle_selection_techniques_in_circular_knitting_machines.
5. http://nptel.ac.in/courses/116102008/9.
6. http://article.sapub.org/10.5923.j.textile.20140304.03.html.
7. http://www.google.com.pg/patents/US3539782.
8. http://www.protechna.de/en/rundstrickmaschinen/needle_sensor_4022.
9. http://www.tritex.co.uk/dsm.htm.
10. https://www.tib.eu/en/search/id/tema%3ATEMA19990900129/Quality-control-in-circular-knitting-by-monitoring/?tx_tibsearch_search%5Bsearchspace%5D=tn.
11. http://textilelearner.blogspot.in/2011/05/defination-and-properties-of-warp_8342.html.
12. http://textilelearner.blogspot.in/2011/05/defination-and-properties-of-warp_8342.html.
13. https://wn.com/working_principle_of_flat_weft_knitting_machines.
14. http://shodhganga.inflibnet.ac.in/bitstream/10603/102320/9/09_chapter%203.pdf.
15. https://encrypted.google.com/patents/US5862683?cl=un.
16. https://link.springer.com/chapter/10.1007/978-3-662-46341-3_7.
17. http://www.karlmayer.com/en/products/warp-knitting-machines/warping-machines-for-warp-knitting/.
18. http://www.ia.omron.com/data_sheet/cat/zx-t_dsheet_gwe344-e1-04.pdf.
19. http://www.hella.com/hella-za/assets/media_global/HASA_Thermo_Range_Borchure_LRes.pdf.
20. https://books.google.co.in/books?id=zkf7CAAAQBAJ&pg=PA143&lpg=PA143&dq=Warp+Sensing+Unit+in+knitting+machine&source=bl&ots=b8DFf85ISm&sig=Yle_nW20mTdxAEoKzJ4U4CeviLU&hl=en&sa=X&ved=0ahUKEwj0qevm-9jVAhVLto8KHZD-APMQ6AEISTAH#v=onepage&q=Warp%20Sensing%20Unit%20in%20knitting%20machine&f=false.

12

Controls in Testing Instruments

LEARNING OBJECTIVES

- To identify the working principles of different textile testing instruments
- To know how sensors and transducers are used in measuring different textile parameters
- To understand the concept of monitoring of different textile parameters

12.1 Introduction

Textile testing is always an essential part of the production and has become more important in recent years as a result of the new demands imposed upon the products of textiles manufacturers. Advancement in technology has made a great impact on textile testing. If we compare the conventional testing instruments with modern testing instruments, we will find a spectacular change due to application of electronic controls. Although the use of electronic systems has changed in the entire textile industry, the impact on testing instruments is amazing. For example, all the important fiber properties can be determined by feeding fiber samples in a high-volume instrument within a few minutes by any person; otherwise, it can take 5–6 h and extensive skilled work by many operators in ordinary type instruments.

12.2 Fiber Properties

The following fiber properties are measured:

- Fiber length
- Fiber diameter
- Fiber fineness
- Fiber maturity
- Fiber contamination
- Fiber strength
- Trash in cotton

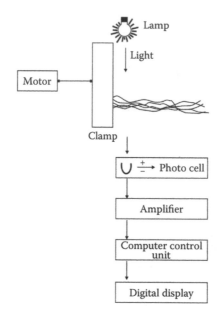

FIGURE 12.1
Optical principle measurement of fiber length.

12.2.1 Fiber Length

In the past, fiber length was found by the sorter method and in some mills, the sorter method is still used. This is time consuming, and skill is required. To get quick results with more information, fibrograph instruments came into the market.

In the manual fibrograph, the fiber sample is analyzed by the photocell principle and a fibrogram is drawn. From the fibrogram, the required fiber length particulars are found.

12.2.2 Principle of Fiber Length Measurement

The block diagram shows the optical measurement of fiber length measurement. A narrow beam of light is allowed to fall on the specimen beard. The specimen beard is then clamped and is moved to and fro by a motor with a traversing spindle unit arrangement (Figure 12.1).

Referring to the figure, as the specimen beard passes, the light is allowed to fall on the specimen. The attenuation of light through the specimen at different areas of the beard is received by a photocell, and signals are fed to the signal processor. The processor converts the signals to fiber length values and through a microprocessor, the fiber parameters such as number of fibers, 2.5% span length, 50% span length, uniformity ratio, and short fiber index are calculated and displayed in a display unit and can be printed out.

12.3 Fiber Diameter Analyzer

In fiber diameter analyzer (FDA) analyzer, a thorough beam reflectance type-optical type of sensor is used to sense the diameter of fiber (see Figure 12.2). The technique is also called Fresnel diffraction. The diameter ranging from 0 to 80, 160, or 240 μm is found. Up to 3000 measurements per minute can be taken.

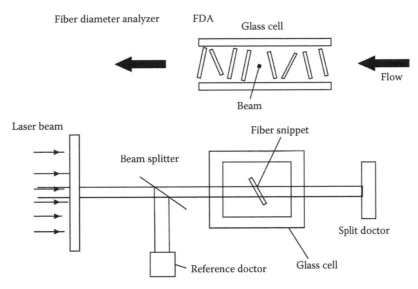

FIGURE 12.2
Fiber diameter analyzer—optical type of sensor.

12.4 Fiber Fineness

Fiber fineness can be measured using the vibroscope method. A schematic diagram of the instrument is shown in Figure 12.3. The weighted specimen is clamped to the vibrator at A and passes over the knife edge K. The clamp and knife edge are connected to a 150 V source so that the specimen is electrically charged. Transverse vibrations of the specimen will therefore induce a charge in a brass screw S situated midway between the clamp and the knife edge and spaced 1 mm from the specimen. The screw thus acts as a transducer;

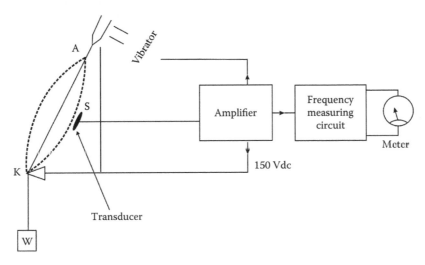

FIGURE 12.3
Vibroscope method.

if the signal from it is amplified suitably and fed back to the vibrator, an oscillatory loop is formed, thus causing the specimen to vibrate at its resonant frequency. The voltage across the vibrator can then be fed into the frequency measuring circuit and the frequency of the oscillation indicated on the meter. From the weight of specimen, wavelength and frequency the fineness of fiber is calculated.

12.5 Fineness and Maturity Testing

Cotton fineness and maturity are important fiber properties in the textile mill lay down. Some yarn and fabric properties associated with fiber fineness and maturity can be defined in various ways. Fineness is expressed as mass per unit length (mtex) and the fundamental measure is cross-sectional perimeter (µm). Maturity involves dye uptake, nep formation, strength and uniformity of yarn, and resistance to surface abrasion. Maturity is expressed as the degree of wall thickening/0.577 (dimensionless), and the fundamental measure is cell wall thickness (µm) (Figure 12.4).

The fineness of cotton is important because yarn made from fine fibers is generally stronger and more uniform than yarn from coarse fibers. Fiber maturity is important because mature fibers, those with well developed cell walls, absorb dye better. Micronaire measurements are considered a combination of fiber fineness and maturity.

The reference method for fineness and maturity is based on the analysis of fineness and maturity by the resistance to air flowing through cleaned cotton that an operator has placed in a short cylinder. This machine is called a fineness and maturity tester (FMT).

The strategy is to use fundamental reference methods (e.g., image analysis) to develop a moderately small set of calibration cottons that can subsequently be used to convert the air pressure readings of the FMT into fineness and maturity values. These instruments are then used to develop a much larger set of cottons with known fineness and maturity

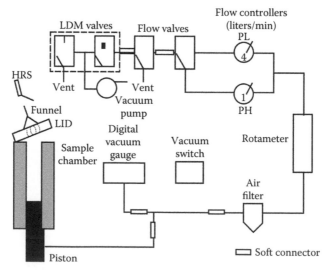

FIGURE 12.4
Upgraded Micromat with HRS and LDM.

values that can then be used to calibrate the near infrared high-volume instrument system. The improvements to both the FMT and the near infrared device have gained recognition, but important gaps must be filled before the technology is widely used. The focus of this research was to develop techniques to minimize the variations that occur when the cotton is mechanically cleaned (prepared) for FMT analysis and when the operator puts the specimen in the sample chamber.

The Micromat tester is being used in the laboratory as a reference method to calibrate fast spectroscopy-based instrumentation to measure fineness and maturity. This instrument is a double compression airflow device that measures the pressure drop of air drawn through a sample that is compressed, during the test, to two different densities. The initial and second stage pressure drops are referred to as *PL* and *PH*, respectively, and are converted to fineness and maturity by appropriate empirical equations.

The most important features of the instruments are the sealing of the air flow system, the installation of a leak detector module (LDM), and the use of physical standards dubbed headspace resistance standards. Calibration is a three-step process. The calibration order is detector, air flow system, and sample chamber volume (the detector is used to calibrate the other instrumental settings and the air flow system is needed to calibrate the chamber volume).

12.5.1 Fiber Contamination Technology

This system works under the principle of the laser signal analysis system.
It measures the following:

- Contamination such as stickiness, seed coat neps, and fiber neps.
- Evaluation of performance of pre-cleaning systems, such as carding, comber, and so on, for determination of their optimum operational settings and efficiency.

The sample in the form of bundles is fed into a self-cleaning micro carding device integrated in fiber contamination technology (FCT) to produce about 10 m of transparent web. The laser beam is passed through the sample, the beam reflectance signals are analyzed, and contaminations like stickiness, seed coat neps, and fiber neps are found.

12.6 High Volume Instrument

The high volume instrument (HVI) gives a certificate of its quality. In this system, a set of automatic fiber-measuring instruments are linked to a common computer, which together characterizes a fiber sample and prints a report on it. Hence in this system, the bundle testing is automated. In Uster high volume instruments, there are three spectrums: HVI spectrum I, II, and III. These three spectrums are linked together to a common computer.

By using these spectrums, the following fiber parameters are found: fiber length, length uniformity, strength, elongation, micronaire, color, and trash.

Length and length uniformity: The length and length uniformity of the fiber are measured by optical methods such as the digital fibrograph.

Strength and elongation: The fiber strength and elongation are found by the constant rate of elongation (CRE) method.

Micronaire: The fineness value is measured by the resistance to air flow principle. Instead of fixed weight as used in other methods, sample of between 9.5 and 10.5 g is used and the instrument itself makes allowances for it.

Color: The color of the fiber is measured by the optical method. Both percentage reflectance and the yellowness are found by this method, which is then combined onto a USDA color grade.

Trash: Trash of the fiber is also measured by the optical method. By using a video camera, the number of trash particles in that sample is counted.

NEP: Nep is measured by the optical method.

In this HVI system, the fiber sample is placed on a chamber, and automatically the sample is selected, measurements are taken, and the above parameters are found. Then the measured signals are passed on to a computer control system and printed out.

12.7 Advanced Fiber Information System

As shown in Figure 12.5 the fiber sample is fed by a fiber feed unit to an individualizer, which separates the fiber sample into individual fibers by an opening and cleaning unit. Each individual fiber is sensed by electro-optical sensor, and signals are fed to computer control unit. The data are displayed in a display unit (Figures 12.5 through 12.7).

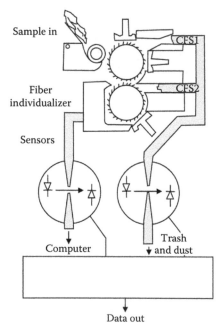

FIGURE 12.5
Schematics diagram of operating principle AFIS.

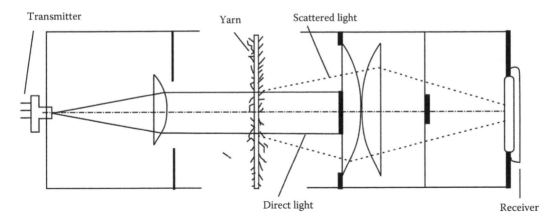

FIGURE 12.6
Scanning principle of AFIS.

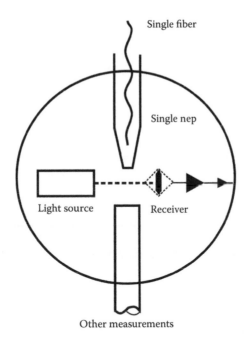

FIGURE 12.7
Single fiber analysis.

In this instrument, there are two modules, namely:

1. *For finding out number of neps and size of neps*: By using an optical sensor, the number of neps and size of neps are measured and the values are evaluated by microcomputer.
2. *For finding length and diameter of fiber*: By using the optical method, the length and diameter are found.

12.8 Auto Sorter

In the past, the yarn count was determined by using Knowles balance, Beesley balance, quadrant balance, and Stubbs's balance, which involve mechanical action but give direct value.

The auto sorter is used to find the count of yarn or linear density and weight per unit area of a fabric. The sorter consists of an electronic balance with computer interface and an optional printer to print out the results. With the computer interface, we can get a graphical representation of the results in the form of histograms and Strokes diagrams for fast and easy verification. Here the maximum possible measuring range is 200 ktex and the weight per unit area of 20000 gsm.

12.9 Electronic Twist Tester

In this twist tester, the twist in the yarn is electronically analyzed by removing the twist in the yarn automatically. The instrument is provided with the following features.

- Digital reading suitable for both single and twisted yarns (S and Z) up to 9999 torsions.
- High sensitivity pretensioning device with built-in electric clamp for the seizing of yarn.
- Optical tracer and warning light to signal the starting and stopping "zero."

The adjustable distance between clamps ranges from 0 to 50 cm and from 0 to 20 inches. The specimen is fixed between the two jaws and rotated by a motor. After untwisting and twisting, the mechanical rotation of the jaw is converted into signals, which are amplified and finally converted into TPI/TPM. This is shown in a display unit.

12.10 Yarn Evenness Measuring Instruments

There are three important methods of measuring evenness of yarns: They are (i) Photoelectric method (ii) Capacitance method (iii) Infrared sensing method

12.10.1 Photoelectric Method

In this method, depending on the changes in the intensities of light, evenness is measured. As shown in Figure 12.8, the light is emitted through the yarn and the rays are received by a light receiver. From the received rays, signals are converted and amplified, and through suitable computer controls, the evenness of the yarn is measured.

12.10.2 Capacitance Method

The Uster evenness tester works under the capacitance method. The principle schematic diagram of evenness tester is shown in the Figure 12.8. Two oscillators A and B have equal frequencies when there is no material in the measuring capacitor C. When the two frequencies are superimposed, the difference in frequency is zero.

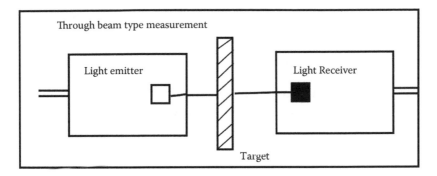

FIGURE 12.8
Yarn evenness measuring principle.

The presence of material in the capacitor causes its capacity to change and so alter the frequency of oscillator A. There will then be a difference between the two frequencies, which varies according to the amount of material between the capacitor plates. Suitable circuit D translates these frequency signals, which can perform or indicate the following:

- The variation in meter M
- Driving the pen of the recorder to draw a variation chart

The results are fed into the integrator, which indicates the average irregularity either as percentage mean deviation or coefficient of variation. Modern evenness testers can indicate both U percentage as well as CV percentage.

12.10.2.1 Choice of Measuring Indicator

In order to achieve a change in capacity, which is linearly related to the amount of material between the capacitor plates, the thickness of the material relative to the size of the capacitor must not exceed certain limits. In the Uster instructions manual, a table states which capacitor must be used for the different hanks and counts.

12.10.2.2 Normal and Inert Testing

When the Uster is used in the normal mode, the trace records all the variations, namely, short-, medium-, and long-term variation. The results are similar to those obtained by the cutting and weighing method.

12.10.2.3 The Imperfections Indicator

Signals from the unit are fed to the imperfection indicator, which simultaneously measures neps, thick places, and thin places. Counts of all three types of faults are made and the results are shown on separate counters. It is possible to adjust the sensitivity of the system so that a chosen size of fault is counted and smaller faults are ignored.

Presently third generation and fourth generation of evenness testers are available, which can provide extraordinarily detailed information within a short amount of time. The use of the microcomputer improved the accuracy of the installation and the reliability of the data obtained. Further, various checking and adjustments involved are eliminated from the routine work of operation. In Uster Tester-3 (UT-3), the instruction of the operator is undertaken via the video screen. The Keiosokki Evenness Tester (KET-80 B) is a digital evenness tester applying the latest electronic technology.

The following information is available with a complete set of KET-80 B.

- Diagram of weight variation per unit length in a sample material
- Mean deviation U % or coefficient of variation CV %
- Relative yarn count, average of a sample material
- Number of thin places, thick places, and neps
- Deviation rate, DR %
- CV (L) %; CV % at a set length L
- Spectrogram

The block diagram shows the control system used in the Uster evenness tester. The signals received from an infrared sensing unit and capacitance sensor are converted into signals by a converter. The computer control unit receives the signal from the converter, and the data display unit displays the yarn mass variations, imperfections, and hairiness value.

12.10.3 Infrared Sensing Method

The infrared sensing unit is the third method of measuring the evenness of the yarns (Figure 12.9).

12.10.3.1 Keisokkis Laserspot

Keisokki Koggo Co., Japan, has developed an instrument, laserspot (LST), using laser technology for measuring hairiness and evenness simultaneously. It works under the principle of measurement of Fresnel's diffraction using a semiconductor laser beam. Laserspot separates the interference at a right angle to the yarn from the direction of yarn running by means of a spatial filter and then detects the singles in proportion to the thickness and hairiness respectively. Since the hairs are supposed to be at right angles

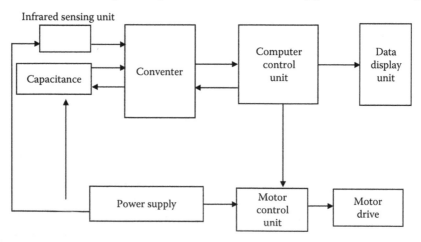

FIGURE 12.9
Keiosokki evenness testing principle.

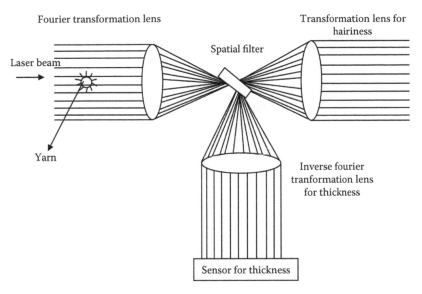

FIGURE 12.10
Laser technology for measuring hairiness measurement.

to the main yarn body, the interference fringes caused by the main yarn body and hairs are separated into two components, one for measuring hairiness and other for measuring thickness.

When irradiating the coherent parallel rays (laser beam) with identical wave length (λ) onto the linear sample material such as yarn, the interference fringes that are proportional to the thickness (d) appear at right angle to the yarn (Fresnel diffraction).

Let the angle α be the expansion of interference fringes.

$\text{Sin } \alpha = \lambda/d$ From the above-mentioned formula, the thickness (d) can be determined by measuring the expansion of interference at right angle of the yarn. At this time, the hairiness exits at a right angle to the yarn, and it does not affect the interference fringe of the yarn. Hairiness also causes the interference fringes, but they appear like diffusion along with the direction of the yarn running, which is proportional to the amount of hairiness (number of hairiness × length). The change of the thickness influences the direction of only the right angle and has no influence on the direction of the yarn running (Figure 12.10).

The conventional optical hairiness tester is influenced by the variations of yarn thickness due to the utilization of the diffused reflection by the light. Laserspot is free from the influence of the yarn thickness because of the spatial filter, which enables laserspot to measure the hairiness alone. The figure shows the above-mentioned principle schematically.

12.10.4 Variation in Thickness under Compression Method

In the compression method, the evenness is measured by measuring thickness variations. Two rollers sense the thickness mechanically and give the variations of the thickness of the sliver (Figure 12.11).

The displacement is sensed by an LVDT. According to displacement, the LVDT converts the displacement into signals. An amplifier then amplifies the signals and through suitable computer control, the evenness is measured.

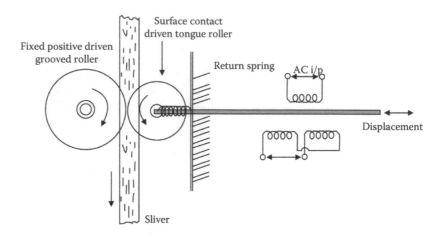

FIGURE 12.11
Evenness measuring by measuring thickness variations.

12.10.5 Classimat

Classimat classifies the yarn faults according to the variation in the linear density and length of the fault. It works under the principle of capacitance type. As the yarn is passed through capacitor plates, it is analyzed for linear density variations and length variations according to the lengths and with the help of software; faults are drawn on a chart. The block diagram explains clearly about the operational diagram of Uster Classimat-II fitted in automatic cone winding machines.

The uncleared yarn is passed through a capacitor head for measuring the yarn cross section and length. The yarn faults are measured into 23 categories according to the fault length and cross-sectional area as shown in the Figure 12.12. The classified faults are counted and then recorded. The faults are analyzed and required clearing setting limits are set in the clearer. This is based mainly on the quality of yarn required by the yarn manufacturer. Then the clearer installation is set. According to the setting, the EYC works in the machine and the yarn is cleared (Figure 12.12).

In the diagram, it is also seen that from quality consideration, measures to be undertaken in the spinning mill are considered to get good quality yarn. The faults are again subdivided into the following categories.

Short slub channel (N):

- *Length of fault*: No preset limitations
- *Diameter of fault*: 4–9 times the normal diameter

Short slub channel (S):

- *Length of fault*: 1–6 cm
- *Diameter of fault*: 1.8–4.8 times the normal yarn diameter

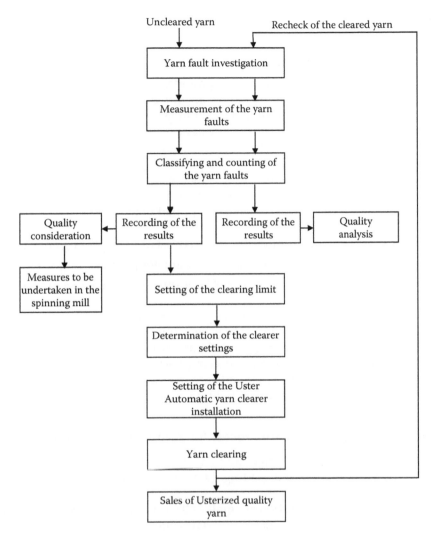

FIGURE 12.12
Working principle of Uster Classimat.

Spinners double channel (D):

- *Length of fault*: 20 cm minimum
- *Diameter of fault*: 0.2–0.9 times increase in the normal yarn diameter

Thin place channel (T):

- *Length of fault*: 20 cm minimum
- *Diameter of fault*: 0.3–0.9 times decrease in the normal yarn diameter

12.11 Electronic Inspection Board

The general appearance of yarn is very important, and practical yarn property determines its quality. The electronic inspection board provides very detailed, useful information about the appearance of the yarn, unlike the conventional method of comparing the yarn board with ASTM standard and grading the yarn subjectively. In this instrument, there are hardware and software systems, which are designed for the analysis of yarn appearance and grade. The obtained effects can be stored in the computer for future reference. This instrument is manufactured by Lawson–Hemphill of the United States. The important feature of electronic inspection board (EIB) is its flexibility to produce a different visual appearance and corresponding grades for tested yarn electronically.

There are three different software packages supplied with the EIB.

1. Electronic inspection board mode for the analysis of yarn grading appearance
2. Yarn profiler mode for generating the yarn diameter profiles
3. Production mode for the quality control of large quantities of packages in textile mills

The EIB is a transport system in which the yarn is passed through an optical (digital) camera that does not drift and has a very high resolution of 3.5 microns that continuously measures the yarn diameter at very high speeds, providing high resolution. At 100 m/min, the diameter is recorded every 0.5 mm, allowing very small variation in the yarn to be measured. The yarn diameter measurements are received by the computer, and the software generates important information including the yarn appearance.

12.11.1 Electronic Inspection Board Process for Yarn Appearance Grade Evaluation

The optical technology of the EIB offers many new advantages over conventional testing relative to the measuring faults and grading the yarn, including:

- It has a minimum fault detection as small as 0.5 mm in length and high resolution of 3.5 microns, the highest available in today's market.
- It is not affected by moisture, color, or blends that contain electrostatic constants.
- It has the ability to measure the effect of twist on diameter and appearance in fabric.

EIB is able to combine various measurements to generate a grade consistent with the human grading techniques with inspection boards and can analyze different yarns in different types of fabrics to identify the best end use for appearance. The most important calculation is the EIB appearance letter grade (i.e., A, B, C, D). It is easy to understand how the level of each of the components (diameter, average, CV % diameter, and events) or combinations of them can give bad appearance. For example, introduction of two yarns with different diameters because of different twist into a fabric would cause objectionable fabric appearance. Likewise, a yarn having a high diameter variation with high CV %, having many defects, would produce bad fabric. In the printouts, the defected areas are shown clearly at the top of the diagram and fault classifications are made by length. This will bring the yarn spinner a long way in delivering yarns with consistent appearance day to day, month to month, and year to year (Figure 12.13).

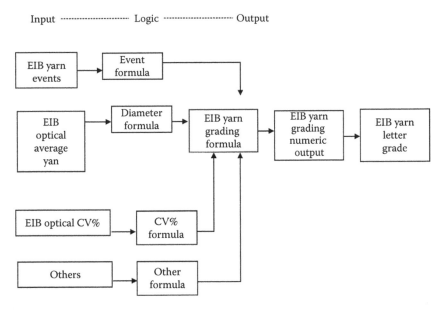

FIGURE 12.13
Electronic inspection board processing principle.

12.12 Hairiness Measurement—Photoelectric Method

The hairiness of yarn characterizes the number of projecting and freely moving fiber ends or fiber loops. The hairiness of spun yarns is considered to influence the appearance, weaving performance, and overall quality of the products. Due to recent trends demanding improved quality, product diversification, increased processing speed, and decreased man hours, good-quality spun yarns with minimum hairs are becoming essential.

In the photoelectric method, light is directed out of the yarn from its side to form a shadow of the yarn including its hairs. The yarn light and shade as well as magnitude are converted into electric signals for analysis. This method is defective in that the measurement results are considerably affected by the yarn count and unevenness, causing the range of the measurement values to be indefinite. A variation of this method has been introduced in Japan in which the shadow of the hairs in the ring concentric with the cross section of the yarn is converted into electric current.

12.12.1 Photoelectric Counting Method

In this method, photoelectric elements evaluate hairiness by counting the number of hairs along the axis of the yarn that are longer than the specified hair length (Figure 12.14).

Referring to the figure, the hair is raised electrostatically by running the yarn between a pair of high voltage electrodes. The resulting enlarged image of the hair is made by using a lens system, and the number of hairs is counted at the distance of 3 mm (or 6 mm) from the yarn by moving the photoelectric tube vertically. Some manufacturers change the distance to 1 mm.

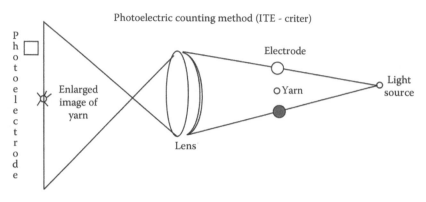

FIGURE 12.14
Photoelectric counting method of hairiness measurement.

12.13 Tensile Strength Testing of Textiles

The strength characteristics are still a primary test as a guide to compare the performance of various processes in quality control and determining the end use behavior of most of the textile products. The range of instruments that are available for testing the strength of yarn is quite wide. Single thread testers operate on the principle of the pendulum lever types. Some of these instruments are relatively simple, some complicated, some have recording devices, and some are automatic. The strain gauge transducer is most popular due to its versatility in all types of textile materials.

A strain gauge is a passive transducer, which converts a mechanical displacement into changes of resistance of the strain wire. The basic principle of the strain gauge wire is that a resistance wire stretches within its elastic limit and shows an increase in resistance when elongated. When attached to the surface of the structure by means of elastic cement, the variation in the resistance of the wire will follow the change in surface dimensions of the structure due to stress formation.

A strain gauge transducer is made of a thin wire. In Figure 12.15, the upper jaw J_1 is attached to the free end of the beam and the lower jaw J_2 has a controlled vertical movement through a screw mechanism. By moving J_2 downward, a tensile force is developed in the specimen, which causes the free end of the beam to be deflected. The effects of the deflection are used to measure the magnitude of the load on the specimen. When the beam bends, the length of the upper face AB increases and the length of the lower face decreases. These changes in length are proportional to the applied load. Between the two outer faces, there is a neutral plane NL whose length remains unchanged.

When a piece of resistance wire is stretched, its electrical resistance increases; conversely, if it contracts, its resistance decreases. Further, the change in the resistance value is proportional to the change in length. Consider a resistance wire R firmly bonded by cement to the face AB of the beam so that elongation of AB will produce an elongation of R. Thus, a load applied at the end BC causes a change in the length of the resistance wire, and the change in the value of the resistance is proportional to the magnitude of the load. It is now necessary to convert this change in resistance into a visual recording of the load. This conversion is usually achieved by means of a Wheatstone bridge.

Four resistances are employed in the loading system, two resistances on the upper and two resistances on the lower surfaces. They are connected in the form of a Wheatstone

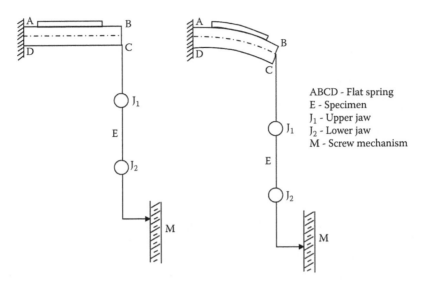

FIGURE 12.15
Working principle of strain gauge transducer.

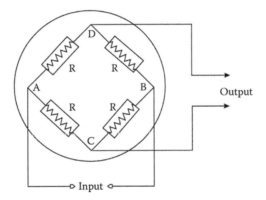

FIGURE 12.16
Wheatstone bridge principle of tensile tester.

bridge as in Figure 12.16. With a beam undeflected, no voltage will be developed across CD when the voltage is applied across the AB. The bridge is said to be balanced. When a load is applied to the beam, the deflection causes changes in the values of the resistance and a voltage is produced across the CD. Its value is proportional to the load. This voltage output is fed to suitable electronic circuits, which is finally converted the voltage to load value and recorded by a pen recording instrument.

12.13.1 Breaking Force Measurement

The figure shows a pin, B, connected to a beam, A, which is fixed at one end. There are two sensors F_1 at the top of the beam and two sensors F_2 at the bottom of the beam. The spacing between the two sensors is equal if the beam is horizontal. The sensors are connected to oscillators O_1 and O_2.

Referring to the Figures 12.17 and 12.18, if a force, F, is applied downward, the pin moves down and the beam bends down. So the space between top sensors F_1 increases and the space between F_2 decreases. These sensors measure the frequencies with oscillators O_1 and O_2. Now there is a frequency difference between the two sensors, which is the measure of the force F. This frequency difference is fed to a signal processor to convert into the breaking force required to break the specimen. The more the difference between the sensors, the greater the force. If there is a frequency difference of 1 Hz, this is a breaking force of 0.0000167 N under 50 N measuring head (Figures 12.17 and 12.18).

Referring to the figure, the movable clamp is connected to a chain K and a rotational transducer disc S. Around the circumference of the disk, 1000 evenly distributed markings are arranged. As the movable clamp moves down, the chain moves, which will rotate the disc in the counter-clockwise direction. So the movement of the markings on the disc is measured optically, which produces impulses. These impulses are converted to find the elongation values. Each marking in the disc produces four impulses. Thus, a maximum of 4000 impulses are produced when the disk rotates to one rotation. Also, the maximum distance moved by the movable clamp is 168 cm and 1 impulse refers to 168/4000 = 0.42 mm. So through these impulses, the elongation values are measured. These are fed to the signal processor and the results are stored (Figure 12.19).

The breaking force and elongation values can be stored in the processor, and the results can be printed out.

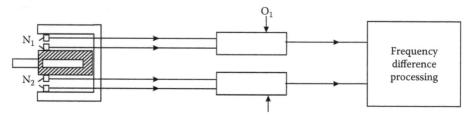

FIGURE 12.17
Breaking force measuring principle.

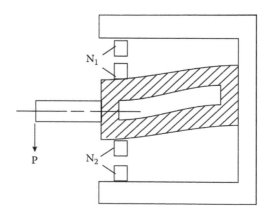

FIGURE 12.18
Elongation measurement.

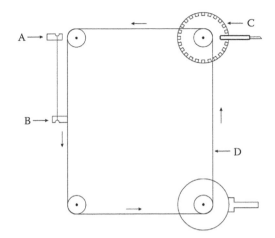

FIGURE 12.19
Breaking force measuring principle.

12.13.2 The Instron Tensile Testing Instrument

The Instron is an American tester that uses the bounded wire type of strain gauge. A general view of the table model is seen in Figure 12.20. In order to accommodate a wide variety of specimens, several interchangeable load cells containing the strain gauge are used in this instrument. In this way, a range from 2 to 100,000 g is possible. The load cell is located centrally in the fixed crossed rail. The upper jaw is suspended from the cell through a universal coupling. The lower jaw is mounted on the traversing crossed rail, which is driven upward and downward by screwed rods on each side. A range of gear changes enables the speed of the crossed rail, that is, the rate of extension of the specimen, to be varied in steps from 0.02 to 50 in./min. This can be seen on the bottom panel of the crossed rail. The load cell output is fed by cable to the control cabinet, which houses the various electronic circuits and the pen recording instrument. The main controls for load range selection, calibration, and so on, are mounted on the front panel below the recording chart.

12.13.3 The Load Weighing System

The load cell contains metal beam in the form of a cantilever. Strain gauges, consisting of a grid of very fine wire, are bounded to the surface of the beam. Four strain gauges are used and are connected in the form of a Wheatstone bridge, each gauge having a nominal resistance of 120 ohm. The block circuit diagram shows a simplified picture of the method of converting the unbalancing of the bridge in to the record of the behavior of the specimen under load.

The Wheatstone bridge network of strain gauge in the load cell is excited from an oscillator at a frequency of 375 c/s. The amplitude of this oscillator is stabilized at a constant value by a separate circuit. An applied load on the cell changes the resistance balance of the bridge, and the resulting signal is amplified by a circuit, which also includes means for initially balancing the bridge (Figure 12.20).

The gain of the amplifier can be varied in steps by a load selector switch, and a calibration control can vary the sensitivity continuously between any of these steps. A zero button shorts out the input to the following amplifier so that the zero position of

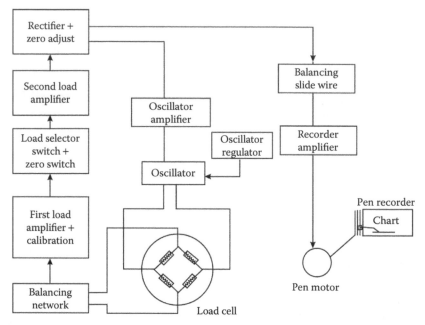

FIGURE 12.20
Instron tensile testing principle.

the recorder may be adjusted without the presence of any load cell signals. The signal is further amplified and then rectified by a circuit, which includes the zero adjustments for the recorder.

From this last circuit, the output is a dc signal that is fed to the potentiometer type of recorder, where it is compared electrically with the voltage across a balancing slide wire. Any unbalance between the two voltages is converted by a vibrator into a 60 c/s ac signal and amplified to operate the pen drive motor, which simultaneously moves the pen until the recorder is again in balance. Hence, the effect of extending the specimen is to change the pen to move across the chart a distance proportional to the tensile force in the specimen. A load-elongation curve is drawn on a chart.

12.14 Optical Trash Analysis Device

An Optical Trash Analysis Device (OPTRA) has been developed jointly by Textile Research Institute, Denkendorf, and Schubert and Salzer to measure the content of disturbing particles, especially the content of trash in the yarn. In the OPTRA device as shown in Figure 12.21, the yarn to be examined is wound up into thread layers on a board with defined distances by a winding device. The video camera takes one picture per measuring cycle of the thread surface thus formed and transmits the data to the computer, where the particles are identified.

In the course of each test, at least 500 pictures are taken to ensure a reliable result for every lot of yarn. Particles of 0.05–5 mm surface area are evaluated, classified in three nominative sizes, and calculated per yarn length of 1000 m. According to these three classes (trash imperfections), the different parameters in the spinning mill having an influence on the content of disturbing particles in the end product can be examined.

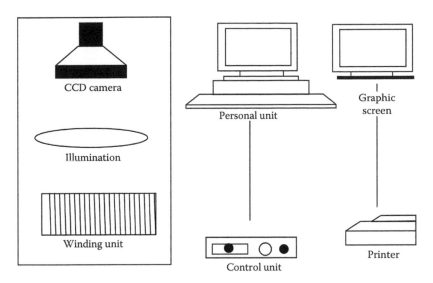

FIGURE 12.21
Optical trash analyzer.

12.15 Summary

The chapter discusses the importance of textile testing instruments. The working concepts of textile testing instruments were discussed in detail. The user will understand the concept of using sensors and transducers in measuring textile parameters.

References

1. http://nptel.ac.in/courses/116102029/12.
2. https://en.wikipedia.org/wiki/Fiber_optic_sensor.
3. https://www.elprocus.com/diffrent-types-of-fiber-optic-sensors/.
4. http://www.micro-epsilon.in/2D_3D/optical-micrometer/fiber-optic-sensors/.
5. http://www.dtic.mil/dtic/tr/fulltext/u2/046318.pdf.
6. http://journals.sagepub.com/doi/abs/10.1177/004051755802800809.
7. https://www.icac.org/tis/fiber_testing/fiber_testing_sept06.pdf.
8. https://textlnfo.wordpress.com/2011/10/27/advanced-fibre-information-system-afis/.
9. https://textilelearner.blogspot.in/2012/03/uster-evenness-tester-or-uster-tester-5.html.
10. http://nptel.ac.in/courses/116102029/65.
11. http://keisokki-thailand.blogspot.in/.
12. http://marantz-electronics.com/technical-articles/What%20is%20AOI%20(No%205).pdf.
13. http://www.brighthubengineering.com/hvac/48653-what-are-strain-gauges-and-how-they-work/.
14. https://www.uster.com/fileadmin/customer/Instruments/Yarn_Testing/Uster_Tensorapid/en_UTR4_TD_Feb14.pdf.
15. http://textileinsight.blogspot.in/2014/08/testing-instruments-for-tensile.html.
16. https://www.miniphysics.com/principles-of-electromagnetic-induction.html.

13

Automation and Control in Chemical Processing

LEARNING OBJECTIVES

- To understand the concept of the control systems used in chemical processing
- To identify the different types of sensors and measuring devices used in chemical processing
- To recognize the importance of measuring various textile parameters in chemical processing

---

13.1 Control Systems in Dyeing Process

Process control is one of the most critical aspects of quality assurance. In textile operations, such as batch dyeing, there are many variables that are under the dyer's control, but many more that are not. Typically, these controllable and uncontrollable factors interact in a very complex way. The dyehouse of the future must feature even better process controls. Traditional manual control methods in textile processes have been automated using microprocessor systems, with corresponding improvement in process repeatability. However, this mode of control utilizes only a minuscule fraction of the total capabilities of modern microprocessor hardware. In an attempt to improve microprocessor utilization, there are two avenues to pursue. The most celebrated of these is the macro scale global linking of information referred to as computer integrated manufacturing (CIM), to which much attention has recently been devoted. In typical CIM implementations, sophisticated user interfaces, attractive graphics, and computer networking capabilities are employed to make information from machines and machine groups available to managers for real time and post process analysis. The focus is on the use of state of the computing hardware and sensors to acquire data from processes, set up common database formats, link islands of automation, and provide management information in a timely manner in a macro or global sense. In CIM, individual workstations are connected through networking paths, and data are stored in a standardized central database, accessible to all workstations. The general structure of such a system is shown in Figure 13.1. One of the most important features of CIM is that it provides data to managers for strategic decision making. This implementation, although useful in certain ways, is incomplete because it fails to focus on optimized hardware and sensor utilization at the micro level, and also because it does not address the underutilization of the microprocessor and other hardware capabilities per set. An important, but far less traveled avenue for enhanced microprocessor utilization is the development of novel control strategies that fully utilize data processing capabilities at the micro level, use improved control models, and employ improved theoretical and empirical process models.

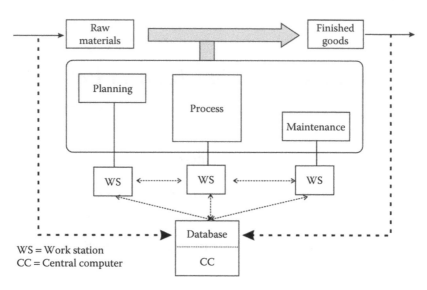

FIGURE 13.1
Structure of a future dyehouse.

The key at this level is not only a better understanding of processes but also a willingness to evaluate known control technologies from other disciplines (e.g., aerospace) and to develop new textile wet processing technology and science such as:

a. Process devices (e.g., valves, pumps)
b. Networking, upload/download capabilities

The state of the art in microprocessor control of textile wet processes is quite advanced, as can be seen from the preceding, but there is still potential for a quantum leap in performance if one is willing to discard certain traditional constraints in textile control concepts. Many of these constraints have become embedded in controller design concepts over time during the evolution of modern textile wet processing control systems. In contrast, literatures has focused on discarding many of these preconceived notions about control in order to evaluate the feasibility of novel systems in terms of performance and overall cost. Methods under evaluation include adaptive, real time, multi-channel control strategies that include sophisticated empirical and theoretical dye models, and that are based on innovative control algorithms.

13.1.1 Novel Control Concepts

Traditional dye process control methods attempt to conform as closely as possible to a specific predetermined process profile (e.g., time, temperature) to achieve correct results. Discipline is emphasized. Uncontrollable variances are accepted and, in some cases, remedied after the fact, for example by shade sorting or dye adds. Several innovative concepts are embodied in this feasibility study. Our approach attempts to control the ultimate product property of interest (in this case, dye shade) by adjusting controllable process parameters in such a way as to arrive at the desired end result. In this approach, the process may or may not be the same each time it is run, but the goal is only to arrive at the correct result. The process may be varied each time it is run to compensate for noncontrollable factors, such as variances in water quality, substrate preparation, and raw materials.

The first and most fundamental departure from the traditional control concept is the use of predictive, result-oriented strategies, as opposed to process conformance strategies. Sophisticated and theoretically sound dyeing models are combined with extensive real-time data acquisition to assess the state of the system and predict process outcome (i.e., final dye shade) about every two minutes during the dyeing. These predictions are the basis for real-time process modifications and departures from nominal process specification. Controllable process parameters are used to offset uncontrollable variances. Another departure from traditional methods is the use of multi/multi control strategies as opposed to the traditional one-to-one approach. For example, in a traditional control algorithm, a standard temperature of 200°F may be the process specification. If temperature deviates from the process specification, the controller will open steam valves to correct. This one-to-one control strategy senses temperature and controls steam. The novel approach does not control temperature for its own sake, but rather predicts the effect of a temperature variation. If no undesirable effect arises from that situation, the controller does nothing. However, if a problem such as an unacceptable dye shade is predicted, then the controller takes action, but not necessarily by opening a steam valve to correct temperature to a nominal value. Rather, action is taken by whatever means will correct the predicted outcome to the desired result (shade) at minimum cost and production time. The best action may be, for example, to add salt or change the pH. Of course, there are constraints built in to prevent the controller from taking absurd actions. For example, rate of temperature change or permissible pH or temperature values may be limited (Figure 13.2).

Central to this concept is the ability to accurately predict a result from the present state of the system. Several prediction methods are used for this. Their presentation is far beyond the scope of this chapter, but they will be described in other publications. Accurate real-time data acquisition is another prerequisite for this approach. The present work was done with a system capable of monitoring real-time values of temperature, rate of rise and

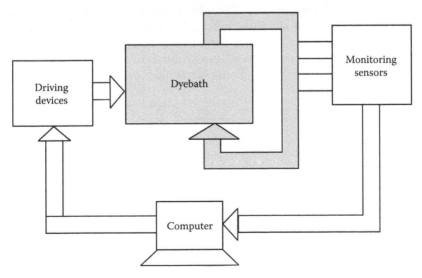

FIGURE 13.2
Structure of a dyeing process control system.

cooling, conductivity, pH, time, and up to three dye concentrations in exhaust dye oaths. This system is comprised of control logic units, dyeing machine, sensors, and interfaces. Another feature of the novel control strategies is their adaptive nature. Most colorists are aware of the necessity of making numerous standard dyeing as data input for laboratory color matching systems. Of course, this would not be feasible in a commercial production setting; therefore, the controller must have some adaptive methods of altering its database and control algorithm, that is, to "learn" from actual production dyeing. Traditional control strategies do not do this and, in fact, dyers are constantly altering dye recipe percentages to adjust to standard. The set point temperature, for example, would rarely if ever be changed for an individual shade. In this novel approach, nominal starting set points for various controllable process parameters (temperature, etc.) are adjusted, according to results of previous dyeing, to optimum values that would produce the desired shade at the lowest cost and minimum production time. By combining the above concepts of real time, adaptive, multi-channel, predictive process control with state of the art computing devices and sensors, outstanding results have been achieved and in fact, as will be shown in the owing examples, dye bath exhaustion can be brought consistently to a desired get value, thus producing excellent shade repeats in batch dyeing. The same novel control principles could apply to any process. We have selected batch dyeing as our example for feasibility evaluations (Figure 13.3).

13.1.2 Novel Control Schemes

Due to the complexity and uncertainty of dyeing processes, they are very difficult to control. Possible control schemes can be divided into two categories: parametric methods and nonparametric methods. The parametric methods require prior knowledge of process model structure and the range of process model parameters. Nonlinear robust control and adaptive control belong in this category. Adaptive control is a technique to handle uncertainties by designing the control algorithm to be self-adapting. There are two basic approaches to adaptive control. The first approach, which is called model-reference adaptive systems (MRAS), attempts to make the l/O behavior of the controlled process identical to that of pre-selected models. The adjustment of parameters can be determined based on gradient methods or stability theory. The other approach is called self-tuning regulator (STR) because it has facilities for tuning its own parameters. Sliding mode control is a very important scheme of nonlinear robust control. By utilizing specific information about the dyeing system that is being controlled, a sliding surface s = 0 is defined,

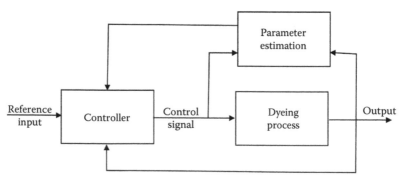

FIGURE 13.3
Structure of adoption.

where s is a function of state variables and time, which defines desirable process dynamics. The control law is defined so that the sliding surface $\mathbf{s} = 0$ becomes "attractive" in the state space, that is, a desirable state to which the system will converge. The nonparametric methods include artificial neural network (ANN) control, fuzzy logic (FL) control, and expert system (ES) control, which is also called knowledge based control or intelligent control. ANN is a massively parallel architecture for information processing, which has results in a new paradigm for learning similar to that in the nervous system. There are several ways to design an ANN controller. Supervised control is a common one, in which an ANN "learns" desired control actions from sensor output.

Training sets can be supplied by system models, other well-designed control systems, or human experts. Neural adaptive control, direct in verse control, and other methods are also used. The primary advantage of an ANN controller is its learning ability, which can continuously improve its control performance. The FL and ES controls are both used to simulate the decision-making activities of an experienced expert. The difference between them is the logic. Classical ES control uses crisp logic. Currently, the two approaches are often used together in many cases. Usually, the control decisions of an expert can be expressed linguistically as a set of heuristic decision rules. These rules are used to build rule bases for FL and ES controllers. Also, certain algorithms are used to convert the rules to quantitative control outputs. The disadvantage of all the nonparametric methods is that the control parameters do not correlate with physically meaningful parameters. Any of the aforementioned schemes can be used for dyeing process control. Two designs of dyeing process control, adaptive control for parametric methods and FL control for nonparametric methods have been evaluated for feasibility and performance in batch dyeing.

13.1.3 Fuzzy Logic Control

The structure of a FL controller is shown in Figure 13.4. It is comprised of three parts: fuzzifier, rule base, and defuzzifier.

A computation of the control action consists of the following stages:

1. Compute current error (E), which is the difference between the ideal output and the measured output, as well as its rate of change (CE).
2. Convert numerical E and CE into fuzzy E and CE.
3. Evaluate the control rules using the FL operations.
4. Compute the deterministic input required to control the process.

The fuzzifier is used to quantify the available measurements E and CE, which may be of limited accuracy, into certain coarse levels. The fuzzifier includes a scaling part and

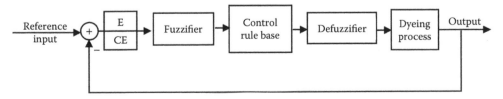

FIGURE 13.4
Structure of a fuzzy logic controller.

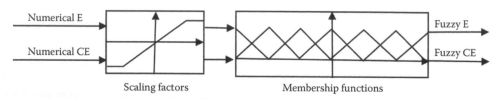

FIGURE 13.5
Structure of the fuzzifier.

a membership function part as shown in Figure 13.5. The scaling factors can be linear or nonlinear; also, the membership functions can have different shapes, such as triangle, bell, trapezoidal, and sinusoidal. Here, a linear scaling factor and triangle membership function are used. The fuzzifier converts numerical E and CE, such as 2.01 and −0.93, into fuzzy E and CE, such as large negative (LN), medium negative (MN), small negative (SN), zero (ZE), small positive (SP), medium positive (MP), and large positive (LP), with grade of membership m(E) and m(CE) from 0 to 1. The grade of membership values are assigned subjectively to define the meaning of the fuzzy values, such as large negative.

The rule base contains the control rules, which are developed heuristically for the particular control task and implemented as a set of fuzzy conditional statements of the form: If E is LN and CE is ZE, then CTRL is MP. This expression defines a fuzzy relationship between error (E) and error rate (CE) and change of process control input (Ctrl) for the particular system state. Also, the rules used are evaluated by using the FL operations, such as union, intersection, and complement of fuzzy sets. The control rules can also be processed graphically as a surface in the three-dimensional space (Figure 13.11). The defuzzifier is the inverse of the fuzzifier. It converts fuzzy process control input obtained through rule evaluation numerical deterministic process control input. Many algorithms can be used here, the center of gravity method being the most popular one.

13.1.4 Auto Jigger Controller System

This is a modified version with user-friendly features to convert any ordinary jiggers to automatic jiggers. It is an ideal substitute to the old mechanical counter type auto jiggers. Operational controls are extremely user friendly. This is a microprocessor-based system with digital indication of cloth on both sides and number of cycles completed. It has auto and manual modes. If there are problems in the system, it can be switched to manual mode. It also has start/stop and set cycle facility. This system can take care of elongation or shrinkage during the process of the cloth. For higher reliability, the sensors are inductive proximity switches, which are chemically resistant, dust-proof, and oil-proof with an IP 56 protection rating. It does not have any contact with the moving parts of jigger, so there are no wear and tear problems. It also has relay for controlling forward-reverse of jigger. This microcontroller takes care of false pulses due to electrical noise in the industrial environment. The system has been tested in a real environment. It can be installed at a visible place. In case of power failure, it indicates the direction of the rotation, so the user can rotate the machine in same direction. Forward counter/reverse counter indicate the rotations of cloth. If there is a power failure, it has the memory saving facility through E^2PROM.

13.1.5 Automatic Dispensing System

The automatic dispensing system for dissolving large quantities of powder, chemicals, and dyes from Color Services has several obvious advantages. The system is devised for weighing, dissolving, and dispensing large quantities of powder, dyestuffs, and chemicals, and meets health, safety, and environmental requirements. The powder dyestuff or chemical can be weighted (5–100 kg) automatically into the dissolving station. This is followed by the controlled and auto-addition of water, up to 70 liters per minute. The mixed solution is automatically dispersed to the tank or the trough of dyeing machine. The dissolving vessel is mounted on a trolley, and after the completion of operation, this vessel and pipelines are washed. The complete operation is controlled by the software designed by the Color Service.

The precision bulk-chemical mixing and dispensing system, from Kinder International Systems, is based on automotive gravimetric devices. It can mix up to 100 constituents of 10–1000 kg. The company has systems for automatic dispensing of finishing chemicals, dyestuffs, and print paste. The basic system is called CDS; however, the software and electronic controlling the dispensing are continuously developing.

A fully automatic dispensing system for package dyeing is another development toward dyehouse automation. In 1986, Technorama introduced Dosorama WSD, an automatic machine for dispensing solid and liquid products (dyes, auxiliaries, and chemicals). The package dyeing automatic dispensing system, from Color Service, integrates lab with production. This is used for powder dye and chemical dosing to package dyeing unit. Dyes and chemicals are dispensed to an intermediate preparation tank, where dyebath is prepared. This is subsequently sent to the dyeing machine using a multiple distributor. In the dispensing of solutions, the materials are mixed, according to the recipe, by means of three-way valves and Teflon pipelines, and the solutions are automatically delivered to the dyeing machine. These machines are particularly advantageous for the dyeing units that have to meet the requirements of short run production and delivery deadlines running the dyeing unit 24 hours, seven days a week with an increased operational control.

Logic Art Automation has introduced the Tube-Free Automatic Laboratory Dispenser to produce a dyeing recipe in shade matching. This dispenser can be used in labs of dyeing, continuous printing, and chemistry. The dispenser consists of several solution bottles, each with one injector. The injector dispenses solution by volume and weighing by scale. The recipe is fed through a PC, and the program is easy to read, learn, and operate. The dispenser is built with a module design and has 48 dispensing positions with simple connection, operation stability, and easy maintenance.

The chemical dispensing and distribution system, introduced by Adaptive Control Solutions, can extract the chemical, through a pump, from the supplier's containers. This system does not need the intermediate storage vessels for chemicals. A high accuracy flow-meter ensures the precise dispensing of chemicals to their destinations. Each dispensation can be followed by a hot or cold water supply. The system is useable for batch and continuous operation. An important part of this system is the control by an adaptive touch-screen industrial PC that can be operated in stand-alone mode, or can be connected to a third-party recipe-formulation system. Such connection permits the creation of an integrated dispensing system that can further reduce the order receiving and processing time. The history of each dispensation is created, which can be accessed by the external system.

13.1.6 Online pH Measurement and Control

13.1.6.1 Introduction

These systems have been used in a wide range of monitoring and control applications where pH is vital to the operation. This Technical Note sets out to describe the background to these various monitoring and control schemes, but for those who are unfamiliar with the fundamentals of pH, it is recommended that the Technical Note "pH Measurement" is studied first.

13.1.6.2 pH Monitoring

In its simplest application, pH measurement of an industrial sample provides an indication of alkalinity or acidity. A permanent record of the pH of a sample would be of greater benefit in many applications, and so an output signal is normally provided from the pH amplifier to a paper-chart recorder, satisfying most requirements. In more sophisticated schemes, input may be made to a controller, data logger, or computerized control system.

13.1.6.3 pH Control

In most cases, changes in pH signify the need for a change in the quantity of a reagent being fed into the sample. Very often the pH measuring equipment can be used as the primary element of a closed-loop automatic control system, directly controlling pumps and valves. Control loops are found in a wide variety of applications, from industrial effluent treatment to ore processing in the uranium mining industry. Known applications include:

- Water treatment for potable water
- Boiler systems, baking, brewing, preserving, soft drinks, and nutrient film soilless cultivation, as well as in the chemical, electroplating, detergents, and dyeing industries (Figure 13.6).

Automatic pH control is most simply achieved by on-off switching of solenoid valves using the alarm/control contacts on the pH transmitter. This method is frequently used for the

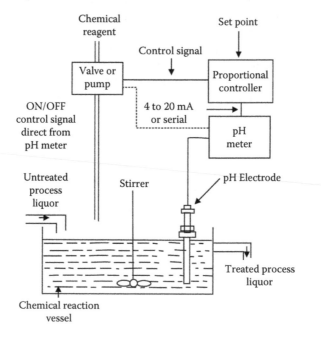

FIGURE 13.6
Basic control scheme.

treatment of effluents (lime slurry addition to neutralize an acid effluent, for example), as it combines simplicity with low cost. But there are some disadvantages. Thorough mixing of the sample is necessary, and care must be taken in siting the pH electrode such that it measures a truly representative sample. The system cannot compensate for large load changes or extreme changes in pH, and system overshoot time responses may become unacceptable long in these cases. Reaction time and electrode positioning must also be augmented to reduce overshoot from excessive reagent addition. When an on/off control system is used, the resultant pH value will rarely be steady, but the mean pH level of the treated liquor should be within specified limits.

Applications are often in effluent systems where holding or buffering tanks can reduce extreme changes in pH and varying loads in the effluent. A proportional control system provides a signal that varies in proportion to the deviation of the measured value from a set point. This signal is used to vary the opening of a suitable valve, or the speed or stroke of a pump. Unfortunately, proportional control by itself gives an "offset" in the resulting pH if there are load changes (such as flow rate) in the system. An additional control function called integral action control (or automatic reset) is usually incorporated in the controller (Figure 13.7).

This function repeats the effect of the proportional control action to eliminate the offset. The period of repetition (integral action time) is adjustable on the controller. Proportional + integral (P + I) schemes require careful setting up procedures, but once commissioned

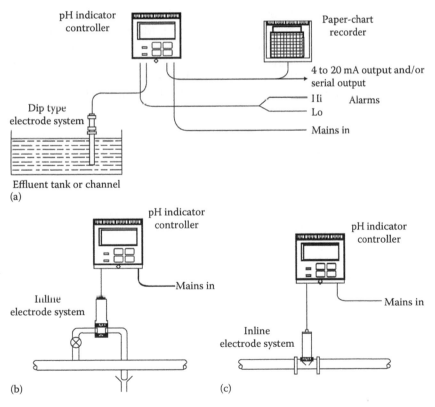

FIGURE 13.7
pH controlling devices: (a) dip system, (b) flow system in bypass line, and (c) inline system.

provide a very steady pH in the treated sample. Consequently, they are most often used in process control schemes. In practice, controllers usually have integral functions and, less commonly, derivative (or rate of change control) functions. These serve to compensate the system for irregular operation. pH control schemes most often employ P + I control schemes and only the derivative (D) function is usually switched out. Once set up, the system will compensate for offsets due to load changes and will give a smooth controlled pH value, without excessive hunting, that is, oscillations between the two extreme limits.

The effectiveness of any control is dependent on the response of the system, and this is primarily affected by the response times or lags in the plant and measuring apparatus. The most significant of these are reaction lag and distance-velocity lag. The former is the time taken for chemical reaction to be completed following reagent addition; strong acid-strong base processes are virtually instantaneous, for example, while at the other extremeTreatment of acids using carbonates requires reaction times of 10–15 min to eliminate generated carbon dioxide from the system. Distance-velocity lags comprise the time taken for the treated sample to reach the electrode system, which may be some distance from the point of reagent addition. The placing of pH probes and selection of dosing points must be carefully planned to ensure that representative samples are measured by the probes.

13.1.6.4 On–Off Control—Two Step

Operation of a solenoid valve by one of the control contacts of an industrial pH meter provides the simplest form of automatic pH control. Many sample streams are of reasonably steady pH and flow rate, and may thus be known to be always either acid or alkaline. Acid samples, for instance, require neutralization with an alkali, often lime slurry. One control contact on the pH meter can be set to achieve the required value for neutralization by opening a valve to admit alkali to the sample when the pH falls below the set value.

13.1.6.5 On–Off Control—Three Step

As a logical progression from two-step control, three-step control uses two solenoid valves to increase the dose rate when a sample pH value is far from the required value. Three-step control of an alkaline effluent, for example, might be arranged so that sample pH values between the set point, say 6.5 pH and 8 pH would cause one solenoid valve to be opened (by the low alarm contact). At 8 pH a second valve would open to dose more acid into the sample, controlled by the high contact (Figure 13.8).

Advantages of three-step control over two-step control are as follows:

- At pH values far from required value, greater flow rate of reagent gives rapid neutralization or pH adjustment.
- Nearer the required pH value, a smaller valve can be used to give finer adjustment to the pH. Advantages over more sophisticated control schemes are those of simplicity, and the lack of complex controllers.

A disadvantage is that the dosing reagent is added at a fixed rate, so that under varying load conditions, undershoot and overshoot would occur.

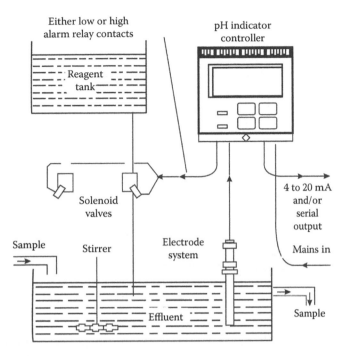

FIGURE 13.8
Three-step control of pH.

13.1.6.6 On–Off Control of Two Reagents

A simple neutralization system for effluent treatment can be produced using two solenoid valves, one controlling acid addition and the other alkali addition. On-off control would be effected by the control contacts—the low control switching the alkali dosing valve, and the high control for the acid dosing valve.

This system works satisfactorily if the dead band is large, and "overshoot" due to the plant load variations is small. The system is virtually unresponsive to load changes, which in the case of step load changes would cause the final pH to overshoot, and there would be a tendency for the reagents to neutralize each other rather than the process stream. Interlocks to prevent both valves operating together help to prevent reagent wastage (Figure 13.9).

13.1.6.7 Proportional, Integral, and Derivative Control

Accurate control of pH (or other parameters) within time limits can be most readily achieved using proportional, integral, and derivative control of valve position or pump stroke. In use, the required control point is set on the controller. Valve position or pump stroke is increased or decreased in proportion to the deviation from the set point. Actuator movement for a given error is governed by the proportional band setting; the proportional band is defined as the percentage change of full-scale deflection on the pH meter that would cause the pump or valve to operate through its full range, for example, at a setting of 25%, a change of 25% (f.s.d.) would cause the valve or pump to move through its full range; the proportional band is thus in effect a sensitivity control for the valve or pump.

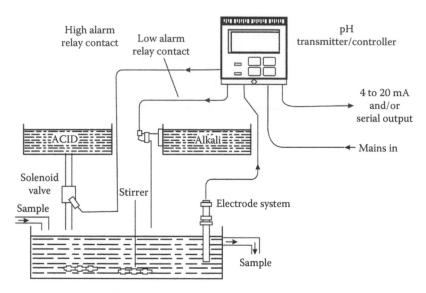

FIGURE 13.9
On–off control of acid and alkali.

In practice, the controller will be set up to deal with the prevailing conditions. When these change, an offset between the measured value and the required value occurs. If proportional control alone is used, the system will normally stabilize eventually, but at a point that is offset from the required value. The addition of integral control removes the offset. If the integral action time is too short, the process will oscillate. Further addition of a derivative term to the control action reduces the time needed to stabilize.

On start-up, if the process is changing rapidly, there could be an unacceptable overshoot past the set point value. The approach band can be used to bring in the derivative term before the proportional band is reached. This has the effect of slowing down the rate of rise. However, if the rate of rise is very slow, the introduction of the derivative term can be delayed. For applications in which large load changes (e.g., process steam composition and flow rate) are encountered, stroke adjustment alone may be insufficient to adequately control reagent dose rates. In these cases, pumps with both speed control and stroke control are preferred. The speed is controlled automatically by a flow meter adjacent to the addition point, increasing in proportion to sample flow. Stroke control is used as a trim on the pump output, controlled as before by a proportional, integral, and derivative (PID) controller (Figure 13.10).

13.1.6.8 pH Control in the Dyeing of Polyamide

Uptake of acid and metal-complex dyes by polyamide is temperature-dependent. The fiber takes dye at temperatures above the glass transition point because the segments in the polymer chains become more mobile at higher temperatures. The fibers thus open and allow the dye, which is attracted to the positive charge on the amino end groups in the fiber, to penetrate the polyamide. It is then bonded to the fiber through intermolecular forces. At the end of the dyeing process, a thermodynamic equilibrium is established between the dye dissolved in the liquor and the dye that has diffused into the fiber. The relationship between the dye concentration in the liquor and on the fiber is roughly described by the well-known Langmuir and Nernst isotherms. A finite time is required

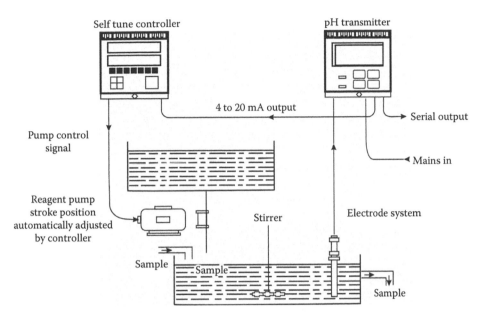

FIGURE 13.10
Typical installation of pH system for P & I control.

to achieve this state of equilibrium. The time required depends on the substrate, the dyes, and the process parameters. This process can be described by the laws of dyestuff diffusion. Applying these physical chemical principles in conjunction with practical experience shows the dyeing process.

13.1.6.8.1 Systematic Optimization

There are two established methods of dyeing polyamide with acid and metal-complex dyes: the constant-pH process and the pH-sliding process. While the constant-pH process ensures that the dye exhausts onto the substrate by raising the temperature, the pH-sliding process achieves exhaustion by a combination of raising the temperature and reducing the pH. In the constant-pH method of dyeing polyamide (Polyamide S process), the aim is to produce dyeing that are as level as possible from the outset and have good reproducibility. Level application of dyes, particularly at the start of the dyeing process, can be ensured if all the dyes exhaust onto the substrate uniformly. This is what is referred to as their combination behavior. It can be achieved in two ways.

The first is to select suitable dyes with good combinability. If this is not possible on coloristic grounds or because of the fastness properties required, a second method can improve the combination behavior of some dyes by using an auxiliary with affinity for the dye. The optimum concentration of such auxiliaries depends on the type of dye, dye concentration, and the auxiliaries used. The combination behavior of the dyes may deteriorate if the amount of auxiliary exceeds the optimum level.

To ensure good levelness, the dyebath must contain enough dye to allow sufficient migration. However, if bath exhaustion is too low, too much dye is wasted. As well as raising dyestuff costs, which are relatively low compared with the overall process costs, this alters the final shade and thus impairs reproducibility. A low pH increases bath exhaustion and reduces migration, while an increase in the concentration of an auxiliary with affinity for the dye has the opposite effect. The pH therefore needs to be optimized to ensure optimum bath exhaustion. At the same time, the strike rate (i.e., kinetic properties) plays a major role in regulating the dyeing process. During the heating-up phase, the dyes exhaust onto the fiber in a given temperature range. This is known as the critical temperature range. The beginning and end of this range are indicated by TStart and TEnd (TStart must not be confused with the temperature at the start of the dyeing process). The optimum heating gradient depends primarily on the unit used. To ensure good penetration of the fiber and thus stabilize the shade and improve fastness, a diffusion phase is required after exhaustion of the dyes. The time required for this depends on the substrate and diffusion properties of the dyes. In practice, the diffusion period is often unnecessarily long. This raises process costs and can damage the goods. Optimum conditions can be calculated using DyStar's Optidye™ N computer program, which includes both the underlying theory and the necessary dyestuff data.

13.1.6.8.2 The Optidye N Program

The Optidye N program contains the formula required for systematic optimization of polyamide dyeing. It contains data on the properties of DyStar products and a number of auxiliaries. Users enter details of the recipe, dyeing units, and substrate for each batch. The program then calculates the optimum dyeing profile, the pH required to achieve average bath exhaustion of approximately 95%, and the auxiliary concentration required to ensure optimum compatibility of the dyes. Alongside the recipe data, details of the substrate are very important. Two parameters are required to describe the dyeing properties of polyamide to be dyed in the constant-pH process. The V value shows how quickly a standard dye exhausts onto the substrate. The fiber saturation value (SF value) shows the maximum amount of dye that can exhaust onto the substrate. Any dyehouse laboratory can determine these values by carrying out a few dyeing. Values determined in bulk conditions vary between 1 and 3 for SF and 0.2 and 5 for V. For a typical recipe with Telon A dyes, the following exhaustion ranges may be calculated on the basis of the SF and V values given above. For V = 0.2, a range of $T^{Start} = 74°C$ and $T^{End} = 102°C$ is calculated. To save time, the bath can be heated to 74°C as fast as the equipment allows, without any risk of unlevelness. By contrast, for V = 5, the final temperature $T^{End} = 31°C$, while T^{Start} would be below the normal water temperature (theoretically below 0°C). Consequently, heating the bath too quickly at the start would entail a risk of unlevelness. A dwell time at the start of the process (e.g., 15 min at 20°C) can reduce this risk but not eliminate it entirely. Moreover, a higher starting pH would be required. In other words, a pH-sliding process would have to be used. Using a pH-sliding process, the bath could be heated rapidly from 31°C. These examples show how effective optimization of the dyeing profile can be. The next section looks at the progress made on the basis of these theoretical findings.

13.1.6.8.3 Extending the Polyamide S Process

As we have seen, it is necessary to regulate bath exhaustion to ensure the correct level of migration. In the constant-pH process, the aim is to achieve final bath exhaustion of around 95%. However, to increase reproducibility, bath exhaustion of nearly 100% would be ideal. This can be achieved by reducing the pH. Similarly, if the V value of the substrate is very high and the strike rate cannot be controlled solely via temperature, a modified version of the constant-pH process is required. The pH-sliding process was developed from the constant-pH process specifically for critical shades and problematic units and substrates.

This process uses a combination of temperature rises and a reduction in the pH during dyeing to control exhaustion of the dye.

Under standard dyeing conditions (pH 7), the amino end group and the carboxyl group are protonated. In other words, viewed from a distance, this material has a positive electrical charge. A negatively charged dyestuff molecule is therefore attracted to the PA and tries to react with it. Raising the pH—in other words, reducing the concentration of H^+ ions—deprotonates the carboxyl group, leading to a negative charge. The fiber thus takes on a neutral or negative charge, and the dyestuff molecules are not so keen to react with it. Therefore, pH plays a major role in the exhaustion of the dye from the liquor onto the surface of the fiber and in the diffusion of the dye in the fiber. The optimum pH range required for the pH-sliding process depends to some extent on end groups in the fiber. Practical trials have shown that a pH^S that generates bath exhaustion of around 70% is most suitable.

Since the reproducibility of a dyeing is best at 100% bath exhaustion, wherever possible the pH should be reduced to a level where this is achieved. pH^{End} thus depends on the substrate, dyeing recipe, and auxiliaries used. This value can be calculated using the Optidye N program. It is about 3 points below pH^{Start} and 0.5 points below the pH used for the constant-pH process. To optimize the pH-sliding process, we look at three different ranges in which the pH can be reduced. Depending on requirements (i.e., the level of difficulty), three different points are recommended for the addition of the acid. In the pH process as the most rapid method, the pH is reduced as soon as the temperature reaches the start of the critical range. This method is suitable for relatively uncritical conditions. The dyeing time at T^{max} is then equivalent to the diffusion time in the constant-pH process. The critical temperature range can be passed through faster because the pH at the start of the process is above the level used in the constant-pH method, allowing for more level dye uptake.

In the second method illustrated here (the "universal" method in Figure 13.4), which has proved reliable in practical trials, the pH is reduced when the temperature reaches the end of the critical range. This generally ensures an optimum balance between dyeing time and reliability. The third method, which is shown in Figure 13.4 as the most reliable method, should be used only for very difficult shades such as turquoise. In this method, the pH is not reduced until the temperature reaches T^{max}. The low level of bath exhaustion at the higher initial pH value is utilized for migration of the dyes. Compared with the constant-pH method, the dyeing time at maximum temperature is increased by the length of time required for dosing. Regrettably, the most suitable of these three methods can only be determined empirically at present, as no mathematical formulae are available. Alongside the migration properties of the dye, other major influences are the equipment used and the composition and pH-dependence of the polyamide.

13.1.6.8.4 Setting the pH

Changing the pH during the dyeing process is a method that has been used since the 1970s—to dye carpets on winch becks, for example. Methods such as the Telomat and Dosacid processes are well established. Since the pH is an inverse logarithm of the concentration of H+ ions, it is often difficult to regulate automatically. Methods used to control the pH during dyeing include pH buffers, acid or alkali donors, and automatic control and regulation units. While a pH buffer maintains a constant, predefined pH, acid and alkali donors alter the pH as the temperature rises. Automatic pH measuring and control units can be used for the constant pH and pH-sliding methods. Moreover, they are often used to monitor the pH and thus control the process. However, in the past, such units have not been stable enough to become established. A modern pH-control unit is expected to meet the following requirements:

- Robust technology with low maintenance requirements
- User-friendliness
- Automatic calibration
- Long-lasting electrodes
- Simple connection to dyeing units

The pH sensor (electrode) is built into a bypass parallel to the liquor pump. To prolong the life of the electrodes, they are only placed in the liquor to take measurements. The bypass has a back cooling system that cools the liquor to 80°C. This is also designed to prevent wear of the electrodes. In the mobile unit, the acid or alkali is added via a dosing pump. If several pH-FiT units are installed in the same plant, it makes sense to pump the chemicals in a closed-system circuit with a dosing valve for each dyeing unit.

To demonstrate how the Polyamide S process can be used to optimize the pH-sliding process in conjunction with a pH-FiT unit, Figure 13.3 shows exhaust samples taken from the dyebath after beam dyeing of a polyamide taffeta fabric with Supranol and Isolan dyes. To achieve an optimum dyeing profile, the starting temperature for the critical range was calculated at 48°C, while the final temperature was calculated at 88°C. Since the dyeing involved a critical shade with high-molecular dyes, the reduction in pH was selected to ensure maximum migration (i.e., the pH was reduced after the critical temperature range). The optimum pH range in this case is 8.5–5.6.

A statistical evaluation was made of the reproducibility of bulk dyeing performed using the pH-sliding method and the Polyamide S process on the basis of approximately 300 batches dyed on the jet. After optimization of the process, the optimized and non-optimized batches were compared. Process times were reduced by 27% per batch, mainly due to a 65% reduction in the need to correct faulty dyeing. The use of acetic acid instead of an acetate buffer cut chemical costs by 48%. Altering the chemicals used also reduced effluent contamination, for example, but lowered the COD.

13.1.6.8.5 The Compudye System

Optimizing the pH-sliding process through the Polyamide S process is only one aspect of our approach to modern dyeing methods. Following successful application of the Optidye N PC program, where certain data such as the recipe have to be entered manually, it is far more effective for routine work to automate both these inputs and the transmission of the dyeing parameters calculated to the control units.

Dyehouses often have a host computer system on which the servers and PCs depend and which is responsible for coordinating routine tasks. The Optidye N program may use data already in the system. Alternatively, the host system may require data from the Optidye N program. Once the dyeing recipe has been calculated, data on the recipe, substrate, and equipment are transmitted to the Optidye N program to optimize the recipe. In the simplest case, this takes place manually. In a fully automated dyehouse, the data would be transferred automatically via a network. The Optidye N program would then pass the relevant optimized recipe parameters on to the control unit on the dyeing machine—which needs data like T^{Start}, T^{End}, heating-up rate, pH range, pH reduction time, and so on,—and to the color kitchen, which needs data on the dyes required and the optimized additions of auxiliaries.

Compudye is the cost-effective solution for complete control and visualization of all types of textile dyeing and finishing machines. The active color display shows a windows user interface that requires little training. The SEDOMAT 5000's powerful "open concept" with internal programmable logic controller (PLC), Profibus DP field bus, and remote input/output allows connectivity with most available peripherals. This future oriented concept enables a modular approach to machine automation. The SM5000 can easily be integrated into a total CIM-concept. Control of machines and functions of all types is guaranteed by the modular and free programmable PLC concept including a link with a fully automatic dye kitchen for the preparation of dyestuff and chemicals.

13.1.6.8.6 Batch Processes

Batch processes control all batch dyeing machines such as yarn, jet, jigger, and inch, as well as control of temperature, time, speed, flow, differential pressure, pH, circulation control, and dosing.

13.1.6.8.7 Continuous Processes

Continuous processes control all continuous machines such as singeing machines, bleaching ranges, mercerizing machines, washing machines, dryers, continuous dyeing machines, stenters, shrinking machines, decating machines, calenders, raising and emery machines, and so on. They control speed, residual moisture, exhaust air humidity, fabric temperature, dwell time, shrinking and stretching, pH, and fabric width. By simultaneously executing two batches and several functions, the next batch can be prepared while the current batch is still active. This increases the machine efficiency. The functionalities of the continuous process are

- Windows user interface for a simple and intuitive operation.
- Entry and management of processes.
- Animated VGA mimic of the machine improves the process overview without providing a machine diagram on the panel.
- PCMCIA memory card for data transfer and backup.
- Interface to SEDOMASTER central system.
- Entry of processes.
- Scheduling of batches on the machines.
- Graphic visualization of the process values.
- Flexible report generator.
- Several languages and character sets (e.g., Chinese) are supported.
- Allows logging of all process data, a must for ISO9000 certification.
- Storage and graphic evaluation of process data and events (e.g., operator calls, consumption, alarms, fabric length, stops).

13.1.6.8.8 Internal PLC

All intelligence for the complete control of a textile finishing machine is integrated in the SEDOMAT 5000. The process know-how (process editing and execution) as well as the machine know-how (PLC program) are available in the SEDOMAT 5000. There is no longer any need for a separate, external PLC, thus reducing the hardware requirements. There are two ways of programming the internal PLC software:

- *PLCPROG compatible PLC*: All PLC programs written for SEDOMAT 2000 or 3500 will also run on the SEDOMAT 5000. So there is no need to rewrite existing PLC programs.
- *IEC1131 compliant PLC*: This soft-PLC offers the following advantages:
  - Windows programming interface
  - Standard programming languages, variables, and operators
  - Faster implementation
  - I/O independent programming
  - Arithmetic calculations
  - Unlimited number of timer and counter etc. by using the principle of instances
  - Powerful library and debugging functionality
  - Multitasking
  - Event driven

13.1.6.8.9 Profibus DP Field Bus

PROFIBUS DP is a worldwide accepted standard field bus for the connection of inputs, outputs, frequency inverters, PLC components, sensors and valves, remote displays, and so on. This standard industry bus has been chosen as part of our open connectivity strategy. The field bus concept significantly reduces the panel building overhead by eliminating a large part of the wiring. Worldwide, thousands of PROFIBUS DP-ready peripherals are available, providing an enormous choice of components. This flexibility and freedom of choice will have a positive effect on performance and price of the machine automation.

13.1.7 Indigo Dyeing

The indigo-dyeing-pilot analyzes the critical concentrations in the dye-bath, compares to the set values, adjusts the dosing commands, and includes a display that informs you about the essential process parameters. Automatic titration is the basic way to value the concentrations, it is a method of analyzing and has been proofed over a longer period of time and you can rely on. When the yarn varies, the dosing is automatically adjusted.

The indigo dyeing pilot keeps the concentration of hydrosulfit dithionit) and the alkali (pH-value) constant and outputs the indigo concentration in g/l for the purpose of control. The display shows either the diagram of the pH value or the curve of the redox titration and at least the essence (concentrations) and more data. The indigo dyeing pilot lets you perform your process efficiently.

An automatic redox titration guides the way. A well-defined portion of the dyebath is presented and a reagent is given into it step by step until the final point is reached. The redox titration is based on the reaction of a reducing agent like leukoindigo and hydrosulfit (dithionit) with a matching oxidation agent like Kaliumhexacyanoferrat. This redox reaction is similar to the dyeing, where reduced indigo went on the fiber, and it is fixed through the oxidation with oxygen out of the air.

H&R's redox titration measures the amount of indigo able to dye active, that is, the reduced form, his correct name is leukoindigo. The display shows its concentration in g/l. The used reagent solutions are simple to prepare, and their durability is at minimum a week. Also, the titration is a simple method without exotic sensors. In use are standard probes well known in your laboratory (calibration, cleaning, storing).

The indigo concentration is displayed and characterizes the condition of the dyeing machine, interesting when the fabric often changes. A higher concentration shows a lower affinity to the yarn and a lower indigo concentration shows a higher affinity to the yarn. The dosing controller keeps the addition of indigo constant at, for example, 1.8% (18 g/Kg). The dyeing pilot keeps the indigo concentration in the dyebath constant through adjusting the dosing amount for the indigo. The display shows the amount of indigo in relation to the weight of the fabric. The yarn with a high affinity to dyestuff looks darker and vice versa. Before starting your lot, you can decide the way you want to dye on the controller.

The concentration of hydrosulfit (Na-dithionit) found at the final point of titration, that is more less than you inject to the machine. If the concentration of hydrosulfit is too low, then the dyeing is at high risk (deepness and shade of the color) or if the concentration is too high, the costs of hydrosulfit increase as along with the pollution of waste water. The controller keeps the concentration of hydrosulfit constant by driving the Hethon dosing system. After the lot changes, you must not change anything unless the set value for the hydrosulfit concentration is constant at, for example, 1.5 g/l, but you can change the set value at any time with the keyboard

The reduced agents can deposit as particles in the ground sludge of the bath. The particles in the ground sludge are found in the titration curve. Their average concentration is displayed.

13.1.7.1 The pH Value Guides the Way...

The pH value is proportional to the concentration of the alkali, or chemically seen to the concentration of OH^- ions. At high pH values, the di-salt of the leukoindigo arises and promotes the introduction of the indigo to the center of the yarn, that is, full penetration dying through the profile of the yarn. At lower pH values (below pH = 11) the mono-salt of the leukoindigo arises. The substantivity is higher and the leukoindigo sit down on the outer layers of the yarn, that is, ring dyeing through the profile of the yarn. This ring dying seems deeper and darker, although the same amount of dyestuff was inside the dye bath. It is easy to see the consequences if the pH value is not kept near the given set value.

The indigo dyeing pilot continuously measures the pH value. The dosing amount of the alkali pump is adjusted accordingly to the difference of the actual to the set value. An automatic check of the pH-probe determines the state of the probe (durability, sensitivity, zero point, etc.). The check is made during rinsing of the armature. If the parameters are out of tolerance, an alarm sounds, and the probe can be replaced in time.

Rely, or better control of direct online measuring of the machine speed, should be constant but is not. Crawling or stopping is inevitable. The dyeing pilot reads the machine speed, controls the dosing accordingly, and displays the information. Salt is important for the dyeing, as well as for the indigo. It substantively increases and influences the shade of the color. The indigo dyeing pilot measures this value. The signals of the probes are a function of the temperature. They must be corrected to the basic temperature.

Therefore, the temperature is continuously measured, and the controller calculates the correction and makes the adoption automatically.

13.1.8 Automatic Control of the Dyeing the Dosing of the Agents

The dosing must be exactly adapted to the machine. Examples for different dosing techniques are:

- Stock vat indigo with increased concentration of reductions
- Stock vat indigo and chemical auxiliaries separately stored
- Stock vat indigo and injection of hydrosulfit as powder and injection of caustic soda

To keep the above-mentioned concentrations constant, we need three independent sources for the dosing

1. Dyestuff (stock vat indigo or indigo solution BASF 20%)
2. Reductions (hydrosulfit in solution or as powder)
3. Alkali (normally caustic soda 50% or 50°Be)

Proportional dosing in relation to the weight of the fabric (constant amount per Kg warp) is important. The dyeing pilot must be aware that the free hydrosulfit disappears while being in contact with the air. The fine adjustment is done during the feedback process, because every component has a different consumption.

13.1.8.1 The Smart Gray Cells of the Controller

The controller is a personal computer in industrial standard with a Pentium processor (Intel), H&R Model TICCO 3. The operating system is Windows NT, for this purpose a more stable platform than Windows 95 or 98 (Microsoft). The structure of the software is modular, so modifications can be done without damaging well-running parts. The main focus is on so called algorithms for closed-loop control, which output a dosing command to the pumps, conveyor screws, valves, and so on based on the concentrations and the boundary conditions. These algorithms for closed-loop control influence the quality of the process and the evenness of the concentrations. A problem is the dead time inside the machine, that is, the time from the beginning of a change in flow rates until the response through titration. During this dead time, plausible parameters are needed to avoid the swinging of the system.

Another important task is preparing the rough data of the sensors and extracting the searched concentrations out of the titration curves. Of course, the calibration of the sensors is supported by the controller's software. When changing the lot, the controller needs information such as yarn number, thread count (weft), depth of shade, and machine speed. A lot that had run before can be called simply out of a small database.

The main panel contains all input and output information, interfaces, and analog/digital transducers. Service, especially for the sensors, is easy because there is enough space, you don't need special equipment, and the sensors don't contain embedded intelligence. All electronic components are inside the water protected area, which includes the controller also. The controller is ready to connect it to a Windows network. At every connected terminal, necessary software presupposed, the data of the last lot and the current lot can be watched. This way, remote access works over a telephone line and the collected data is stored over half a year. Although the system has a big variety of tasks, the handling is still simple through user-friendly support of the software.

13.1.9 Automation in Dyehouse

In addition to package dyeing, wool tops are also dyed in the same machine system. Carrier for yarn packages with 5, 10, 20, 40, or 80 spindles each are used, which can be coupled to one another in various ways or treated individually in the dyeing machine systems available. The corresponding carriers for tops are suitable for 2, 4, or 16 and can be loaded with 4 bumped tops in each case. In the dye kitchen, one preparation tank per dyeing machine is provided as well as appropriate dye-dissolving vessels in a range of different sizes.

The delivery of textile auxiliaries from their respective supply tanks is fully automated (MPS-L) with tank containers being provided for common liquid chemicals (e.g., acetic and formic acids). In order to ensure a high degree of reproducibility in dyeing, intensive laboratory preparations with a data color dispensing system are necessary. The entire transport system and yarn storage are fully automated, which includes the loading and unloading of material carriers. The presses used are for the press-packing of yarn packages with appropriate positioning of the top plates without manual intervention. For accurate dyeing, the net weight of each batch is determined exactly and recorded data transmitted to the dye kitchen and dyeing machine system ready for appropriate order allocation. The liquor ratio used for dyeing is approx. 7:1, which has a very favorable effect on the dyeing process in terms of dyeing costs for energy and auxiliaries.

The realization of this fully automated yarn dyehouse was made possible by providing overall control with an OrgaTEX system, which determines, controls, and regulates all functions of the yarn dyehouse. Besides the loading and allocation of dyeing machines, this system also includes dyeing machine control, management and control of the dye kitchen, integration of the dyehouse laboratory and the entire allocation of orders with the necessary movements of material within the dyehouse, and the simultaneous acquisition of all dyeing data. This dyehouse is, of course, linked to an effluent treatment plant with integrated heat recovery so that, among other things, cooling water can be returned directly, and the corresponding heat energy from discharged effluent is recovered by means of a heat exchanger.

This company operates a batchwise piece dyeing plant that is mainly engaged in processing cotton circular knits and their blends for the underwear market. The dyehouse is equipped with both atmospheric and HT "roto-stream" dyeing machines for the dyeing and bleaching processes.

The entire system is again controlled by means of an OrgaTEX computer. Besides the management of dyeing recipes and the control of information to the dyeing machines, this system controls both the dye store and the integrated paternoster as well as the corresponding dye-dissolving station (MPS-D). The dye, which is weighed in powder form, is fed to a special dissolving system MPS-D, and, depending on the particular dye class used, it is completely dissolved with the minimum quantities of water and temperatures required in accordance with specific programs, then transferred to the dyeing machine.

An MPS system that doses all the chemicals individually is available for chemical dosing. The feed pipe is cleaned with the minimum quantity of water after each chemical addition until all the chemicals necessary for the dyeing process have been transferred to the preparation tank. These pipes are cleaned in each case with compressed air. For the solid chemical dissolving station (MPS-S), four supply tanks have been provided for salt, hydrosulphite, sugar, and soda. These solid chemicals are delivered by tanker or in sacks and are fed to the dissolving station through a pipework system dried with hot air.

In the solid chemical dissolving station, salt is dissolved at a mixture ratio of 1:1 and the other solids at a ratio of 1:2. A prerequisite here is the availability of a constant quantity of water for preparation. The respective solid chemicals are dispensed gravimetrically. The resultant mash is then fed to an individual dyeing machine by special pumps with a final rinsing charge followed by air impingement to clean the pipework systems. The fully automated delivery of necessary chemicals by the computer system also requires a correct decision with regards to the particular piece dyeing machine to be used. Here, the type of material and fibers involved, in this case knitgoods, is decisive. The effect of automation should bring a corresponding reduction in the total consumption of dyehouse water. Of course, the design advantages of the machine such as the arrangement of the reel, the filling of the storage chamber by means of a plaiter, as well as the use of a rotating drum as a fabric storage chamber to facilitate tension-free fabric transport also need to be taken into consideration. The control system must be capable of analyzing all processing techniques and store the production data in such a way that fault detection is also possible at a later stage.

The use of new rinsing systems including, among others, continuous rinsing with exact quantities of water (superwash rinsing) assisted by a 100% addition tank, is also crucial here. The possibility also exists to drain the liquor with a liquor pump under valve control in order to accelerate the draining process. The 100% addition tank can also be used during bath changes for pre-preparation of the next process baths. The so-called CCR (combined cooling and rinsing) procedure, the use of cooling water in the cooling phase with simultaneous return as rinsing water, can also be integrated within the closed machine system.

A further point that must not be neglected here is the use of a suitable filter system. For a fully-automated piece dyeing operation, this can supply two different systems. A so-called self-cleaning filter is able to filter out the lint that accumulates in the main circulation and remove it by controlled separation via a rotating system. With another filter system, which is integrated into a so-called bypass circulation, it is possible to filter out lint continuously via a second flow and remove this lint from the dyeing machine system at every bath change.

13.1.10 Plant Manager System for Dyeing and Finishing

The IT plant manager is applicable to all the dyeing and finishing processes. The following IT plant manager modules will assist the daily procedures in the dyeing and finishing mill: *Production organization*: Using the IT-Architect application, the IT plant manager becomes the knowledge repository of the company. All information related to production sources can be defined in the system's database. The following production processes of dyeing and finishing can be defined in the system:

- Batch dyeing
- Continuous dyeing
- Singeing
- Washing
- Bleaching
- Mercerizing
- Calendering

- Raising
- Fixation
- Emerging
- Shearing
- Printing
- Final Inspection

13.2 Control Systems in Textile Finishing Machinery

Finishing is the final process for textile materials to make them attractive and user friendly. Public demand regarding the quality and appearance of manufactured textiles has always been the controlling factor in the final sales. The modern trend is toward the better presentation of manufactured textile materials. The better finished textile material sells faster and at better prices. The modern development of control systems in machinery, apart from finishing chemicals and recipe, has helped considerably in improving the marketability of textile materials.

13.2.1 Stenters

- The stenter is a preshrinkage machine used mainly for synthetic and its blends (heat setting).
- The main function of the stenter is to stretch the fabric width-wise and to recover the uniform width.
- The stentering machine is also used for the following operations:
- To dry the fabric
- Curing treatment for some special finishes such as resin finishing and water repellent finishes etc.
- To give soft finishes to the fabric
- To stretch the fabric to required width
- There is a control over the width and length-wise stretching during drying, which is not so in cylinder drying.

Typical monitoring and control at a stenter frame include the following:

- PLC heatset plus with
  - Dwell time control
  - Residual moisture control
- Air temperatures
- Moisture retention
- Stretch/shrinkage
- Width monitoring and control

- Exhaust humidity
- Fabric weight
- Down time reports and deviation reports

The controller can also be used as a reporting station for plant production floor monitoring.

13.2.1.1 Control of the Fabric Temperature

The temperature of fabric is an essential process parameter in a large number of production processes during and immediately after thermal treatments (e.g., dryer). The measuring computer processes all the values of a traveling infrared temperature measuring camera across the width of the web.

More and more measurement of the fabric temperature after pad patch is important to guarantee reproducible dyeing results and to avoid shade variation over length and width of the fabric. The fixation speed of reactive dyes depends strongly on the dwelling temperature and should be known to calculate minimum and maximum fixation times. Experience from the practice in monitoring the fabric temperature shows significant influence and improvement of the shade continuity; this was underestimated in the past.

13.2.1.1.1 Sensors and Control Devices at a Dryer—Measurement
of Fabric Temperature Inside a Dryer

13.2.1.1.1.1 Preheater Sensors   Heat treatment processes (e.g., drying, fixing, condensing, vulcanizing, cross-linking, and shrinkage) can be optimally achieved if temperature patterns of products being heated in the dryer can be measured accurately and continually. To measure the temperatures of the products in the dryer, sensors have proved to work most successfully. The sensors can be used inside a heat treatment machine (e.g., dryer) up to a temperature of 400°C. Several sensors are mounted and distributed over the length and width of the machine. Up to now, the sensors have been used successfully in the textile industry at stenter frames, multi-layer stenters, and hot flues (Figures 13.11 and 13.12).

13.2.1.1.1.2 Dwell Time Control and Process Viewing   For the automatic control of the stenter frame, Pleva provides two different controllers:

1. The PLC heatset with a maximum of eight sensors
2. The climatic heatset as a control and process viewing system for up to 120 sensors and many additional functions

13.2.1.1.1.3 The PLC Heatset Plus   The PLC heatset plus is comprised of a programmable logic control system and an operator's panel with color monitor and keyboard. The PLC scans in the data and speed using up to eight sensors. It controls the speed of the machine in order to attain the required temperature dwell time or a given product temperature in the last processing zone of the dryer (temperature control).

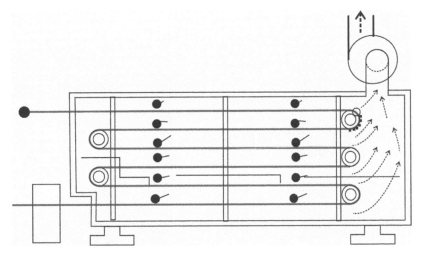

FIGURE 13.11
Drying units.

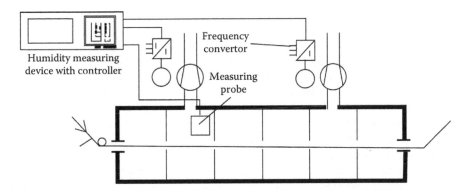

FIGURE 13.12
Heat treatment processes.

The manner of operation is entered into the system at the operator's panel:

- Data sensor measurements and displays without control
- Dwell time control by means of setting values for the dwell time and the dwell temperature
- Temperature control with the setting values for the product temperature at the last zone of the dryer

All relevant data, such as the above settings and the actual values of the temperature of the product, are displayed on the integrated color monitor. Dwell time or product temperature and speed of the last 4 h also can be displayed as a line graph diagram. Sensors for exhaust humidity and residual moisture can be connected optionally. The relevant control loops are already installed.

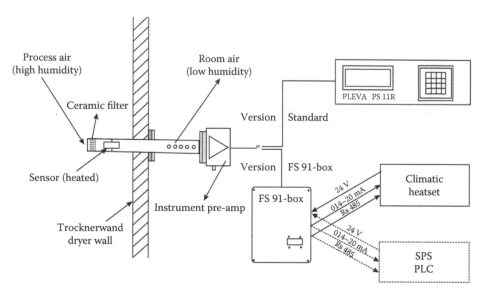

FIGURE 13.13
Controls in heating systems.

The advantages of dwell time control are that without information about the temperature patterns of the product, the product may well be left in the dryer too long, in order to be on the safe side. Using a dwell time controller, these excess periods in the dryer are avoided. Increases in production of 30% and more are feasible. At the same time, the product is not exposed for excessively long periods to high temperatures (Figure 13.13).

The new heatset plus process viewing and dwell time controller is available for more demanding tasks; in its basic version, it can process up to 30 sensors. It features a plurality of additional functions that are not provided by the PLC heatset plus. The climatic shows the actual data and provides trend charts according to length and time. Additional sensors for residual moisture and exhaust humidity can be integrated.

13.2.1.2 Residual Moisture after Dryer

The measuring and control of the residual moisture constant of fabrics are of considerable importance in both technological and economic aspects. In usual practice, the goods are over dried to a higher or lower extent. This will cause a rise in costs. The goal is therefore to control the speed of the dryer (or if not possible, then the heat) to achieve the moisture regain (e.g., 7–8 at cotton) after the drying process.

Residual moisture meters are used as well at humidifying processes of dry fabrics before sanforizing, decating, or compacting. In these cases, the residual moisture is measured before or after the humidifier and controlled to the required value (e.g., 12%–14% at sanforizing of cotton).

13.2.1.2.1 Residual Moisture Meter

The residual moisture meter is a contact measurement of running fabrics. The residual moisture meter is based on measurement of electrical resistance. This increases exponentially as the residual moisture decreases. The residual moisture meter allows

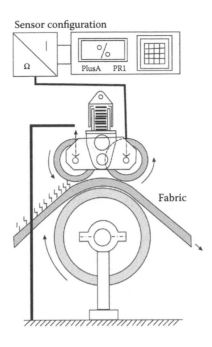

FIGURE 13.14
Residual moisture meter.

lower residual moisture contents to be measured than was previously the case (measuring range 10^{14}–10^{15} Ohms, which means, for example, at cotton: 0.9%–1.5%, at synthetics: 0.1%–5%). The electrostatic charges occurring during such measurements are discharged by the residual moisture meter itself, meaning that they do not impair measurement (Figure 13.14).

The contactless measurement is based on the microwave absorption by water. The residual moisture meter therefore allows measurement of terry towels, high plied products (velvet, cord, etc.,), and delicate printed goods. In addition, it can be used at higher residual moisture values to measure and control the residual moisture after IR-predryers where there is 20%–40% residual moisture.

13.2.1.2.2 Exhaust Humidity

Controlling the humidity of the exhaust air is a standard established procedure today. The energy savings can be 10%–30%. The installation can be usually written off in a few months. Controlling the humidity of the exhaust air also maintains constant atmospheric conditions. The quality of the dried material is maintained at a high humidity level in the exhaust air, and at the end of the drying operation there is less yellowing of the material and improved handle.

The sensor consists of a measuring probe and evaluation electronics. It contains two electrodes, one of which is exposed of the air, the other to the environmental air of the room. In line with the progressive absorption of humidity by the dry air, the sensor generates the signals, which in connection with the preamplifier are presented to the evaluation electronics calculating the absolute humidity in gram of water per Kg dry air or volume % in the dryer.

13.3 Special Purpose Drying Machine and Felt Finishing Range

The special purpose drying machine and felt finishing ranges are ideally suitable for creaseless drying of all kinds of fabrics from poplin to bottom weight (wears). The machine is equipped with a couple of tension bars, and an adjustable tension arrangement is provided. This tension arrangement can also be modernized by using a control system. The control system:

- Controls the uniform pressure of the pneumatically operated mangle by using a pressure sensor on the mangle
- Can also control the water level on the trough of the mangle by using a water level controller

13.3.1 Predryer

- We can control the temperature of the drying cylinder by using a thermostat.
- We can control the pressure of the steam in the drying cylinder by using a pressure gauge. If a steam pressure in the cylinder that senses the pressure gauge is varied, then adjust an electrically operated steam solenoid valve through a control unit.

13.3.2 Belt Stretcher

- The belt stretcher is equipped with suitable fabric in feed devices consisting of either a feeler switch acting on reversible couplings or infrared sensors acting on feed systems ideally suited for high speed.
- The machine is equipped with a two-point ac frequency control drive. On request, a four-point drive can also be offered with precise tension controls at all stages.
- An automatic tension control device for felt can also be installed.
- A suitable electronic device is incorporated to get precise and accurate synchronization of various units.

13.3.3 Compressive Shrinkage Unit for Tubular Knitted Fabrics

The compacting process of cotton knitted fabrics is able to import dimensional stability that is resistant to various washing tests. In this way, garments made from the treated process will not suffer abnormal shrinkage after initial domestic washing.

Compacting of fabrics has therefore become an essential condition for obtaining dimensional stability and for satisfying varies demands of garment makers and the market. The compressive shrinking (compacting) principle is based on the difference in the speed of the fabric between its entry point into the compactor and the area in which it is subjected to the shrinking process.

13.3.4 Controls in Compressive Shrinkage Unit

The controls provided in the Compressive Shrinkage Unit are

- Motor driven introduction unit for ensuring minimum fabric tension
- Inverter transmission to control the speed of the machine
- Automatic positioning of the felt centering system by using a control system
- Synchronization of the motor and electronic load cell tension control
- Automatic control of the fabric width by using a tacho meter and servo meter
- Motor driven Teflon shoes controlled by a PLC for improved reliability of the process
- Automatic control of the steam pressure and temperature also by using an electrically operated solenoid valve

13.3.5 Air Relax Dryer

The conseptin relax drying is to form better quality drying without tension and excellent shrinkage values in tubular or open width knitted form. Fabric overfed in the drying chamber is in a fully relaxed condition to ensure excellent dimensional stability and soft and pleasant handle, increasing the value of the final product. A relax dryer normally built in two rows of the nozzle arrangement allows high shrinkage and good quality drying of the highest production capacity, reducing energy costs. It has its entry and exit on the same side, and it is easy to control the whole machine by a single operator. The control system is located nearer to the machine. The operator can control the machine very efficiently. The dryer has a suspension jet nozzle system, and the gab between the upper and lower nozzle can be adjusted by a motor up to 90 mm in height. Adjustment is made automatically by operating a push button. The complete upper line of nozzles is designed to move upward. Thus, the construction becomes a totally integrated adjustment system for the up and down movement of nozzles. By this, different qualities of fabrics can be dried and optimum capacity can be obtained. Circulating air turbines are provided in each section and are designed specially to operate with maximum efficiency. The exit section has a special conveyor system in order to avoid static loads of delicate fabrics. The compact machine layout and arrangement of equipment and accessories are easily accessible while the machine works.

13.4 Digital Textile Printing Ink Technologies

The rapidly evolving world of digitally printed textiles is a reflection of several unique and contrasting business models that create challenges, threats, and opportunities for the future of the textile printing market. The traditional textile industry looks at it from their traditional mass production business models and complains it is too slow and too expensive relative to the conventional screen printing technologies most commonly used in the market today. For these companies, digital textile printing has proven to be a tremendous cost savings in sampling only, while conventional methods best fit their mass production needs.

With textile seminars and exhibits being presented on an increasingly frequent basis at wide format printing and graphics trade shows, it is apparent that other industries are looking at this technology with an eye toward what it can do, as opposed to what it can't do. Since these industries already support short run and customization business models as a reflection of their technology driven businesses, textiles simply represent a new market to which they can sell their excellent command of printing technology as well as their ability to produce short run production with quick turnaround—a business practice that is foreign to the conventional printing industry as a result of the analog technology on which it is founded.

If the textile industry waits for the technology to evolve in order to adopt it as a production tool, then they will have lost many opportunities to new players. If, on the other hand, they can re-engineer their businesses to support the growing consumer demand for customized product, then they can lessen the probability of market erosions.

The wide format printers are not without their challenges to the new market opportunities. While most have mastered the basics of printing on paper, vinyl, and even plastics, printing on textiles that vary in fiber content, weight, thickness, ink absorbency, and yarn size, that must be washable, light fast, crock resistant and wearable, and require multiple ink sets can present a whole new set of challenges, if not at least a learning curve. Satisfying a textile industry that is accustomed to the color accuracy that spot color offers is yet another challenge. For the growing number of wide format printers, graphic artists, and entrepreneurs that see the market opportunities, we offer a primer in digital textile printing. The four principle technologies considered here are as follows:

Chemistry	Fibers	Processing
Acid Dyes	Silk, Nylon, Wool	Steam/Wash
Disperse Dyes (Sublimation)	Polyester	Hi-temp. Steam/Wash (Heat)
Reactive Dyes	Cotton/Cotton Poly	Steam/Wash
Pigments	All Fibers	Dry Heat

Reactive dyes provide the ink maker with the least difficulties in that they are water-soluble dyes. Purification of the dyes and formulation into inks designed to work well in the print head of choice has been achieved with relative success, and several products are available that will function successfully for samples and strike offs. In cases where new dyes have been synthesized specifically for inkjet, care must be taken that their fastness meets the end use needs where functional performance is required. Specially developed dyes may be offered in place of traditional textile dyes in order to achieve higher process color gamut and to allow for inkjet friendly performance. If the printed fabric is to be used for more than sampling, tests should be run to determine the dye's suitability.

Disperse dyes and pigments present a more difficult set of problems for the ink maker. Both exist in water as dispersions of small particles. These inks must be prepared with a high degree of expertise so that the particles will not settle or agglomerate (flocculate). Very few good examples of these types of inks have been proven in the market.

13.4.1 Process Color

Textile printing is primarily a spot color process, and most inkjet printer implementations utilize process color. These two approaches to color differ in that the colorants used to color the textile are premixed in the case of spot color, and mixed on the fabric in the case of process color. Process color printing is generally composed of black, cyan, magenta, and yellow inks that are mixed in varying proportions by jetting droplets onto the fabric to create the colors in between. The colors achievable by mixing only four colors of the same chemistry cannot come close to the colors obtainable in spot color printing from a well-selected set of "mother" colors numbering between 10 and 12. In an attempt produce more colors with process color printing, either dilute four-color process inks or up to eight different colors have been used. While this results in improvements, it still does not reach the combination of color correctness and functionality of spot color printing in color critical applications. Another technique to try to simulate the colors between the colors in process color is "dithering." This is an attempt at achieving more colors by choosing a "super pixel." It is made up of a block of four or more inkjet drop printing locations, which are treated as a single location. Varying the number of drops of each process color printed within this "super pixel" allows for a simulated gray scale in the printing. An unwanted side effect of this technique is a reduction in sharpness and frequently an unevenness in color. In many prints, this is objectionable. Another difficulty with process color printing is the "gray" contribution to the color. The mixing of two or more inks with very different colors causes this. As a rule of thumb, the larger the difference in color between two colors (the hue angle), the larger the gray component in the resulting color. The gray component dulls down the color, making it less attractive in many applications.

13.4.1.1 Quality and Productivity

Additional points should be considered in the effort to achieve the desired color, resolution, quality, and productivity. From the printer standpoint, spot color can be printed at adequate resolution with accurate color at the lowest potential number of inkjet nozzles, largest ink drops, and simplest software solution. This is particularly important to the role the ink must play since the larger the drops and the fewer the nozzles, the easier it is to provide reliable ink. To achieve good color match with process color, very small drops from very small nozzles must be used. To achieve reasonable throughput, the frequency of drop firing and the number of nozzles must be high. High frequency, larger numbers of nozzles, and very small nozzles place a very difficult burden on the ink designer. In comparison with office or wide format inkjets on the market today, the amounts of ink required to be fired from a high speed, high quality textile inkjet printer are several orders of magnitude higher. No inkjet inks have yet been commercially demonstrated that provide defect-free performance at the rate necessary for short run productions.

13.4.2 Advantages of Digital Printing

Digital printing requires minimal press setup and has multicolor registration built in to its system. This eliminates many of the time-consuming front-end processes and permits quick response and just-in time print delivery. Digital processes can vary every print "on-the-fly," that is, while production printing, providing variable data, personalization, and customization. Most digital printing technologies are noncontact printing, which permits printing of substrates without touching or disturbing them. This eliminates image distortion encountered in some analog processes such as screen printing. It also does not require as aggressive substrate hold down methods, which can distort or damage some substrates such as fabrics.

Digital technologies can print proofing, sample, and short runs more cost effectively than analog methods. Digital color printing processes offer a range of color processes including 3-color process (CYM), 4-color process (CYMK), and 5, 6, 7, and 8 extended gamut color options in addition to some spot colors. These match growing market demand for full color. Most digital print processing requires less or no color overlap or trapping. Digital printing does not use film masters, stencils, screens, or plates. It requires much less space for archiving text and images than analog printing methods require. Generally, digital printing uses less hazardous chemicals, produces less waste, and results in less negative environmental impact than analog technologies. Digital printing employs sophisticated color matching and calibration technology to produce accurate process color matching. Digital web printers can print images limited only by the width of fabric and the length of the bolt or roll. They can print panoramas and are not restricted to repeat patterns. Digital files are usually easier and quicker to edit and modify than analog photographic images.

Designers, artists, photographers, architects, and draftspeople are increasingly creating and reproducing their work digitally. Digital processing has replaced optical and manual methods for typesetting and page composition. Telecommunication has largely converted to digital processing. One can use the same digital files for electronic media, such as Internet, CD-ROM, video and TV, print media, and multimedia. One can readily convert analog images and text to digital with scanning and optical character reading (OCR) software. Digital files are easy to transport and communicate. One can send a digital file to any digital printer on the planet within seconds. This permits distribution of design to many locations for quick response printing. Industries are adopting digitally generated and communicated art and print copy.

13.4.2.1 Advantages of Analog Printing

1. Analog print technologies print many multiple copies quickly and inexpensively.
2. Offset lithography and gravure produce very high resolution and image quality.
3. Analog printing usually does not require expensive coated substrate to print satisfactory images as most digital printing does.
4. Analog inks do not require the high degree of refinement and small particle pigment sizes that most digital printings do. Most analog inks cost less than digital ink.
5. Analog screen printing provides a wide range of single pass ink deposition thicknesses.
6. Screen printing can print opaque inks that cover dark substrate surfaces.

13.4.3 Digital Printing Technologies

Digital printing encompasses many technologies. These include various forms of inkjet, thermography, electrophotography and electrostatic printing, ionography, magnetography, and digital photographic imaging and developing. None of these require a physical master but instead rely on digital data to create images.

13.4.3.1 Hybrid Digital-Analog Printing Technologies

Both analog and digital printing methods have advantages that the other lacks. Numerous opportunities exist for combining the strengths of each to garner the best of both worlds. The digital takeover of prepress analog operations illustrates this example. Prepress requires the generation of a single master, which is best generated digitally. Once created, analog printing can reproduce large numbers of it cost effectively. Digital can print variable information in a print job, while conventional prints the unchanging elements. Other marriages are also possible to use the best of both.

T-shirt printing device employ the Image continuous inkjet heads and UV curable water-based pigmented inks. This first direct digital garment printer could image on cotton, linen, rayon, silk, wool, polyester, polyamides, Lycra, and sponge. The printed images altered the character of the fabric's hand so slightly as to be indistinguishable. In addition to the UV inks, Toxot, the research and development arm of Imaje, developed a water-based, pigment-loaded, thermally cured inkjet color ink that exhibits greater color density, wash fastness, and adhesion than its water soluble UV curable predecessor. The Embleme team established that the Imaje heads and UV ink could print textiles. Continuous inkjet heads can cost about $5000 per head, piezoelectric heads range from about $30–$3000, and thermal inkjet heads about $20. Continuous heads offer very high reliability and resistance to failure commensurate with their cost. Continuous inkjet heads are hand assembled while the production of thermal inkjets and the Epson multi-layer Stylus systems are automated. A few tens of thousands of continuous inkjet print heads satisfy worldwide demand each year while companies such as Hewlett Packard, Canon and Lexmark produce drop-on-demand inkjets by the millions.

13.4.4 Continuous Multi-Level Deflected Inkjet

Large-scale production of many drop-on-demand print heads keep their costs very low. Their reliability is considerably less than continuous printers. This makes CIJ effective for high volume, long run printing. Embleme produced commercially acceptable T-shirt prints with a set of four CYMK scanning Imaje/Toxot inkjets operating at 120 dpi. Theoretically, a configuration involving dedicated arrays of Imaje/Toxot continuous inkjets could print bolt textile at a rate of 600 running meters per hour at 180 dpi. For 1.5 m wide bolts of fabric, this would yield productions rates of 900 square meters, or about 9000 square feet, per hour. (Toxot is currently constructing a 120 dpi 6-color multilevel CIJ array capable of printing two meter widths of floor vinyl at production rates of 2400 square meters per hour.) Though still falling far short of rotary screen printings cruising production rates of 60 to 90 meters per minute or 3600 to 5400 meters per hour, it is considerably faster than drop-on-demand inkjet printing capability. In addition, rotary screen printers usually print electrometric and other dimensionally unstable fabrics at slower speeds of about 20 meters per minute. For print jobs which could accept its resolution, continuous inkjet could compete with rotary screen printing for short to medium production runs, electrometric fabrics, and for on-demand printing. The advantages of multilevel continuous inkjets are their speed, their ability to cover a larger bandwidth print area with one pass, reliable operation, long print head life over thermal or piezo drop-on-demand printers, proven track record, and available chemistry. The disadvantages of multilevel continuous inkjets are that they initially cost more than drop-on-demand printers, they currently operate at resolutions lower than most drop-on-demand printers, they are limited by their requirement for inks with extremely low viscosity between 3 and 6 cp, and they use electrical conductivity that usually involves the addition of soluble salts. In general, the initial cost of CIJ heads currently prohibits their use for low-volume applications. Binary CIJ and multilevel CIJ systems currently offer the most cost effective and reliable means to digitally print larger volumes.

13.4.5 Piezoelectric Shear Mode

Shear mode print heads use an electric field perpendicular to the polarization of the piezo-electric PZT driver. Electric charge causes a shearing action in the distortion of the PZT piezo plates against the ink causing ink to eject from the nozzle opening in drops. 3D Systems manufactures prototype building devices that print dimensional prototype models using Spectra shear mode print heads. These print systems melt thermoplastic resins that Spectra heads shoot layer upon layer to form models. Spectra heads have the advantage of high reliability, proven performance, robust capability, wide ink choice, and ink processing in the 20–25 cP (centipoises) range, which is relatively high for piezoelectric inkjet. Spectra manufactures a number of head versions made of materials varying from sintered graphite to stainless steel. Spectra Inc. has advanced shear mode with the use of CNC machined sintered polycrystalline graphitic carbon as the structural print head base, the placement of a filter between the piezo pumping chamber and the nozzle, and the edge shooting placement of the piezoelectric transducer. The shear mode action makes it possible to achieve tightly packed assembly of many jets in a print head with just one piece of piezoelectric plate. Although these heads have the disadvantage of high sticker price, print heads developed with shear mode technology can deliver lower cost per jet at higher speeds with superior jet uniformity for jetting various inks on a wide variety of substrates. Water-based inks corrode these electrodes unless they are passivated, that is, coated to prevent corrosion. Without passivation, these heads must use non-corroding, solvent-based inks. The others have successfully passivated the print head electrodes to permit the printing of water-based inks. The Xaar piezoelectric print technology is called shared wall because each of the piezoelectric activated membrane walls is shared with its neighboring ink channel. This means that only every other chamber can fire simultaneously. In actuality, one can only fire every third chamber due to the possibility of accidental droplet generation from the chambers immediately adjacent. By adjusting the angular orientation of the print heads to the direction of substrate movement, one can achieve a workable pattern of ink deposition to compensate for this head firing limitation. The Nu-kote/MIT version of the Xaar print heads can print either 200 or 360 dpi depending on the angle of orientation (Figure 13.15).

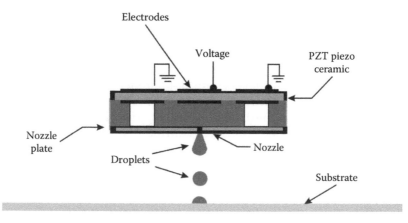

FIGURE 13.15
Piezoelectric shear mode.

13.5 Summary

At the end of the chapter, the user will understand the concept of automation used in chemical processing. The importance of measuring the textile parameters was discussed in detail. The different types of plant manager system in the chemical processing are explained.

References

1. http://nopr.niscair.res.in/bitstream/123456789/19290/1/IJFTR%2021%281%29%2057-63.pdf.
2. https://ahmedabad.indiabizclub.com/catalog/246788~gas%20singening%20machine%20super%20singe%20~ahmedabad.
3. http://www.indiantextilejournal.com/News.aspx?nId=RzcecNI3U0nyMLelLSpqsQ==&NewsType=Swastik%E2%80%99s-gas-singeing-machine-India-Sector.
4. http://www.fibre2fashion.com/osthoff/technology.asp.
5. http://kuesters-calico.com/singeing/.
6. http://textilelearner.blogspot.in/2012/03/what-is-singeing-process-of-singeing.html.
7. http://ptj.com.pk/Web%202003/10-2003/art-sheikh.htm.
8. https://www.energystar.gov/sites/default/files/buildings/tools/EE_Guidebook_for_Textile_industry.pdf.
9. http://www.swastiktextile.com/maxi_jig_electronic.html.
10. http://nptel.ac.in/courses/116102016/39.
11. http://library.aceondo.net/ebooks/Home_Economics/Handbook_of_Textile_and_Industrial_Dyeing_Vol_1_(Woodhead,_2011).pdf.
12. https://www.scribd.com/doc/203871786/Chemical-Technology-in-Pre-treatment.
13. http://www.indiantextilejournal.com/articles/FAdetails.asp?id=148.
14. https://www.scribd.com/document/32938937/02348.
15. http://infohouse.p2ric.org/ref/13/12378.pdf.
16. http://citeseerx.ist.psu.edu/viewdoc/download?doi=10.1.1.593.6115&rep=rep1&type=pdf.
17. http://shodhganga.inflibnet.ac.in/bitstream/10603/50973/6/chapter%203.pdf.
18. http://textilescommittee.nic.in/writereaddata/files/publication/Pro11.pdf.
19. http://www.srmuniv.ac.in/sites/default/files/files/PAPERINDUSTRY.pdf.
20. https://www.researchgate.net/profile/Nikos_Karacapilidis/publication/222478850_Production_planning_and_control_in_textile_industry_A_case_study/links/00b495176d36163da8000000/Production-planning-and-control-in-textile-industry-A-case-study.pdf.
21. https://www.slideshare.net/sheshir/erp-software-for-textile.
22. http://www.techexchange.com/library/A%20Primer%20in%20Digital%20Textile%20Printing.pdf.
23. http://www.inkworldmagazine.com/issues/2017-07-01/view_features/the-digital-textile-market/.
24. http://www.imaging.org/site/IST/Resources/Imaging_Tutorials/Progress_and_Trends_in_Ink-Jet_Printing_Technology/IST/Resources/Tutorials/Inkjet.aspx?hkey=4af47800-9584-4480-be8d-45fc3ee53e86.

13.9 Summary

14

Automation in Garments

LEARNING OBJECTIVES

- To identify the machinery used in garment technology
- To describe the sensors, transducers, and measuring devices used in garment machines
- To recognize the importance of automation in a garment industry
- To understand the concept of interfacing a machine with a computer

14.1 Introduction

The apparel industry, in order to be more competitive, is driving toward mass customization. This move toward made-to-measure apparel requires underlying technology to facilitate acquiring human body measurements and extracting appropriate critical measurements so that patterns can be altered for the customer. Traditionally, tailors have taken these measurements themselves for their own pattern altering methods. For this reason, tailors' measurements are notoriously inconsistent when compared with other tailors.

The first stage in the manufacture of garments is the cutting of the materials into the necessary pattern shapes. These are then joined together by means of seams to create three-dimensional garments. When a single garment is cut out, the garment pattern is attached to one or two plies of the fabric in a way that allows for any special requirements such as marching of the design of the fabric. The garment parts are then cut out with hand shears or electric cutters on dies. Where large quantities of a garment style must be cut, a lay is created that consists of many plies of fabric spread one above the other. From this must then be cut all the garment pieces for all the sizes that have been planned to cut from that lay. The pattern shapes for these garments may be drawn on a paper marker placed on top of the lay, or information as to their shape and position may be held within a computer, to be plotted similarly on a paper marker or used to drive an automatic cutter. Depending on the method of cutting that is used, there is not always a need for the pattern shapes and positions to be physically drawn on a paper marker, but whether the marker is drawn out or not, a marker plan must be made in which the pattern pieces are closely interlocked to achieve minimum fabric usage. In spreading the fabric to form a lay, the plies of fabric in the lay will be nominally the same length as the marker plan (Figure 14.1).

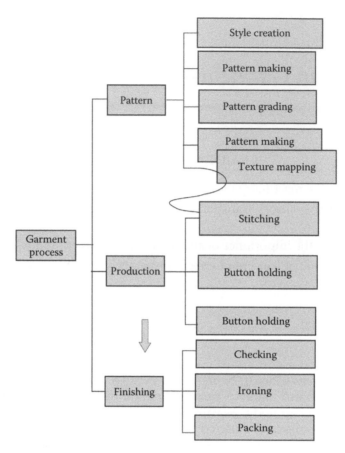

FIGURE 14.1
Sequence of the garment process.

14.2 Automated Fabric Inspection

Fabric inspection has proven to be one of the most difficult of all textile processes to automate. It has taken decades for computer and scanning technology to develop to the extent that practical, consistent, and reasonably user-friendly systems could be produced. In this article, we will look at three automated optical fabric inspection systems: BarcoVision's Cyclops, Elbit Vision System's I-Tex, and Zellweger Uster's Fabriscan. Today's automated fabric inspection systems are based on adaptive, neural networks. They can learn. So instead of going through complex programming routines, the users are able to simply scan a short length of good quality fabric to show the inspection system what to expect. This coupled with specialized computer processors that have the computing power of several hundred Pentium chips makes these systems viable. They are designed to find and catalog defects in a wide variety of fabrics including greige fabrics, sheeting, apparel fabrics, upholstery fabrics, industrial fabrics, tire cord, finished fabrics, piece-dyed fabrics, and denim. They cannot currently inspect fabrics with very large and complex patterns with the exception of the EVS Prin-Tex system, which is specifically designed to inspect and monitor the production of rotary print fabrics.

14.2.1 Strengths and Weaknesses of the Visual Fabric Inspection

In the fabric inspection, a trained person can inspect all types of fabrics, find and correctly identify basically all defects, and divide them into the corresponding classes. This shows once again: Man is the measure of all things. However, he or she has weaknesses that we have to understand to really be able to assess the performance of the visual fabric inspection. These weaknesses make a consistent, objective assessment of the fabric quality difficult. The reason lies in human nature.

The highest level of concentration is maintained only for a period of 20–30 min. After that, a person will tire continuously. Moreover, the highest concentration will only be achieved if the fabric is interesting enough. It is interesting if it contains approx. 200 events per hour. If no defects are observed over a period of 20 s, then the concentration decreases sharply (sleep mode) and only very distinctive defects will be detected.

Even under test conditions, as many as 30% of all defects are not detected. That stems from the fact that the human eye, at a distance of one meter from the fabric, can only see a circular area of 18 mm in diameter clearly and therefore never inspects the entire fabric.

Even in a well-run operation, the reproducibility of a visual inspection will rarely be over 50%. The problem is that a detected characteristic, depending on its appearance and shape, may or may not be a defect. The difficulty of this decision is increased by two weaknesses: On the one hand, there is still no possibility of calibrating the human eye and, on the other hand, these decisions are always dependent on the respective mood of the person, which we all know is never the same. Decisions of this kind will therefore always be made in a subjective manner.

In summary, the visual fabric inspection shows the following strengths and weaknesses:

- Inspection of all types of fabrics and the possibility of detecting all types of faults
- Low productivity
- Inadequate reproducibility
- Inconstant inspection results

14.2.2 Requirements for the Automatic Fabric Inspection

The requirements are as usual: faster, smaller, better, and on top of that, the new product should be available at low cost. If we focus on the economic viability, then we only have to make concessions with regard to the application range. In all other important points, we can make much higher demands than the visual inspection would be able to meet by making full use of the technical advantages. We lay particular emphasis on the reproducible, objective assessment of the fabric defects and a high defect detection rate. Only if these requirements are met will it be possible to set a new standard for fabric defects (Figure 14.2).

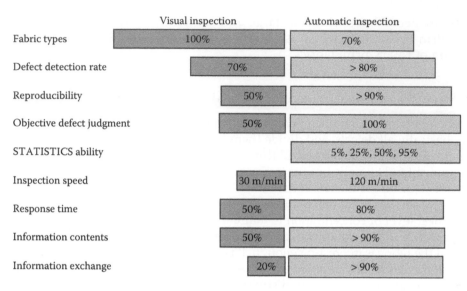

FIGURE 14.2
Demand on an automatic fabric inspection.

The demand for a higher inspection speed is not made arbitrarily. With the speed of 120 m/min, it is possible to integrate into existing production machines. This reduces the operational requirements and the reaction time in the event of technical problems in the production process. Of course, the troubleshooting is to be supported by exact data and images of the defects (Figure 14.3).

The cloth passes over a two-part illumination module that is designed for an inspection in either reflected or transmitted light. The choice of the illumination type is dependent on

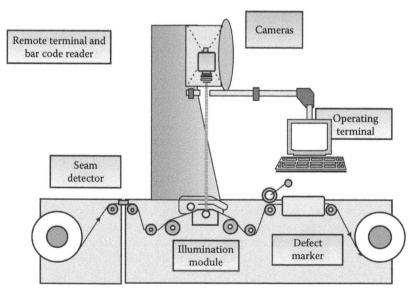

FIGURE 14.3
Automatic fabric inspection system.

the density of the fabric, the special types of defects, or the textile process stage in which the inspection is carried out. Above the light source, there are three to six or, in special cases, up to eight CCD high-resolution line scan cameras depending on the inspection width. The system is therefore designed to inspect the usual fabric widths of 160–330 cm (up to 440 cm). The cameras scan the fabric continuously for deviations. In this process, the fabric is inspected with the resolution that is achieved by an inspection person at a distance of one meter to the fabric.

The inspection system is operated from an operating terminal, where article-specific inspection parameters are set and the necessary piece data are entered or read with a bar code reader. The various reports are also called up via the operating terminal. After the inspection, the detected defects can be displayed on the screen for a quick and easy visual analysis. When the seam sensor detects the end of a piece or an article, the inspection system terminates the running inspection and automatically starts the next one. This requires that the sequence of the pieces to be inspected has previously been specified.

14.2.2.1 The Inspection Process

After the learning phase, the system is ready for the inspection. The inspection progress can be observed during the process. Every defect is graphically displayed with its size and position. In addition, a table informs the user on the exact position, the classification result, and the type of defect. This is only preliminary information, not a report.

14.2.2.2 The Reports

In all, there are seven different reports, which in various ways support the increasingly demanding quality management task in a textile mill. Some reports provide information on the quality while others mainly support the process optimization.

Aside from the measured length and width of the fabric, the *Standard Report* also informs on the absolute number of defects in the piece and per 100 m. The exact position and the size are indicated as well as the classification according to FABRICLASS and the type of defect. If a defect appears to be particularly interesting, a click on that defect is all it takes to display its actual image. The *Position Report* provides a quick overview of the defects in a piece. Especially the frequency, size, and position of the defects are visible at a glance. Together with the images of the defects, this report and the following one support above all the process optimization.

In the *Defect Type Report*, the user can see whether a certain type of defect occurs too often. In addition, the layout has been designed to clearly show whether a particular type of defect occurs with a higher-than-average frequency within the piece.

The FABRICLASS report provides information on the frequency of the characteristics in the individual classes. The list shows the disturbing and nondisturbing characteristics according to their contrast and length. The frequency of the nondisturbing characteristics becomes interesting when their number increases to such an extent that the characteristic appearance of the fabric is changed.

14.2.2.3 New Generation in Fault Detection

The visoelas-tec is designed for inspecting, measuring, and rolling highly elastic fabrics and incorporates novel measuring and tensioning technology. Fault detection has become an important element of the fabric production process. Production costs have to be controlled, and there is no room for the waste generated by undetected faults. The ability to recognize faults and stop production as soon as possible after the fault begins is of growing importance for producers watching their profit margins.

Indisputable consistency of fault diagnosis and detection, together with automatic report generation, are also important features for those mills looking to save time and reduce labor costs. The equipment that is available ranges from online, real-time detection and analysis systems to stand-alone on or off-line equipment that requires a greater degree of operator interaction. The Minifit is its latest introduction for fault detection on elastic and non-elastic tapes, ribbons, and braids. Inspection speeds of 200–400 m/min. are possible.

The Minifit offers advantages to users that include personnel-independent inspection of both sides of a tape, with reproducible results and individual adjustment of fault tolerances. At the heart of the system, light is projected on to the tape being inspected. The intensity of the light beam that is reflected by the tape is measured by a photocell and interpreted within a millisecond. The Minifit has two of these sensor units, which work in unison. When preset tolerances are breached, a fault stop is induced in the system. It is the design of these sensor units that allows the Minifit to run at such fast speeds.

As standard, the Minifit includes an integrated disentangling and feed device, pre-feeder drums for tape-tension control, inspection unit with one pair of white sensors for 4 mm tape width (front and back surfaces), and a transportation motor for tape unload into box.

14.2.3 Inspecting Elastics

The visoelas-tec is designed for the inspection, measurement, and rolling of highly elastic fabrics. It is able to perform calibrated measuring without any adjustment for elasticity grades 1–5 (firm to slightly elastic) and also for highly elastic fabrics up to an elasticity grade K of $60 \times 10\text{-}2$ N/m2 (PTB approval mark 1.3/00.14). It is also suitable for pile fabrics.

The fabric is guided to the measuring unit by means of a conveyor belt with crosswise-arranged slats, which allow the fabric to form small loops between the slats. This means that it is conveyed in a completely relaxed state. Fabric tension is further reduced by a specially designed composition of drive and dancer rollers.

14.2.4 I-Tex System

The I-Tex system is based on the automatic fabric inspection systems that use image-understanding algorithms, CCD cameras, and intensive computing capacity to imitate the human visual system. The I-Tex system automatically inspects and grades textile and technical fabrics, detecting diverse spinning, weaving, dyeing, finishing, and coating defects on any unicolor fabric. I-Tex can be used as a stand-alone inspection unit or positioned online. There are three models within the I-Tex family. The I-Tex 100 is used for inspection of greige and technical fabrics; I-Tex 200 is used on unicolor dyed and finished fabrics while the I-Tex 2000 is a computer-vision inspection system that automatically detects, maps, and memorizes weaving and finishing defects on any unicolor dyed and finished fabric. The I-Tex 2000 is capable of inspecting a wide range of fabrics, including home furnishings, technical, denim, and apparel. It is based on acquired defect data and user-defined grading criteria to automatically grade and classify the inspected fabric.

The objective of cutting room technology is the cutting of garment parts accurately and economically and in sufficient volume to keep the sewing room supplied with work. The three processes involved are:

- The planning, and if appropriate, the drawing and reproduction of the marker
- The spreading of the fabric to form a lay
- The cutting of the fabric

14.3 Automatic Pattern Making System

It is useful to break marker making down into marker planning, or the replacement of pattern pieces to meet technical requirements and the needs of the material economy, and marker utilization, which may include drawing the marker plan directly onto fabric, drawing it onto a paper marker by pen or automatic plotter, or, where the cutting method allows it, recording pattern piece information on the paper marker on the fabric without actually drawing pattern lines on it. Provision may have to be made for the same marker plan to be used many times.

14.3.1 The Requirements of Marker Planning

The industry has always paid great attention to marker planning, because when the cutting room cuts cloth, it spends around half the company's turnover. Any reduction in the amount of cloth used per garment leads to increased profit. Marker planning is a conceptualizing, intuitive, open, and creative process, in contrast to putting together a jigsaw puzzle, which is an analytical, step-by-step, and closed process.

In order to plan efficiently, it is necessary to visualize the marker as a whole, to see it at a glance. The planned proceeds by first positioning the larger pattern pieces in a relationship that looks promising and then fitting the smaller pieces into the gaps. Since most of the pieces are irregular and often tend to be carrot-shaped, one skill lies in discovering those edges that fit together most neatly and placing side by side across the marker those pieces that fill the width most nearly.

These relate to:

1. The nature of the fabric and the desired result in the finished garment
2. The requirements of quality in cutting
3. The requirements of production planning
4. Pattern alignment in relation to the grain of the fabric
5. Symmetry and asymmetry

14.3.2 The Design Characteristic of the Finished Garment

1. For the majority of cutting situations where a knife blade is used, the placements of the pattern pieces in the marker must allow freedom of knife movement and not restrict the path of the knife so that it leads to inaccurate cutting. A blade, which has width, cannot turn a perfect right angle in middle of a pattern piece, and space must always be allowed for a knife to turn such corners. Also, in practice, a curved part of a pattern such as a sleeve head, when placed abutting a straight edge, leads

to either a shallow gouge in the straight edge or the crown of the curve being straightened. The amount of space that must be left will depend on the actual cutting method employed.

2. A pattern count must always be made at the completion of the planning of a marker to check that the complete menu of patterns has been included. This is not just a formality when, for instance, a 12-garment trouser marker, where each garment may have 16 pattern pieces, signifies a complete marker of 192 pattern pieces.

3. Correct labeling of cut garment parts is essential if, in sorting and bundling a multi-size lay after cutting, operators are to identify correctly the parts that make up whole garment sizes. It is the responsibility of the marker planner to code every pattern piece with its size as the marker is planned.

14.3.3 Computerized Marker Planning

This method is normally part of an integrated system that includes digitizing or scanning of full size patterns into the computer, facilities for pattern into the computer, facilities for pattern adaptation, and, by inputting appropriate grade rules, the means to generate all the sizes required. Planning uses a virtual display unit with keyboard, tablet and data pen, puck, or mouse.

Automatic marker planning involves calling up data defining the placement of pieces in markers previously planned and selecting from a series that marker conformation which gives the highest marker efficiency.

All the pattern pieces are displayed in miniature at the top of the screen. In the middle of the screen are two horizontal lines defining the marker width and a vertical line at the left representing the beginning of the marker. At the bottom of the screen is a written marker identification, with marker length and efficiency constantly updated during the planning process.

A data pen (or puck or mouse), tablet, and the computer keyboard are used to manipulate the pattern pieces. When using a data pen, the pen is touched on the surface of the tablet and it registers a position on the grid, which shows as a position on the screen. A combination of movements of the pen and commands via the keyboard enable pattern pieces to be moved about the screen and positioned in the marker. The computer will provide an accurate piece count, calculate a marker plan efficiency percentage, and total the length of pattern peripheries.

14.3.4 Optitex Marker Making

Eastman offers several software packages for specific product applications. These packages are noteworthy for their ability to take full advantage of the benefits offered by Eastman's automated cutting systems. Studies have shown that customers moving from manual to computerized pattern design realize from 3% to 8% fabric savings and a two- to three-times increase in efficiency. The Optitex CAD systems are developed and supported by Scanvec Ltd., a leading provider of CAD/CAM software. Contact Eastman for a free demo disk to see how Optitex software can revolutionize your business.

Eastman's OptiMark interactive marking enables both experienced and novice marker makers to achieve tighter, more accurate markers in less time. Pieces may be placed flipped, rotated, or tilted, depending on each piece's preset restrictions. SGS's automatic and interactive marking quickly generates nested layouts while minimizing material waste. SGS's OptiMark Module is designed to maximize productivity and minimize labor and material costs.

Sophisticated stripe and plaid matching features incorporate scanned bitmap images of materials on screen during the marking process for optimizing pattern layout on fabric repeats. Users can optimize cut sequence for an entire nest all at once or by individual piece. You may also modify cutting direction of individual pieces or use the "Shared Lines" feature for common line cutting on automatic cutting equipment.

14.3.5 Pattern Design System

Pattern design system (PDS) encompasses a wide variety of features for the art of pattern engineering. Whether working with simple geometric shapes or sophisticated designs, the simple yet powerful PDS tools enable the user to draft or modify existing patterns quickly, easily, and accurately. Toolbar icons representing frequently used design techniques can be accessed with one click of the mouse—there is no need to memorize difficult commands or navigate through endless menus.

Users can create various types of notches, darts, pleats, and even the most complex seams in no time. Computer-accurate seams sew together perfectly. PDS improves organization and makes it possible to quickly modify styles for manufacturing and fabric savings. Moveable toolbars and dialog boxes allow the design engineer to use several related functions simultaneously. For example, the "Compare Length" dialog box is used to compare line lengths of multiple pattern segments, while the "Insert Toolbar" is used to add pleats, seam allowance, drill holes, notches, and text (Figure 14.4).

Grading, one of the most time-consuming aspects in the design process, can now be accomplished in seconds. OptiGrade provides a standard fit and makes turnaround time virtually instantaneous. OptiGrade includes features that make short work of time consuming manual grading challenges such as split parts and notches. Sizes are easy to recognize using SGS, OptiGrade's built-in color coding. Grade rules can be made for single

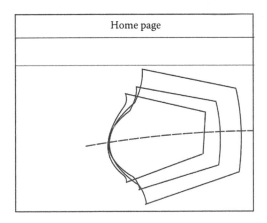

FIGURE 14.4
Graded pattern.

points or for entire pieces, while entire grade libraries are exportable into other applications such as Microsoft's Excel program. Angle grading simplifies grading on difficult points, while complicated grading on notches and split parts is simplified with built-in dialog boxes.

14.3.5.1 Numonics Accugrid Digitizers

The digitizer allows the user to input all hard patterns (made from paper, cardboard, fabric, plastic, or other material) into the computer. All pattern geometry, internals, and text can be input into the computer via digitizing. Accugrid digitizing tablets are easy to set up, easy to use, and reliable. They can be ordered with either 16- or 4-button corded or cordless mouse/pointer. A pen stylus is also included.

The thin Accugrid tablet contains a wide range of features including:

- User-friendly set-up menu: no dip switches to set
- On-table soft keys for easy configuration switching

  Strong single pedestal design supports even the large 44-in. × 60-in. tablet. Each stand features a tilt angle of 0°–80°. Vertical lift of 35–49 in. allows the user to adjust the height of the stand for maximum comfort.

- Manual lift and height adjustment
- Motorized height adjustment; manual tilt adjustment
- Dual motors control height and tilt angle settings

14.4 Body Measurement System

The apparel industry, in order to be more competitive, is driving toward mass customization. This move toward made-to-measure apparel requires underlying technology to facilitate acquiring human body measurements and extracting appropriate critical measurements so that patterns can be altered for the customer. Traditionally, tailors have taken these measurements themselves for their own pattern altering methods. For this reason, tailors' measurements are notoriously inconsistent when compared with other tailors. An accurate data set of the surface of the body is needed in order to develop consistent body measurements.

Several technologies have been employed by researchers for body measurement including 2D video silhouette images, laser based scanning, and white light phase measurement. In order to get a full three-dimensional representation of the body without the use and cost of lasers, the structured light and PMP application is well suited for body measurement because of the short acquisition time, accuracy, and relatively low cost.

14.4.1 System Design

The Body Measurement System (BMS) was designed to achieve unique coverage requirements of the human body. In order to scan the majority of the population, the scanning volume was designed to be 1.1 m wide by 1.0 m thick by 2.0 m high. The BMS has two frontal views with a 60° included angle and a straight on back view as shown in Figure 14.1. With this configuration, there is overlap between views in areas where detail is needed and minimal overlap on outer edge regions where surfaces are smooth. In order to get adequate height coverage, there are six views in total: three upper views and three lower views. The system design uses six stationary surface sensors that encompass the body. The sensors are stationary, so each must capture an area segment of the surface. The area segments from the sensors are combined to form an integrated surface that covers the critical areas of the body that are needed for making apparel.

14.4.2 Sensor Design

Each sensor consists of a projector and an area sensing camera, thus forming a vertical triangulation with the object or body. The camera and projector are separated by a baseline to form the necessary geometry for mapping points onto the surface of the body. The projector contains a two-dimensional patterned grating, which is projected onto the body. The pattern varies in intensity sinusoidal in one direction, and is invariant in the perpendicular direction. Both coarse and fine grating patterns are employed. The projected pattern is imaged by an area array charge-coupled device (CCD) camera.

14.4.3 System Software Design

The software program was developed in the Microsoft Developer Studio Visual C++ under the Windows NT platform. OpenGL was utilized to create the graphics display tools. The software performs the following functions: graphical user interface, controlling the acquisition sequence, acquiring and storing image buffers, processing acquired images and calculating resulting data points, and displaying graphical output (Figure 14.5).

14.4.4 Theory of Operation

Phase Shifting The PMP method involves shifting the grating preset distances in the direction of the varying phase and capturing images at each position. A total of four images is taken for each sensor, each with the same amount of phase shift of the projected sinusoidal

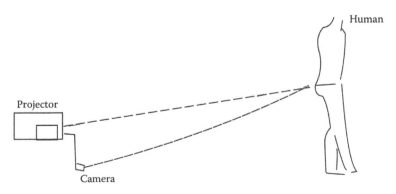

FIGURE 14.5
Triangulation between projector camera and target subject.

pattern. Using the four images of the scene, the phase at each pixel can be determined. The phase is then used to calculate the three-dimensional data points.

14.4.5 Image Acquisition

As mentioned earlier, four images per sensor per grating are acquired. Since there are six sensors in the BMS, a total of 48 images are captured. In order to minimize the capture time, the computer acquires the images using a round robin acquisition scheme. In this manner, all of the images of the first grating position are captured in succession. After each sensor's image is acquired, the computer starts shifting the respective grating. When all six sensor images are acquired, the sequence is restarted for the next grating position. This process continues until all 24 images have been acquired for the fine grating. The gratings are then moved to project the coarse grating. The round robin sequence is repeated to acquire 24 images for the coarse grating. In this manner the critical portion of the scan, which occurs during the fine grating acquisition, is captured within 2 s.

14.4.6 Scanning Results

The intermediate output of the PMP process is a data cloud for each of the six views. The individual views are combined by knowing the exact orientation of each view with respect to one another. Their orientation is derived by scanning a calibration object of known size and orientation. This is known as system calibration. These data points are the raw calculated points without any smoothing or other post-processing.

14.4.7 Measurement Extraction

With the raw scan data acquired, a wealth of 3D geometric information is available that can be used to make clothing for the scanned individual with a level of fit that would be difficult to achieve with a manual measurement process. However, a data extraction step is necessary to get the key measurements that can be used to alter the clothing pattern. This process is a fully automated computer process. The automation is desirable because the time required for a computer operator to extract the information using an interactive data analysis tool can be as great or greater than the manual process using measurement tapes.

14.4.8 Actual Scan—Raw Data

Filter, Smooth, Fill, and Compress the Data The raw scan data is further processed into a proprietary format that has several advantages over the raw form of the scan data. This proprietary format results from a sequence of processes including:

- Data filtering, which removes any stray points
- Segmentation of the body into individual limbs (arms, legs, torso)
- Smoothing, which removes low level noise in the scan data
- Filling, which closes any small gaps in the scan data
- Compression, on the order of 100:1, to achieve a very "light" yet fully defining data set

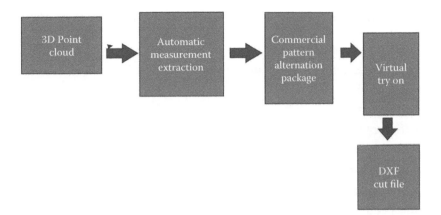

FIGURE 14.6
Made to measure apparel process summary.

The primary advantage of using this "processed" scan data is that it allows for the creation of measurement extraction algorithms that are relatively more robust, repeatable, and accurate as compared to algorithms that operate on the data in its raw form. This in turn allows for the measurement extraction process to be automated, that is, to occur without operator intervention. Some of the measurements are highlighted on the processed image and are shown as darkened lines superimposed on the processed data (Figure 14.6).

14.4.9 Automatic Pattern Alteration Using Commercial Apparel CAD, and the Virtual Try-On

Numerous apparel computer aided design (CAD) packages have made-to-measure or pattern alteration functions that can be used in concert with the scanner-based measurements to create a custom pattern for the customer. The measurements are formatted specifically for the target apparel CAD package so that the input of the data to the system and the output of the made-to-measure pattern, either to a fabric cutter or to a plotter, occurs automatically. Additionally, the possibility now exists for the customer to view the fit and appearance of the made-to-measure garment as a computer simulation (a virtual try-on). The garment is draped on their 3D scan image to approve the fit and appearance of the garment before it is made.

14.5 Automatic Fabric Spreading Machine

The objective of spreading is to place the number of plies of fabric that the production planning process has dictated, to the length of the marker plan, in the colors required, correctly aligned as to length and width, and without tension.

A study of spreading must include the following considerations:

- The requirements of the spreading process
- Methods of spreading
- The nature of fabric packages

The requirements of the spreading process include the following

- Shade sorting of cloth pieces
- Correct ply direction and adequate lay stability
- Alignment of plies
- Correct ply tension
- Elimination of fabric faults
- Elimination of static electricity
- Avoidance of distortion in the spread

14.5.1 Cradle Feed Spreading System

Eastman's CR spreaders feature a programmable microprocessor controller and a cradle-feed design that provides superior tension control, fast and easy roll loading, and automatic threading. Digital push buttons make the CR spreaders easy to program. The multi-program computer control allows storage of up to 30 programmed spreads. To set the spread length, the operator simply drives the machine to the start and end points of the spread, pushes a button at each end, and then enters the number of plies required on the keypad. Loading and threading is just as easy. The cradle tilts back, the exhausted roll's core is removed, and the next roll dumped in. The operator presses a button, and the machine threads itself.

The CR spreaders can be configured to spread face-to-face, face-up one way, and face-to-face cutting at both ends. The spreaders can spread virtually any material from Lycra® and stretchy materials to canvas and denim. The tension control adjusts automatically to eliminate stretching of materials. CR spreaders include the following features

- Multi-Program Computer-Control
- Counter Roller Guidance System
- Servo Drive Cradle
- Preset Counter
- Multi-Program Computer-Control
- Counter Roller Guidance System
- Servo Drive Cradle
- Preset Counter
- Automatic Tilt Cradle Load Feature
- Table Safety Switch
- Infrared Edge Control
- Easy Feed Windscreen
- Motorized Elevator System
- Manual Speed Control Throttle
- Six Wheel Table Drive

The face-to-face spreading of rolls weighs up to 1200-lb with diameters of up to 36-inches (48-inches optional). With top speeds in excess of 110 yards per minute, System IV is capable of spreading more than 5000-yards per hour depending upon spread length and material. Setting the System IV's spread length takes very little time. Sensors built into the catchers tell the machine when to slow down and reverse; therefore, changing spread length is as fast and easy as relocating the moveable catcher. With top speeds in excess of 110-yards per minute, System IV is capable of spreading more than 5000-yards per hour depending upon spread length and material.

- Setting the System IV's spread length takes virtually no time. Sensors built into the catchers tell the machine when to slow down and reverse; therefore, changing spread length is as fast and easy as relocating the moveable catcher.

14.6 Automatic Fabric Cutting Process

The objective of cutting is to separate fabric parts as replicas of the pattern pieces in the marker plan. In achieving this objective, certain requirements must be fulfilled.

14.6.1 Precision of Cut

Garments cannot be assembled satisfactorily, and they may not fit the body correctly, if they have not been cut accurately to the pattern shape. The ease with which accuracy is achieved depends on the method of cutting employed and in some cases on the marker planning and marker making, as described earlier. In manual cutting using a knife, accuracy of cut, given good line definition, depends on appropriate, well-maintained cutting knives and on the skill and motivation of the cutter. In both die cutting and computer-controlled cutting, the achievement of accuracy comes from the equipment.

14.6.1.1 Clean Edges

The raw edge of the fabric should not show fraying or snagging. Such defects come from an imperfectly sharpened knife.

14.6.1.2 Unscorched, Unfused Edges

The build-up of heat in the knife blade comes from the friction of the blade passing through the fabric. This, in extreme cases, leads to scorching of the fabric, and, more frequently, to the fusing of the raw edges of thermoplastic fiber fabrics, such as those containing polyamide or polyester. The cutter cannot separate individual plies from the pile of cut parts. Forced separation causes snagged edges, and, in any case, the hard edge is uncomfortable in wear.

14.6.1.3 Support of the Lay

The cutting system must provide the means not only to support the fabric but also to allow the blade to penetrate the lowest ply of a spread and sever all the fibers.

14.6.1.4 Consistent Cutting

The cutting system should not be limited in the height of plies it will cut, because of progressive deterioration in cutting quality.

14.6.2 Cutting and Spreading System

This cutting machine enables one operator to spread fabrics up to 144 in. wide and 32 ft in length with efficiency and accuracy. With the Blue Jay, the fabric is located at the end of each table. Each ply is pulled from the end of the table to the desired spread length and cut off. Successive plies are stacked with a high degree of accuracy up to a maximum height of 9 in.

The Blue Jay can handle large-diameter rolls or thick materials. Depending on the fabric, the Blue Jay can spread several ply each time it travels down the table. It processes vinyl, canvas, laminates, poly blends, and a variety of industrial fabrics.

The Blue Jay consists of three components:

The puller: The puller is the moveable pneumatic clamp, which clamps the ply to be spread and pulls it down the table to the desired length.

The cutting attachment: The Blue Jay's cutting attachment features a powerful Eastman 5-inch round knife, which is automatically driven across the width of the fabric in a guided track. Single or multiple plies up to one inch total thickness can be cut.

The let-off: Optional material let-off devices are available, including the fully automatic Power Cradle and various roll stands for single or multiple rolls.

14.6.3 Continuous Cutting Conveyor System

The conveyor cutting system combines the speed and accuracy of computerized cutting with continuous material feed. This machine offers several important benefits including:

- Spreading time is completely eliminated.
- Markers can be infinitely long, improving fabric utilization due to more efficient piece nesting.
- The fabric normally lost at each end of traditional multi-ply spreads is no longer wasted.
- Plaids, stripes, and other relational fabrics can be matched and continuously cut in real time.

The system offers cutting speeds of up to 50 in. per second and linear throughput of up to 20 in. per second, depending on the pattern's shape and material being cut. The unique micro porous table surface ensures maximum vacuum hold-down across the entire cutting area. To feed the material onto the conveyor belt, Eastman offers a wide variety of material handling options. The EC3 features three tangentially controlled tool holders, which may be fitted with any combination of different sized rotary blades, different angles of straight knife blades, variously shaped notch punches, and a wide range of drill diameters. There is also a pen for marking and piece identification and an optional non-contact inkjet printer for faster marking capability. The features of Continuous Cutting Conveyor System are

- High speed servo motor-controlled cutting gantry with MultiTool cutting system and three computer-controlled tangentially driven cutting tools
- Blade mounts
- X, Y, and Theta axis servo drives with closed-loop feedback and optical encoders
- Two pressure-sensitive emergency stop zones on each side of each carriage
- High-density microdrilled cutting belt
- Pentium processor-controlled computer, 15-inch VGA monitor, graphics card, enhanced keyboard
- Onboard plotter control panel with alphanumeric user interface terminal
- Electronics control cabinet
- Diode laser pointing alignment tool
- Rack-and-pinion table attachment system
- Festooning system with RF-resistant cables and E-chain support
- Two remote emergency stops located at table ends
- Motion control board, multifunction board, SVGA color monitor, graphics card, keyboard, mouse
- Onboard plotter control panel with alphanumeric user interface terminal
- Electronics control cabinet
- Diode laser pointing alignment tool
- Rack-and-pinion table attachment system
- Festooning system with RF-resistant cables and E-chain support
- EasiCut Windows-based plotter motion control software, calibration, and diagnostic software
- Translate design communication software interface converts GBR, DXF, and numerous other files

14.6.4 EasiMatch Software System

The EasiMatch system revolutionizes the previous tedious task of matching stripes, plaids, and florals. Equally important, EasiMatch allows the user to identify bow and skew fabric irregularities and correct them by altering the shape of the pattern in real time, immediately prior to cutting. EasiMatch automatically compensates for deviations from the planned fabric repeat and bowing and skewing in the fabric that would previously require the fabric to be straightened in order to be used. As the fabric is conveyed into the projection area ahead of the cutting zone, the pattern data is projected on top of the actual fabric.

Using a trackball mouse, the operator defines the actual fabric repeat and aligns the data to the material. The bow or skew is then compensated for and the perfectly matched data is passed back to the cutter, where the pieces are cut out with precision. This system works in conjunction with Eastman's single gantry OR dual gantry high-speed single ply continuous cutting conveyors, providing a total system that increases throughput and improves fabric utilization with less labor in shorter time.

The main components of the EasiMatch system are:

- A Windows-based computer and trackball mouse.
- Software, which includes Windows operating system and EasiMatch software.
- An overhead projection system for projecting the pattern pieces and bow and skew alignment tools onto the fabric. At 2200 Lumen, the EasiMatch system features the brightest projector on the market. And the EasiMatch system offers a number of mounting options, from the standard projection tower to a ceiling mount.

14.6.5 EasiCut Software System

The latest version of EasiCut software is more graphical in nature and provides the user with information regarding which tools are needed for each programmed cutting pattern. The improved program allows users to completely view the tool path, increasing accuracy during the entire cutting phase. The software is easy to configure and calibrate and will produce the desired outcome faster and more efficiently. Additional security features have been added to provide three separate management levels to prevent the software from being reprogrammed without proper approval.

14.6.5.1 Powerful Variable Speed Motor

The newly developed original variable speed motor has realized many powerful functions. Motor speed is adjustable depending on the features of cloth. The cutter can cut thick material at high speed position and fusible material at low speed position.

Drive handle
- Fits the hand, less vibrations, and is easy to operate

Motor speed indicator
- LED lights indicate the motor speed, which can be easily adjusted by the speed control dial

Other well-known features of km cutters are maintained, such as automatic sharpener capable of sharpening the knife quickly, low resistance stand, and the smooth, low base plate.

14.6.6 Conveyorized Cutting System

The gantry conveyor cutting system combines the speed and accuracy of computerized cutting with continuous material feed and, with two separately controlled cutting gantries, processes two different cut files simultaneously as material flows down the vacuum conveyor. The system features two cutting heads, which provide enhanced throughput and more tool options.

This system offers several important benefits including:

- Spreading time is completely eliminated.
- Markers can be infinitely long, improving fabric utilization due to more efficient piece nesting.
- The fabric normally lost at each end of traditional multi-ply spreads is no longer wasted.
- Plaids, stripes, and other relational fabrics can be matched and continuously cut in real time.

The unique microporous table surface ensures maximum vacuum hold-down across the entire cutting area. To feed the material onto the conveyor belt, Eastman offers a wide variety of material handling options. It features six tangentially controlled tool holders, which may be fitted with any combination of different sized rotary blades, different angles of straight knife blades, variously shaped notch punches, and a wide range of drill diameters. There is also a pen for marking and piece identification, and an optional non-contact inkjet printer for faster marking capability.

14.6.7 Automatic Labeler Option

For piece identification, the EasiLabel utilizes a separate gantry with print head to print and place a self-adhesive label onto pieces before they are cut. Label data comes directly from the CAD pattern database. The EasiLabel adapts to conveyorized cutting systems, allowing you to identify pattern pieces and cut "on the fly" at the same time.

14.7 Sewing

The dominant process in garment assembly is sewing, still the best way of achieving both strength and flexibility in the seam itself as well as flexibility of manufacturing method. Much of the application of technology to clothing manufacture is concerned with the achievement of satisfactorily sewn seams.

The selection of the correct combination of five factors during manufacturing includes:

- The seam type, which is a particular configuration of fabrics
- The stitch type, which is a particular configuration of thread in the fabric
- The sewing machine feeding mechanism, which moves the fabric past the needle and enables a succession of stitches to be formed
- The needle, which inserts the thread into the fabric
- The thread, which forms the stitch that either holds the fabric together, neatens it, or decorates it

Class 1 (Superimposed Seam): The simplest seam type within the class is formed by superimposing the edge of one piece of material on another. A variety of stitch types can be used on this type of seam, both for joining the fabrics and for neatening the edges or for achieving both simultaneously.

1. Superimposed seams
2. French seams
3. Piped seams
4. Lapped seam
5. Lap felled seam
6. Welted seam

Class 2 (Lapped Seam): The simplest seam type in this class is formed by lapping two pieces of material. It is commonly used is in the joining of panels in sails where a strong seam is achieved by using two or three rows of zig-zag stitching.

Class 3 (Bound Seam): In this class, the seam consists of an edge of material that is bound by another, with the possibility of other components inserted into the binding.

Class 4 (Flat Seams): In this class, seams are referred to as flat seams because the fabric edges do not overlap. They may be butted together without a gap and joined across by a stitch that has two needles sewing into each fabric and covering threads passing back and forth between these needles on both sides of the fabric.

Class 5 (Decorative Stitching): The main use of the steam is for decorative sewing through one or more layers of fabric. These several layers can be folds of the same fabric. The simplest seam in the class has decorative stitching across a garment panel. One row would have little effect, but multi-needle stitching is common.

Class 6 (Edge Neatening): Seam types in this class include those where fabric edges are neatened by means of stitches (as opposed to binding with another or the same fabric) as well as folded hems and edges. The simplest is the fabric edge inside a garment that has been neatened with an overedge stitch.

Class 7: Seams in this class relate to the addition of separate items to the edge of a garment part. They are similar to the lapped seam except that the added component has a definitive edge on both sides.

14.7.1 Stitch Types

The six classes of stitch included in the British Standard are as follows:

- Class 100 chain stitches
- Class 200 stitches organizing as hand stitches
- Class 300 lockstitches
- Class 400 multi-thread chainstitches
- Class 500 over edge chainstitches
- Class 600 covering chainstitches

Class 300: Lockstitches

The stitch types in this class are formed with two or more groups of threads, and two or more groups are interlaced. Loops of one group are passed through the material and are secured by the thread or threads of a second group. One group is normally referred to as the needle threads and the other group as bobbin threads. The interlacing of thread in stitches of this class makes them very secure and difficult to unravel. Straight lockstitch, 301, with a single needle thread and a single bobbin thread, is still the commonest stitch used in the clothing industry. In its single throw, zig-zag version, type 304 is commonly used for attaching trimmings such as lace and elastic where a broad row of stitching but no neatening is needed.

Class 100: Chainstitches

The stitch types in this class are formed from one or more needle threads, and are characterized by intralooping; one or more loops of thread are passed through the material and secured by interloping with a succeeding loop or loops after they are passed through the material.

Class 200: Stitches Originating as Hand Stitches

The stitch types in this class originated as hand stitches and are characterized by a single thread that is passed through the material as a single line of thread. The stitch is secured by the single line of thread passing in and out of the material.

Class 400: Multi-Thread Chainstitches

The stitch types in this class are formed with two or more groups of threads, with the interlooping of the two groups. Loops of one group of threads are passed through the material and are secured by interlacing and interloping with loops of another group.

Class 500: Over Edge Chainstitches

The stitch types in this class are formed with one or more groups of threads, and the loops from at least one group of threads pass around the edge of the material.

Class 600: Covering Chainstitches

Stitch types in this class are formed with three groups of threads, and two of the groups cover both surfaces of the material.

14.7.1.1 Sewing M/C Automation

The machine makers related to the garment manufacturing processes are exerting their efforts for the improvements of the mechanism and the systems of the units. The sewing machine is in the improvement stage in automation and labor saving.

In the lock stitch machines, the automations are:

1. Straight/bar tacking/constant stitching pattern selection
2. Setting stop position needle up/down
3. Slow start
4. Presser foot goes up at m/c stop
5. Presser foot goes up after trimming thread

6. Setting number of stitching

7. Start tacking

8. End tacking

In the basic parameter setting:

1. Direction of the motor

2. M/c code

3. Running delay time

4. Half heeling

5. Fine positioning

6. Trimming timing

7. Wiping time

14.7.2 Electronic Lockstitch Pocket Setter Sewing System

The pocket clamp assembly (folding clamp, inner clamp, center blade, sewing clamp) can be replaced easily in about one minute, without the need for any tools. Adoption of a pneumatic chuck system means that there is no possibility of forgetting to carry out mechanical tasks such as tightening of screws.

14.7.3 Automatic Two-Needle Hemmer Sleeves and Shirt Bottoms

This is an electronically controlled workstation consisting of a conveyorized downturn hemming apparatus with two- or -three needle bottom and/or top cover stitch sewing head, electronic motor, automatic edge trim and cut apart, and self-contained waste disposal. The operator places parts to an edge guide and initiates sewing. The unit will continue sewing as long as parts are placed on the conveyor within a specific distance. The sew cycle will stop if the operator fails to position the next part.

14.7.3.1 Automatic Two-Needle Hemmer for Sleeves

This is an electronically controlled workstation consisting of a conveyorized downturn hemming system with a two or three needle bottom and/or top coverstitch sewing head, electronic motor, automatic edge trimmer, cut apart with stacker, and self-contained waste disposal. The operator places parts to an edge guide and initiates sewing. The unit will continue sewing as long as parts are placed on the conveyor. The sew cycle will automatically stop if the operator fails to continue the loading process. Thread savings is achieved as a result of the machine not sewing when the distance between the parts becomes excessive.

14.7.3.2 Automatic Two-Needle Coverstitch Hemmer for Sleeves and Pockets

This is an electronically controlled workstation consisting of a conveyorized downturn hemming system with two or three needle bottom and/or top coverstitch sewing head, electronic motor, automatic edge trimmer, cut apart with stacker, and self-contained waste disposal. The operator places parts to an edge guide and initiates sewing.

The unit will continue sewing as long as parts are placed on the conveyor. The sew cycle will automatically stop if the operator fails to continue the loading process. Thread savings is achieved as a result of the machine not sewing when the distance between the parts becomes excessive. The operator loads the pocket into the pocket pre-loading system, and the pocket is automatically positioned to the edge guide and transferred to the conveyor belts where it is transported to the sewing head. This is equipped with a stacker that will stack both sleeves and pockets with the flip of a switch.

14.7.4 Automatic Clean Finish Elastic Waistband Station with Fold-in-Half Stacker

This is an electropneumatic clean finish waistband station consisting of a multi-needle sewing machine, motor, stand, spiral elastic guide, pneumatic knife (for cut apart), and variable speed conveyor with electronic metering device and fold-in-half stacker. The operator presents the flat garment to the edge guide and into the spiral folder. An electric eye senses the leading edge and starts the sew cycle. The operator holds the garment against the edge guide until completely into the folder. The electric eye senses the trailing edge and stops the sew cycle, leaving the proper spacing for the next garment to eliminate waste. The operator then picks up the next garment and repeats. The conveyor will carry pieces to the knife for automatic cut apart and then to the rear fold-in-half stacker. This unit sews and cuts garments apart then folds the garment in half for the side seam operation.

14.7.5 Computer-Controlled, Direct-Drive, High-Speed, One-Needle, Lockstitch, and Zigzag Stitching Machine

A liquid crystal display has been adopted. It displays stitch shapes, zigzag width, and standard lines with pictographs and settings all together on one screen to improve operability of the panel. The machine has added capabilities that make the most out of the computer-controlled sewing machine functions, such as program stitching (constant-dimension sewing), cycle sewing, and continuous sewing. With these capabilities, the machine is able to respond to sewing in various processes. The thread tension mechanism, bobbin case, and feed timing have been improved to produce soft-textured seams. For the minute-quantity lubrication type hook, the inner hook is finished with titanium to keep it from becoming hot. The machine requires only a very small quantity of oil. This prevents oil stains on the material and improves durability of the hook.

In addition, the hook configuration is carefully designed to avoid needle breakage and has developed a new high-long arm machine head one size larger than the head for the conventional lockstitch machine. With its wider area under the arm, the machine permits easy handling of the sewing material for improved operability. The shape of the jaw gives the operator a clear view of the area around the needle. The machine comes with a more rigid and well-balanced machine head frame. This, coupled with the direct-drive method, dramatically reduces the noise and vibration of the machine and helps reduce operator fatigue.

14.7.6 Direct-Drive, High-Speed, Needle-Feed, Lockstitch Machine with an Automatic Thread Trimmer

This is a needle-feed type DDL-9000 Series sewing machine that has been highly applauded as the top-of-the-line lockstitch machine with a thread trimmer. It has inherited the advanced features of the DDL-9000 Series, such as the direct-drive system, semi-long arm, and elimination of the oil pan. In addition, it comes with a highly reliable needle feed mechanism. The needle feed mechanism, which is widely recognized as offering outstanding efficiency of feed and effectively preventing uneven material feed, responds to a wide range of applications such as the sewing of outerwear, run stitching of men's shirts, and so on, and the attaching of various parts to garment bodies.

The needle-feed mechanism offers excellent efficiency of feed, and the machine ensures accurate stitch pitches (stitch length) as well as preventing slippage of the upper cloth. The bottom feed's locus has been improved to match the needle feed motion, thus the machine produces beautiful seams without stitch gathering, even when it is used for sewing slippery and difficult-to-feed material or for handling a difficult process.

The needle feed mechanism is a so-called dry type, which does not require oiling. With this mechanism, a highly reliable machine structure, free from oil leakage troubles, has been achieved. The frame, which requires only a minute quantity of lubricating oil, is structured to eliminate oil leakage.

The machine can be easily changed over from a needle feed machine to a bottom feed machine through a simple adjustment and gauge replacement. It is a useful feature for those users who also want to use the machine simply as a regular bottom feed type machine in accordance with applications and processes. The machine saves you the time and trouble of removing the knee-lifter and cover. This means that the machine head can be tilted for cleaning and maintenance without the inconvenience of removing the knee-lifter and cover.

14.7.7 Automatic Short Sleeve Closing System

An electronically controlled sleeve closing station with automatic backlatch is designed to close short sleeve tee shirt sleeves. This unit includes an electronic active edge guiding system, providing the capability to sew either straight or contoured seams. The operator folds the sleeve and presents it to the presser foot. A photo cell senses the beginning edge, drops the presser foot, and begins the sew cycle with an automatic backlatch. The electronic active edge guiding system controls the sleeve during sewing, while the operator prepares the next sleeve. When the seam is completed, the machine stops, the thread chain is cut, and the sleeve is stacked automatically. Stacker selection is available for either single or double stack.

14.7.8 Computer-Controlled Lockstitch Buttonholing Machine

The needle thread tension is controlled by active-tension (electronic thread-tension-control system). The needle thread tension for sewing parallel and bartacking sections of buttonholes can be separately controlled through the operation panel and stored in memory according to various sewing conditions (e.g., type of thread, type of material, and sewing speed).

The machine is able to change the needle thread tensions at the parallel and bartacking sections of the buttonhole to produce a beautiful buttonhole shape. This capability helps greatly in preventing thread breakage. Needle thread tension is activated at the beginning and end of sewing. This prevents unthreading of the needle thread and thread fraying that is likely to occur at the beginning of sewing.

14.7.8.1 Feed Mechanism Using a Stepping Motor

The feed mechanism eliminates differences in stitch pitch between forward and reverse feeds, and achieves constant position stopping at the end of sewing. The feed mechanism allows the machine to sew exact square bartacks and exact round tacks.

The machine comes with an intermittent feed system to eliminate stitch skipping or thread breakage due to needle swaying. The cloth cutting length (buttonhole length) can be easily changed at the operation panel.

The machine is able to store as many as 89 different sewing patterns in memory to permit free selection. Sewing conditions can be easily changed on the operation panel.

Using the operation panel, the material cutting length, over edging width, stitch pitch at the parallel section of buttonholes, lateral-position correction of bartacking width, lateral-position correction of knife groove width, and the clearance provided between the knife groove and bartacks can easily be changed.

14.7.8.2 Bobbin Thread Winder

The bobbin thread winder is built in at the top surface of the machine head. The built-in bobbin winder ensures easy replacement of the bobbin thread. The bobbin winder also incorporates a thread cutting knife for cutting the thread after the completion of bobbin winding, as well as a bobbin thread quantity adjusting function. With its wider area under the arm (300 mm), the machine permits easy handling of the sewing material for improved operability.

14.7.9 Automatic Placket Fusing, Cutting, and Stacking

An adjustable, easy-to-load placket attachment utilizing a fusing machine that has both top and bottom heating elements for faster belt speed, an automatic cutter, stacker, and counter is available. When using a roll feed placket, the operator loads a roll of fusing and a roll of placket material, presses the start button, and the machine continues to run until the material runs out or the desired number of plackets is reached. When using individually cut plackets, the operator loads a roll of fusing and hand loads the plackets. The fused plackets are automatically cut to the correct length, counted, and stacked and the waste is discarded. The plackets are indexed to a holding tray when the desired quantity is reached for each stack. The automatic cutter/stacker requires no additional attention. The machine is equipped as standard with a micro-lifter. It works to constantly float the presser foot above an elastic or other difficult-to-sew materials, thereby helping effectively reduce material slippage as well as damages made by the presser foot on the material.

With the newly introduced direct-drive system, a compact AC servomotor is directly connected to the main shaft. This motor demonstrates quick start-up, upgraded stop accuracy, and excellent responsiveness.

The front cover can be opened/closed by a fingertip control. The number of revolutions at the start-up can be switched between three settings according to the operator's needs (standard, low, and high). The function-setting switch is provided with two new capabilities: reverse-feed and repeat. In addition, the switch has a reset capability to restore the machine to the default settings.

14.7.9.1 Control Panel

With this control panel, two kinds of programs combining as many as 15 steps can be established. This control panel has an expanded range of functions such as teaching sewing and a function used to establish the number of stitches during sewing.

Four different kinds of pattern sewing are possible: automatic reverse stitching, constant-dimension sewing, rectangle-shape sewing, and multi-layer stitching. As many as 19 stitches can be established for automatic reverse stitching (single or double). "Without control panel" can also be specified.

14.7.10 High-Speed, Over Lock/Safety Stitch Machine

This general-purpose advanced machine responds to various kinds of sewing materials and processes. The new and powerful over lock/safety stitch machine has been developed to offer increased reliability and ease-of-use, while upgrading seam quality at higher sewing speeds. It responds to a wider range of materials and processes to finish high-quality, soft-feeling seams. With the operating noise having been reduced and increased durability ensured, the cost-effective MO-6900S Series is a leading machine in the new era.

The machine comes with a needle-thread take-up mechanism as well as a looper thread take-up mechanism, to offer upgraded responsiveness from light- to heavy-weight materials with a lower applied tension. It achieves well-tensed, soft-feeling seams that flexibly correspond to the elasticity of the material at the maximum sewing speed of 7000 rpm. The newly adopted upper and lower needle bar bushings of the needle bar mechanism improve both the durability and reliability of the needle bar unit. The machine comes standard with a cartridge type oil filter and cooling fan for further improved reliability and durability.

The machine incorporates various mechanisms as standard, such as a differential-feed micro-adjustment mechanism and an external adjustment mechanism for adjusting the feed dog inclination as well as increasing the differential feed ratio, which can be easily adjusted to finish seams that perfectly match the material to be used. Comfortable operation is all but guaranteed by a wider area around the needle entry, the adoption as standard of a micro-lifter feature that offers improved responsiveness to materials and provides the operator with upgraded operability, and by the reduction of operating noise and vibration, which has been achieved by designing an optimally balanced machine. Gauges and devices used with the existing JUKI machines are also interchangeable with no additional machining. This eliminates both the waste of resources and extra costs.

The servo motor is provided with many different standard functions, including speed-control, soft-start, automatic reverse stitching with the specified number of stitches, constant-dimension stitching, and overlapped (multi-layered) stitching. These functions ensure sewing performance optimally suited to the material and process.

14.8 Finishing Process

By far the most important area of garment construction where an alternative process has significantly taken over from sewing is in the attachment of interlinings. When interlinings are sewn in, it can be difficult on parts such as collars to avoid wrinkling of the interlining inside the collar and pucker around the edge.

The fusible interlining consists of a base cloth, which may be similar to that used for a sew-in interlining, and which carried on its surface a thermoplastic adhesive resin, usually in the form of small dots, which will melt when heated to a specific temperature.

14.8.1 Fusing Interlining

Previously, garments requiring tailored finishes could only be made by highly skilled professionals, but now, with the use of the fusing process, anyone can produce such a garment. Using the fusing process, a good shape can be cheaply produced and will remain in shape for long periods of time.

14.8.1.1 Making Sewing Easier and Increasing Production

- Because of the speed of industrial sewing machines, the material must be in perfect shape before sewing so that the machine operator does not have to try to reshape the piece before or during the sewing time. If, before sewing, interlining is fused onto the material, it keeps its shape, therefore saving time and labor.
- Previously, the process of tailoring had to be done by hand and was only a job for skilled workers. Now, with the fusing process, anybody can create the same effect as a skilled tailor, with only one layer of interlining and no previous skills.

Retaining Shape and Improving materials appearance

- The use of interlining helps the garment material's appearance while at the same time retaining the garment's shape. With the development of interlinings and better fusing press machines, the permanent fusing process was developed. With this process, garments keep their shape no matter how often they are worn or washed.

Making a Functional, Lasting, Easy to Wear Product

- Using the permanent press technique, everyone from the producers and designers to the consumers is satisfied. Production workers find the garments easier to sew, and the clothing designers can achieve shape and long-term performance, while consumers get a good quality product that is easy to care for, looks good, and is easy to wear. The basic aim of pressing is to make the garment look better for longer periods of time, while still being comfortable. Interlining reduces the occurrence of stretching, creasing, and wear.

14.8.2 Permanent Fusing and Temporary Fusing

Interlining can be divided into two groups: temporary and permanent. The purpose of temporary fusing is to make sewing easier and to reinforce the stitching. The fusing intensity does not need to be strong as it is merely to prevent puckering during stitching. Permanent fusing is used when the shape and style of the garment are intact. Therefore, after washing and long periods of use, the interlining must stay fused. If the fusing temperature, pressure, and timing are not correctly set, the intensity of the fusing will differ. Before fusing, the type of material and interlining being used and the desired intensity of the bond must be considered.

- Permanent fusing must stay fused after washing or dry cleaning.
- Temporary fusing is simply to make sewing easier and can become separated after the garment is completed.

14.8.2.1 Continuous Fusing Machine

The highly flexible fusing machine fills the void between low production, platen presses and the larger, more expensive conveyorized fusing machines. A standard "return-to-operator" conveyor allows pressure up to 7.5 kg/cm², making this an ideal fusing machine for dress shirt collars and cuffs. Top and bottom heat with separate control, precise speed control, and a fully adjustable pressure system assures quality fusing at an economical price. This new SC series is designed with safety and operator comfort in mind. This model meets or exceeds worldwide standards in safety and efficiency.

14.8.2.2 High Pressure Fusing Machine for Collars, Cuffs, and Plackets

An extra large platen allows full front pressing and fusing as well as collars and cuffs. Rapid response heating elements, high thermal reserve platen, and up to 30 PSI of platen pressure assure consistent quality. A reciprocating tray system allows parts to be unloaded and reloaded during the fusing cycle.

14.8.3 Pressing

Pressing makes a large contribution to the finished appearance of garments and thus their attractiveness at the point of sale.

1. To smooth away unwanted creases and crush marks
2. To make creases where the design of the garment requires them
3. To mold the garment to the contour of the body
4. To prepare garments for further sewing
5. To refinish the fabric after manufacturing the garment

The means of pressing are heat, moisture (usually as steam), and pressure, single or in combination. These means deform or reform fibers, yarns, and fabrics in order to achieve the effect intended by the designer.

14.8.3.1 Press to Finish

A well-pressed garment is just as important as a well-made garment when it comes to presentation; that is why this range of under pressing equipment has been designed to be of the highest standard, delivering professional, finished results from easily installed and maintained equipment.

14.8.3.2 Pressing System with Automatic Segmented Frames

With the models, Mentasti made obsolete the method of pressing using full frames, with its undoubted limitations. The patented system of adjustable split frames is very versatile and allows accurate adjustment to each garment dimension, drastically reducing the constant changing of frames for each size and style is inevitable with full frames solves the delicate problem of glazing. Particularly in the crucial area between body and shoulder the use of a full frame can lead to annoying marks.

Moreover, there is the possibility of pressing without ironing thanks to the adjustable steam system and drying by vacuum suction, which eliminates the marking problems caused by collars, buttons, and so on. Because of the inclined angle of the beds, the 02 operator keeps a perfect working position and is not disturbed by steam flow. The result is a greater, more consistent work rate through the day. Besides, the design planning of the 02 was founded entirely on achieving the greatest productivity of any other method, combined with a higher finished garment quality and maximum operating efficiency.

A complete industrial steam ironing unit made for design rooms, tailors' workshops, ironing services, alteration shops, small clothing manufactures, and so on comprising of a heated vacuum table, pressure steam generator and stream iron. The pressing station is designed as a compact industrial ironing unit with features similar to those of much bigger tables and generators. It is aimed at users who require professional finished results from compact equipment that is easy to set-up and use. It requires no special installation, uses standard tap water, and is powered by two standard 13-amp plugs.

14.9 Automatic Material Transport

14.9.1 Garment Storage with a Simple Hook Release from Horizontal to Vertical Position

An elegant but simple idea lets your present slick rail trolleys and static storage rails host a convertible auxiliary carrier that works both horizontally and vertically.

Moveable hooks and a universal/reversible carrier design provide horizontal storage for slick rail movement and examining, ready to convert immediately to vertical storage.

Simply lift and release one hook, and the hangers with their garments cascade effortlessly in controlled groups into a dense but secure vertical stack that makes excellent use of available floor to hanger space for storage or transport—in a plant or on a trailer. Eliminate "stringing" in delivery trailers and shipping containers. This system reduces tedious multiple hanger handling and allows the same carrier to perform multiple jobs in the factory or distribution center, the travel trailer/container, and the receiving store location.

14.9.2 Packaging

This task can involve folding and packaging the garments in a bag or a box. We looked at several operations for packaging men's dress shirts, and special considerations for these packaging stations will be described. Important features to consider include:

- The work surface
- Input/output
- Support surface
- Accessories

A good shirt-folding table that we saw had been adjusted to an appropriate height for the packer by placing wooden spacers under the legs. The packer could reach all items at the back of the able without an extended reach. The tabletop had small recessed areas close to the front of the table to hold small, frequently used items. We also saw a good automated bagging station. It was at an appropriate height for the worker. The bagging operation was semi-automated; air was used to blow the bag open. The bag was automatically sealed, and it slid down a ramp into a box.

A good output method is to place the packaged garment on a shelf directly beside the operator. This shelf feeds the garments directly to the next operator in the line. Another good method is to place packaged garments into a box that the garments fit into perfectly. This allows the packer to put the garment into the box and not have to arrange it neatly by hand. Conveyors can be used to transport full boxes, which greatly reduces the amount of lifting required. The packers placing garments into very large cardboard boxes should be able to easily reach over the side of the box. The overhead rails should not be over-full and require excessive reaching.

14.10 Summary

On completion of the chapter, the user will understand the machineries used in the garment industry. The working and electronic components used in the machinery is explained in detail. The automation involved in the garment machinery is explained to understand the working of the machine.

References

1. http://textilelearner.blogspot.in/2012/02/process-sequence-of-garments.html.
2. http://www.indiantextilejournal.com/articles/FAdetails.asp?id=2131.
3. https://www.semanticscholar.org/paper/Automated-Fabric-Inspection-Assessing-The-Current-Dockery/afa0557d22db313025ff5c23885d1f59f3af9a91.
4. http://www.indiantextilejournal.com/articles/FAdetails.asp?id=4664.
5. http://shodhganga.inflibnet.ac.in/bitstream/10603/48961/6/06_chapter1.pdf.
6. http://nopr.niscair.res.in/bitstream/123456789/2018/1/IJFTR%2033%283%29%20288-303.pdf.
7. https://pdfs.semanticscholar.org/142b/f3896d18934a49adc1330d85139c3a733964.pdf.
8. http://www.professorfashion.com/marker-making.html.
9. http://textileclothinginfo.blogspot.in/2015/02/marker-making.html.
10. https://www.scribd.com/document/230780959/Objective-of-Marker-Planning-and-Marker-Making.
11. http://www.indiantextilejournal.com/articles/FAdetails.asp?id=2684.
12. https://www.ercim.eu/publication/Ercim_News/enw39/mtom.html.
13. http://proceedings.spiedigitallibrary.org/proceeding.aspx?articleid=920335.
14. http://www.bodyscan.human.cornell.edu/scenefe80.html.
15. https://textileapex.blogspot.in/2014/03/fabric-spreading-objects-requirements.html.
16. http://autogarment.com/automatic-fabric-spreading-machine/.
17. https://www.tukatech.com/automatic-fabric-cutter/TUKAcut.
18. https://fashion2apparel.blogspot.in/2016/12/types-fabric-cutting-machines.html.
19. http://sewdelicious.com.au/2012/09/different-types-of-seams.html.
20. http://www.coatsindustrial.com/en/information-hub/apparel-expertise/seam-types.
21. http://www.hendersonsewing.com/page.asp?p_key=b1f3d4f2d10346119ebe96cc0ad409a2.
22. http://www.nickosew.com/images/AAC%20211E.pdf.

23. http://www.shelikestosew.com/best-computerized-sewing-machine-reviews-will-give-awesome-sewing-power/.
24. https://smartmrt.com/product/computer-controlled-high-speed-lockstitch-buttonholing-machine-lbh-1790a/.
25. http://textilelearner.blogspot.in/2014/03/garments-finishing-process.html.
26. http://fashion2apparel.blogspot.in/2017/01/methods-equipments-garment-pressing.html.
27. https://www.slideshare.net/lilybhagat3/packaging-and-labeling-of-apparel-and-textiles.

15

CAD/CAM Solutions for Textiles

LEARNING OBJECTIVES

- To explain the concept of CAD/CAM usage in the textile industry
- To understand the design procedure using CAD
- To comprehend the importance of CAM in the garment industry

15.1 Introduction

Computer-aided design (CAD) is industry-specific design system using the computer as a tool. CAD is used to design anything from an aircraft to knitwear. Originally, CAD was used in designing high precision machinery and slowly it found its way into other industries. In the 1970s, it made an entry in the textile and apparel industry. Most companies abroad have now integrated some form of CAD into their design and production process.

Of apparel manufacturers

65% use CAD to create colorways

60% use CAD to create printed fabric design

48% use CAD to create merchandising presentation

41% use CAD to create knitwear designs

Design choices and visual possibilities can be infinite if the designer is given the time and freedom to be creative and to experiment using the computer. Today automation is not only used for substituting labor but is also adopted for improving quality and producing quantity in lesser time. However, a CAD system is only as good (or as bad) as the designer working on it. The computer only speeds up the process of repeat making, color changing, motif manipulation, and so on. It is actually the Computer Aided Manufacturing (CAM) aspect of CAD that will help reduce lead time.

15.2 Textile Design Systems

Woven textiles are used by designers and merchandisers for fabrics for home furnishing and for men-women-children wear. Most fabrics, whether yarn dyes, plain weaves, jacquards, or dobbies, can be designed and in fact are invariably used abroad using a CAD system for textiles. Similarly, embroideries are also developed at CAD workstations (Figures 15.1 and 15.2).

15.2.1 Knitted Fabrics

Some systems specialize in knitwear production, and the final knitted design can be viewed on screen with indication of all stitch formation. For instance, a CAD program will produce a pullover graph that will indicate information on amount of yarn needed by color for each piece. Another example of the new technology is a yarn scanner attached to the computer that scans a thousand meters of yarn and then simulates a knitted/woven fabric on-screen. This simulation will show how the fabric will look like if woven from that yarn.

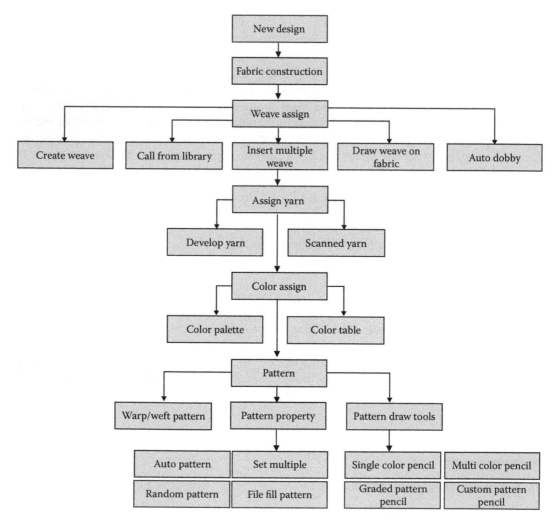

FIGURE 15.1
CAD/CAM solutions in textiles.

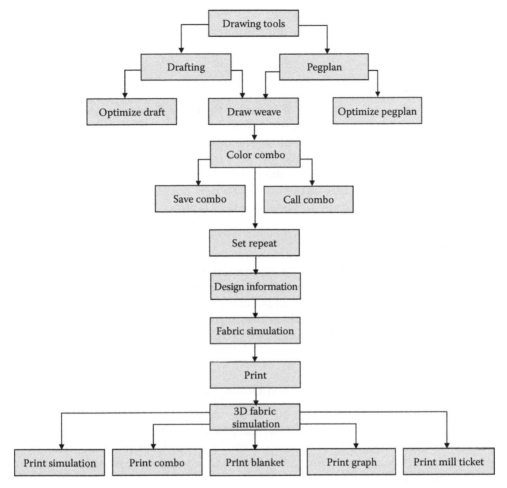

FIGURE 15.2
CAD/CAM applications.

15.2.2 Printed Fabrics

This process involves use of computers in design, development, and manipulation of motif. The motif can then be resized, recolored, rotated, or multiplied depending on the designer's goal. Textures and weave structures can be indicated so that printout either on paper or actual fabric looks very much the way the final product will look. The textile design system can show colorways in an instant rather than taking hours needed for hand painting. New systems are coming that have built-in software to match swatch color to screen color to printer color automatically, so what you see is what you get.

15.2.3 Illustrations/Sketch Pad Systems

These are graphic programs that allow the designer to use pen or stylus on electronic pad or tablet thereby creating freehand images, which are then stored in the computer. The end product is no different from those sketches made on paper with pencil. They have the additional advantage of improvement and manipulation. Different knit and weave simulations can be stored in a library and imposed over these sketches to show texture and dimensions.

15.2.4 Texture Mapping

This technology allows visualization of fabric on the body. Texture mapping is a process by which fabric can be draped over a form in a realistic way. The pattern of the cloth is contoured to match the form underneath it. The designer starts with an image of a model wearing a garment. Each section of the garment is outlined from seamline to seamline. Then a swatch of new fabric created in the textile design system is laid over the area and the computer automatically fills in the area with new color or pattern. The result is the original silhouette worn by original model in a new fabric.

15.2.5 Embroidery Systems

The designs used for embroidery can be incorporated on the fabric for making a garment. For this, special computerized embroidery machines are used. Designers can create their embroidery designs or motifs straight on the computer or can work with scanned images of existing designs. All they need to do is assign color and stitch to different parts of the design. This data is then fed into an embroidery machine with one or multiple heads for stitching.

15.2.6 Design Desk—For Yarn Dyed and Dobby Woven Fabrics

Design Desk is a comprehensive tool kit for the technically discerning and a very simple and powerful CAD system for the visualizer or weaving novice. This system handles weaves, yarns, patterns, and construction parameters to instantly simulate realistic fabric on screen. It can generate true to life simulations and a detailed weaving instruction sheet with yarn consumption and cost calculation.

15.2.6.1 Yarn Development and Management

Designed for the ease of use and strong facilities, the Yarn Development Module creates all types of yarns such as dyed, fancy, melange, slubs, loop, taspa, and so on, or a combination of different fancy effects. Different types of fibers, twist per inch, and direction can be specified to ensure an ideal simulation environment. Simple color controls to research the ideal match and also standard color palettes are accessible. The developed yarns can be visualized in real time for evaluation and interactive changes. Also, the fabric effect of yarn developed can be simulated. The yarn can finally be saved in an efficient yarn management library for easy retrieval and usage with count and cost parameters.

15.2.6.2 Weave Creator

Defining draft, a peg plan can be done numerically or directly drawn on point paper. Simple-to-use cut, copy, paste, and mirror tools are supported. Weave book manages the library of weaves and provides controls for creating complex weave structures by combining weaves. Advanced auto weave generation facility provides for creation of different design effects within the same draft order while changing the lifting plan. An auto weave insertion facility is available for integrating dobby effects directly onto a design in ornamentation or extra warp mode.

15.2.6.3 Design Creation

Design patterns can be numerically specified using the VIDE Direct draw facility, wherein patterns can be specified and scaled online. Weft pattern is adjusted automatically considering the EPI/PPI aspect ratio. Drag and drop operation can be used to specify warp and weft yarns. A simple graphical user interface enables you to easily access all designing control, and design creation is easy. Realistic fabric simulation is generated instantly, combining weaves, design patterns, yarn count, and construction parameters defined. A simulation window can be expanded to develop eight matching for comparative evaluation. Advanced facility generates matching automatically.

15.3 CAD/CAMs Effect on the Jacquard Weaving Industry

15.3.1 Jacquard Design History

Jacquard design has gone through two major phases in the 20 years since the computer was introduced in the mid-1970s. Pre-computer designing process:

- An artist drew onto paper suitable designs.
- A technical (stylist) then translated the artwork into a design that could be woven.
- A technician then punched the cards directly from the technical (stylist's) drawing.

Control points for the above workflow:

- The artwork needs to match the fashion of the day.
- The expertise needed to control the technician's translation.
- The final woven cloth after all the [samples] have been made.

First computerized designing solution:

- An artist drawing on paper; The same as before.
- A technical (stylist) translated the artwork; The same as before.
- The technician who punched the cards now used a scanner to read in the technical design.

The improvement was that the technical design was visible on a monitor and so could be controlled and corrected.
  Control points for the above workflow

1. The artwork needs to match the fashion of the day.
2. The expertise needed to control the technician's translation.

3. The final woven cloth after all the [samples] have been made.

4. The design could now be modified quickly. The first stage of the computerization mainly ran from the mid-1970s to the mid-1980s.

Second computerized designing solution:

a. An artist drawing on paper: the same as before.

b. The artist's sketch was read directly onto the scanner and special software was used to help separate the sketch into a technical drawing. This drawing was controlled by the skilled technician and corrected. The card punching became simpler and more automated.

c. A new process was introduced by Sophis called "Simulation." A simulation is the technical design, with its extra data—weaves and loom information—translated into a picture that is similar to the final woven cloth.

Control points for the above workflow

1. The artwork needs to match the fashion of the day.

2. The technicians translate the design into a weaveable product.

3. Simulation of the final woven cloth.

4. Make the final woven cloth. This stage of the computerization ran from the late 1980s to the mid-1990s. However, it must be recognized that there are many sites where the first stage is still in operation and even sites where the two stages coexist.

15.3.1.1 Industrial and Commercial Trends

Personnel: The industry used to be technically driven. Now it is design driven. This means that more and more companies have opted to take designers from art schools (instead of textile technical schools). As a result, their design skills are good, but they do not have knowledge of the limits as well as the possibilities of jacquard and dobby.

Design life times: The design that sold one million yards per year has become a thing of the past. Today, the runs are shorter and designs are valid for shorter times. A private survey made from data in the region of Kortrijk, Belgium, showed that the number of designs made in 1990 were three times more than in 1980, but the number of yards woven per design was only one third the length.

Head and loom design: The jacquard heads are getting bigger every year. Today it is relatively easy to find a loom with a pattern repeat of 1.4 m. The looms are faster and run with high efficiencies with only one weaver per 20 machines. This means the labor cost per meter goes down, but the capital cost stays high. This is an ideal combination for high labor cost, high skill countries such as in Europe and the United States.

Networking: When the jacquards went electronic, it was only a matter of time before they were networked. It is still a surprise that only a small portion of the total weavers are networked today. The network has two obvious advantages: lower labor cost and less errors. But, the real advantage is that the cost of design change is lower, so the firm can make even shorter run lengths without reducing profits.

Third computerized designing solution is current today and is expected to run into the middle of the next decade. The trend is to have more and more designs while the personnel has less detailed jacquard knowledge. To overcome these two trends, Sophis has moved to a new structure in our software.

The first simplification Sophis made was to have the design problem split up into different areas of expertise and to in such a way that each part works independently from the other but still uses the relevant data from each area. These areas include (a) the technical information: yarns loom information and qualities/constructions weaves and weave families and (b) the design information. This means that the knowledge in each area has only to be in one place, but the information is available to the other people in the design process without them having to understand how to make the files or to have in-depth knowledge.

History has seen mankind expressing art on fabrics using complex weave structures to enliven forms, motifs, and nature to create tapestry, carpets, and furnishings. Millions of interlacements are painstakingly defined to create a weave plan that earlier took months to accomplish. Design Jacquard is a tribute to our predecessors wherein an array of tools is provided for creating artwork interactively, which facilities to creating and attributing weaves to the artwork. It goes further in simulating the fabric and transfers the design to Electronic Jacquard for flawless weaving.

15.3.1.2 The Designer

Designers work in many different ways, and it is important not to limit them. The Sophis system is designed to allow them to move between the traditional design methods to the new way at any time in the process.

15.3.1.2.1 The Idea Phase—Scanning and Editing

The new software from Sophis allows the user to work directly on the scanned image. This means the user can work with designs in 24-bit format (16.7 million colors). In this mode, they are operating purely as artists. The designer can mix and merge designs, scan in pieces and join them together into one design, cut and paste, and use many other design tools. The limits of the conventional jacquard design process are not imposed at this stage and so the creativity of the artist is free to grow.

15.3.1.2.2 The Woven Phase—Translating the Idea into Fabric

- *Preparing the design*: The pattern can be automatically reduced into a limited number of colors and into the correct resolution for weaving. To do this, the program refers to the Quality Files and works out the correct resolution and size that the loom needs. This means that the designer does not need to be bothered with the mathematics.
- *Translating the design*: The design is now automatically converted into a woven article. Again, using the information in the Quality File, the program can automatically choose the weaves and display the pattern as a simulation of the final cloth. We have a special program for drawing right on the simulation.

The artist is effectively sitting at the loom. With the movement of the pen, the artist can change the cards and get a whole new sample of the cloth. On this simulation, the designer is able to change and correct the design, experiment with different weaves, and recolor the fabric.

The major point is that the artist does not have to understand the actual jacquard process. They do not have to understand because the computer is translating the design into cloth in real time, so they are designing cloth as cloth, not as artwork that will later be completely changed in order to be woven.

15.3.1.3 Design Editing in Grid

Design editing allows you to edit the artwork, keeping in perspective the differing warp and weft densities. You can edit the design in grid mode for greater accuracy by using a range of drawing, painting, and editing tools provided. Variable brush size with differing X, Y thickness can be used in tandem with the freehand, geometric, beizer, and advanced editing tools such as auto outline and bandani with multiple undo and redo functions. Color shield facility while editing and copy-paste functions enable protection of intricate motifs. Naturally, all editing features are used in the online editing mode with various repeat types such as Straight, Cross, Mirror X, Y, and X–Y. Navigator allows easy access to different design areas while editing in Zoom/GRID mode to insert hooks/picks in artwork stage.

15.3.1.4 Weave Creation

Single and multi-layered weave structures can be achieved easily. Auto satin and twill generator creates satin weaves with varying repeat and steps. The weave library is provided for better management of weaves, and the simple drag and drop function combines weaves at pre-defined intervals to create fascinating weave structures. Complex weaves can be easily created by combining basic weaves by way of simple assignment to the warp and weft interlacement points and specifying the number of repeats. Thus, an otherwise cumbersome operation is reduced to a couple of clicks. Auto generation of weaves by importing bitmap images adds to the designer's creative abilities.

15.3.1.5 Weave Mapper

Weave mappers facilitates assigning of weaves to different areas of artwork denoted by different colors. A simple mouse click creates the detailed weave plan on a graph. Front and back weaves can be viewed simultaneously for better understanding of the final fabric quality. Extra warp and extra weft weaves can be created to give a special touch to your creation. The number of hooks required for the design can be varied online, and the fabric effect can be viewed. A graph can be viewed in mono, artwork, warp, and weft colors for editing and evaluation. Automatic and manual find float facilities locate warp/weft floats exceeding specified limits. You can assign unique colors for warp and weft floats for ease of control and editing. The floats can be edited using the brush tool.

15.3.1.6 Simulation of Fabrics

This feature aids in presenting the concepts in a realistic way at a click of a button and no extra cost at an internal evaluation or a sales presentation; a good simulation is the acid test of your efforts. Simulation is further enhanced by using yarns created or scanned directly into the library. The yarn editor allows you to work on parameters such as twist, colors, thickness, and so on to generate chenille, fancy, slubs, mélanges, twisted, and all types of yarn effects. You can vary the different parameters of a design or edit the design and observe the result instantaneously. The simulated output can be further tested on a 3D CAD system.

15.4 Computer Aided Manufacturing

Compatible auxiliary modules for electronic jacquards such as Bonas, Staubli, Grosse, and so on enable transfer of design details in electronic form. Easily configurable cast out facility for defining hooks have been reassigned from their normal position in harness ends. These modules have a graphical and easy-to-use interface. The user can define a basic casting sequence, which can be replicated across the harness or user-specified range, with a single click.

Casting files can be saved, enabling management of different harness arrangements within the same shed. Weft selector, cramming, terry, and fringing can be defined in the function file. Variable weft density can also be specified in the electronic function file. Graphs can be printed for manual card punching, or the design information can be transferred to an electronic card punching machine for punching hard card/continuous paper punched cards.

15.4.1 The Software Fundamentals of Fashion Design

When it comes to software for the fashion industry, there are several choices in off-the-shelf software, or software that is readily available to anyone, such as Adobe PhotoShop®, Adobe Illustrator®, or CorelDraw®. Proprietary software, software that is specifically designed for use within a given industry, is often a hybrid that allows the user to use a traditional off-the-shelf software in conjunction with an industry software frequently sold as a plug-in. This kind of software has been adapted to work with the off-the-shelf software and includes features useful to the industry. Regardless of which software application is used to develop apparel and textile designs, there still remains the most fundamental choice of whether you will use a vector-based program or a raster-based program to accomplish the task.

15.4.2 Vector-Based Programs

Vector drawings, also known as object-oriented drawings, are images defined by curves and lines or mathematical formulas. Basically, this means that a vector program stores each image as a series of instructions on how to draw the image. These graphical representations of objects usually consist of line drawings or other primitives such as lines, rectangles, ellipses, arc, spline, and curves. In many cases, the type set is generally simpler and can be highly compressed (made smaller). The most important feature of vector-based images is the resolution or clarity of the drawing. Vector images are resolution-independent and always render at the highest resolution an output device can produce. That means the higher the resolution of the monitor or printer, the sharper the object oriented image will appear.

What this means to the fashion designer is that these drawings are easy to select, color, move, resize (without degradation of the image), reorder, over-lap with other images, access individual objects, and reformat (i.e. change color or fill). Vector graphics are also much smaller files than raster/bitmap files. Vector files can be resized without degrading the file in any way. This is not true of raster images.

Vector images are best used when working with small type and bold, smooth, crisp graphics requiring curves and lines. They are considered the most flexible and use relatively little memory for storage. The downside is that vector based images are not as realistic as raster-based images, which can hold a lot more data. Furthermore, they are known for having a flat versus three-dimensional appearance when compared to a raster-based image.

15.4.3 Raster-Based Programs

Raster images create realistic or real-world images. These types of drawing programs allow the designer to refine details and make dramatic changes with special effects options, and are noted for providing a greater degree of subtly than vector-based graphics.

> *Raster/bitmap*: Best for realistic images such as photographs. These images can be transformed by using image editing filters to create a wide range of special effects and natural looks. These programs work with pixels or bitmapped images that can be enhanced with vector style painting options. Bitmap is a collection of picture elements or dots, also known as pixels. Bitmapped images are resolution-dependent. Basically, this means you must specify a resolution. If you create the image and then change the resolution, you'd grade the image. Scaling up can be a real disaster; scaling the image smaller sometimes yields better results. In fact, the raster image is referred to as a bitmap image because it contains information that is directly mapped to the display grid of x (horizontal) and y (vertical) coordinates.

Bitmapped images are best used with continuous tone images such as photographs and can be modified with great detail because you can manipulate each pixel. If you are scanning a hand drawn image that you plan to modify, you will want to save it as a raster image. Bitmapped images are difficult to modify and to resize, and it can be difficult to freely access objects individually.

Advantages and disadvantages of raster images include:

1. Enlarging: Suffers from aliasing or blurred appearance when enlarged.
2. Reduction of image can result in interpolation or indiscriminate discarding of pixels.
3. Can modify individual pixels or large groups of pixels.
4. Require huge amounts of memory. Usually larger than a vector file; this means they should be compressed to store.
5. Data compression can shrink the size of the pixel data.
6. Slows down the reading, rendering, and printing.

Adobe PhotoShop® is one of the most widely used image editing programs. Another leading painting program for editing and enhancing photos is Corel Painter. Fashion designers love Painter because it simulates natural mediums such as charcoals, chalks, oils, and acrylics to enhance photographic images.

15.4.4 Common File Formats

Each image created will need to be saved in a specific file format that is native to the application. Naming and saving an image along with the file format extension will make it easy for other users of the images to identify the type of drawing it is and what applications may be used to open it. Typically, this extension is added to a file automatically in its own default known as a native format. As in the case of Adobe Illustrator, the file will be saved automatically with an ".ai" extension unless you assign the file another extension.

The challenge arises when the native file is not always readable in another application. This means after you name a file, you have to give the file a special identifying code after its name that will enable you to open the file, no matter what program or platform is used.

15.4.5 Texture Mapping

While designers have historically used CAD software to create sketches, croquis, repeats, patterns, and silhouettes, the latest developments in texture mapping design technology have not only improved the designer's toolset for rendering realistic looking products, but have evolved into web-enabled tools designed for use by the consumer for the development of customized products as well.

Texture mapping, also known as digital draping, is a visualization tool that creates photo-realistic 3D rendering of designs, colors, surfaces, textures, and patterns onto photographs for virtual product generation. It offers life-like representations of digital product samples using scanned-in photographs of furniture, models, or room-sets directly from a CAD system. The process enables test marketing of new products without the need to develop physical samples.

The technique requires the user to identify areas on the photo where a pattern will be applied (often called segments), and to create a series of grid lines that "map" the drape of the fabric across the various segments. Once the photo has been prepared, it may be stored and reused over and over again with alternative designs or textures.

While texture mapping technology has been available to designers for more than a decade and many companies have realized the time and cost benefits, the first generation tools were time consuming and challenging to use. With new releases from several vendors that include increased functionality for both offline and online use, it's time to take another look.

15.5 CIM—Data Communications Standards for Monitoring of Textile Spinning Processes

The yarn spinning industry has undergone dramatic changes over the past 25 years. The need to reduce costs while improving quality and productivity has become a necessity in order to remain competitive in an increasingly global marketplace. To be competitive in today's market, yarn suppliers must do three things:

- Adapt to changing product requirements quickly
- Adapt to changing product volumes quickly
- Continue to provide low price and high-quality products during these adaptations

Improved technologies and automation have increased the spinning industry's use of computerized monitoring and control. This increased computerization has allowed manufacturers' access to data at a rate not available in the past. Unfortunately, this data has only been accessible at a great expense to yarn manufacturers. Often it is spread out across remote systems in differing environments from the plant floor to the corporate office. Several years ago, data residing in these environments seemed mutually exclusive. Today, however, the need to connect these environments and increase the efficiency in which information is assimilated has become a fiscal necessity. The efforts to connect these environments in order to achieve these goals are commonly referred to as implementing computer integrated manufacturing or CIM.

15.5.1 Data Communications

In order to understand data communications in the textile spinning industry, one must first understand the basics and functions of a data communications network, and the topology of a spinning facility.

15.5.2 Network Function

The basic function of a network is to provide communications between the devices on the network. The simplicity of this definition should not distract from the importance of network communications. The network communications are the backbone of the process being monitored or controlled. Without communications, the process monitoring and/or control is lost.

15.5.3 Layered Network Model

The International Standards Organization (ISO) established a framework for standardizing communications systems called the Open Systems Interconnection (OSI) Reference Model. The OSI architecture establishes a set of seven layers used to define the communications process, with specific functions associated with each layer (Figure 15.3). The Layered network model is as shown in the Figure 15.3.

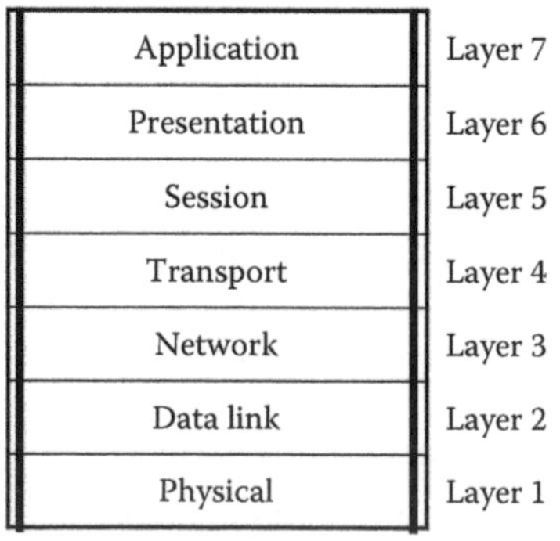

Application	Layer 7
Presentation	Layer 6
Session	Layer 5
Transport	Layer 4
Network	Layer 3
Data link	Layer 2
Physical	Layer 1

FIGURE 15.3
Layered network model.

1. *Physical layer*: The physical layer, at the most basic level, is a set of rules that specifies the electrical and physical connection between devices. The physical layer can be thought of as the network of cables that run through the building that physically connects different devices on a network.

2. *Protocol stack*: The protocol stack is a set of rules responsible for breaking information from one device into packets, routing these packets along the network, and finally reassembling the packets at another device. An example of the protocol stack layer is TCP/IP (Transmission Control Protocol/Internet Protocol), the basis for Internet communications.

3. *Application layer protocol*: While it is the responsibility of the protocol stack to reliably transfer information to and from network devices, it is the application layer protocol's (ALP) responsibility to make sense of the information. The application layer utilizes a data structure to format the information found in the packets and performs the correct task based on the context of this information. Functions performed at this level include file transfers, resource sharing, and database access. Examples of this are TELNET for terminal based connections and File Transfer Protocol (FTP) for electronic file transfer. The ALP can be thought of as common language between network devices.

15.5.4 Analogy of the Three Layer Model and the Telephone System

Consider the telephone system as a data communications network. The phone lines provide the physical medium for transferring data from one phone to another. A phone receives pieces of data from the phone line (in the form of electrical signals) and converts them into another form (sound); the phone executes the functions of the protocol stack. These two pieces alone are not sufficient for effective communication. Consider a call between someone who speaks and understands only German and someone who speaks and understands only English. The lines and telephone can be working properly, but the people are not capable of communicating any useful information due to the lack of a common language. A common ALP implemented in this analogy would be both people speaking the same language (data structure) in order to allow an understanding.

This illustrates the importance to the manufacturing environment of a common ALP. The ALP provides a common language and understanding for the devices on the network. With only a standard physical layer and protocol stack, the devices on a network could not communicate effectively.

15.5.5 Manufacturing Network Topology

The ideal textile manufacturing communication network can be thought of in terms of a hierarchical structure. At the top is a corporate office computer branching down to a work-cell or group of machines and from there, branching to individual machines.

15.5.6 Machine-Level Network

The machine-level network refers to the PLC (or microprocessors), sensors, motors, and so on used to operate a textile machine. At the machine level, the networking systems used are designed for so-called real-time control. These networks consist of direct communication lines between sensors, motors, and so on and control units. For monitoring purposes, these networks are implemented with RS-232, RS-485, Profibus, and/or other proprietary networks.

15.5.7 Work-Cell Network

Groups of one machine type are usually grouped together in a work-cell. The machines in the work-cell are connected to a monitoring computer. The monitoring computer gathers information from the individual machines. These monitoring computers allow operators/technicians to easily monitor many machines from a single location. Historically, the monitoring computers have been fed information through an RS-232/485 type interface. However, the trend in recent days has been toward Ethernet-based connections.

15.5.8 Data Storage System

Along with the monitoring computer, which merely displays data, there is typically a data storage system or data depository. The data depository is a computer with the ability to store or warehouse data; this data can then be used to perform functions such as statistical process control or production planning. Historically, the data depository has been a proprietary system. Currently, however, vendors have begun a conversion to the-shelf database management systems (DBMS). The trend has been toward PCs based DBMSs such as MS-SQLTM, SybaseTM, InformixTM, or OracleTM.

15.5.9 Corporate Office Computers

Traditionally there has been a gap between the equipment on the plant floor and corporate office computers. These have historically been large mainframe computers, which performed the functions related to production planning and the company payroll, and there was not a great need to connect these computers to the equipment on the plant floor.

However, most computers used in corporate offices today are standard off-the-shelf PCs. Due to the gap between these machines and the plant floor, any data about production conditions or quality has been keyed into corporate computers from reports printed at the data depository level. This is tedious and time consuming, and also eliminates the ability to see data until hours or sometimes days after it is collected. Recent advances in plant networks and software have allowed raw data files from a data depository computer to be transferred to corporate office computers using general-purpose applications like FTP. If no physical connection exists, data transfer may be done with floppy disks. But these methods are not the most effective in transmitting data in a readily usable form. In order to aid management in effective decision making, it has become desirable to have these PCs linked to the monitoring systems on the plant floor to collect production data. The connection of these various systems via a plant-wide network is known as systems integration.

15.5.10 Systems Integration

Historically, there has been no standard protocol to connect the separate levels of the communications hierarchy. For example, connecting the work-cell and machine levels has been difficult, as each was thought of as a separate sub-network, each sub-network requiring separate integration work. The integration of these has normally been beyond the ability of in-house information systems (IS) departments within the industry. Therefore, third-party integrators have done this integration work. Connecting these work-cell networks to corporate PCs has been equally difficult and expensive due to the same problems. The efforts to connect these devices or sub-networks will be explained in the section computer integrated manufacturing in the textile industry.

15.5.11 Components of a Communication Standard for Textiles

Successfully connecting the levels of the communications hierarchy has in the past been a difficult and costly endeavor, due to a lack of standards associated with the communications network. For a standard to be useful, it must address the physical and protocol stack, as well as define an ALP. The standard must also address the data that is to be collected by establishing standard names and definitions for individual data elements.

15.5.12 Data Dictionary

The use of different definitions and terms within an industry is a major problem in creating a standard database. In a large multinational environment, it is necessary to ensure that standard language translations are defined and used everywhere. It is also necessary to ensure that units of measure and time are used consistently within the industry.

15.5.13 Physical Layer and Protocol Stack

Textile machinery manufacturers have simplified the task of defining a standard by voluntarily adopting Ethernet and TCP/IP protocols at the physical and protocol stack layers respectively. These are ideal for the standard because these two protocols are also widely used in PCs found in corporate offices.

15.5.14 Application Layer and Common Data Structure

Machinery manufacturers have struggled with defining a common ALP. Samuel A. Moore evaluated many common ALPs and recommended networked Structured Query Language (SQL) be adopted by the industry as a standard. With the existence of a common data dictionary for spinning and an established ALP, the need arises for a common data structure to make integrated communications a reality. Data structure is defined as a common layout or grouping of the different data elements. This grouping is done in a way that provides the most useful information about the spinning processes. If this seems abstract, it will become clear after a discussion of relational databases.

15.5.15 Application of Relational Databases for Monitoring of Textile Processes

The relational database concept has been chosen as the best method for monitoring production and quality data for textile processes. This is due for the most part to the dynamic nature of the data. The variables for spinning processes are continuously changing at a high rate, and the data need only be kept for some fixed (relatively short) period of time then discarded. With data changing so quickly, any database using parent–child relationships, which require pointers to specific records, would be impossible to administer and maintain. Therefore, the most efficient way to access and manipulate the data is to perform operations on the data values themselves. The relational model is the most common and well-established model to facilitate a dynamic application.

15.5.16 Interface to Uster SliverData System

The Uster SliverData monitoring system is utilized at the College of Textiles for monitoring of the in-house spinning equipment and has been used extensively for development purposes during projects. The system is a single-point monitoring system for multiple textile machines. Two versions of this system are available in the pilot spinning facility. The first system, known as a "greybox," requires interface through an RS-232 connection, utilizing Uster's own proprietary command protocol. The second is a Windows NT based system, developed by Uster through work done with the ATMI CIM subcommittee and influence of current market forces. This system allows access via an Ethernet port and supports the SQL/ODBC protocols. To show the advantages obtained through the use of the communications standards, the following section will illustrate the exercise of interfacing with the Uster SliverData system with and without the communications standards.

15.5.17 Interface Using Proprietary Protocol

Without the standard, extracting data involves interfacing with the SliverData system through an RS-232 interface. Once a physical connection has been established, a program can be written (using ANSI-C in this case) utilizing Uster's proprietary command language to extract data. The Uster SliverData system is designed to transfer several different types of data upon request:

- *Full records*: Shift production and quality data
- *Stop records*: Machine stop history
- *Exception records*: Exception records for quality stops
- *Spectrogram records*: Spectrogram data for machines
- *Diagram records*: data for machines

To extract data, a program must first send a series of "hand shaking" commands to the SliverData system. Once these commands are sent and a proper response has been received, the program can then request data. The requested data is then sent by the SliverData system embedded within a long character string. At any point in time, one of five strings can be requested (corresponding to the data types listed previously). These long strings can then be parsed for the individual data elements. The data elements can then be transferred to an RDBMS server, flat file, spreadsheet, or other custom application. In our case, the desired elements are sent to a RDBMS server residing on the network via SQL statements executed through ODBC function calls. To accomplish this interface, the following information must be known:

1. The commands of Uster's proprietary protocol
2. The command sequence of the protocol
3. The relative location of data elements within Uster's data structure (string)
4. The length (and precision for decimal numbers) of each data element

This is a relatively simple exercise for an experienced programmer. However, variations on this exercise must be done for each vendor's monitoring system within a mill, each having its own proprietary command protocol. Also, getting a small set of data elements takes the same amount of work as gathering a complete set, since the long record strings must be parsed to find the desired elements.

15.5.18 Interface Using Standard Protocols

With the standard implemented on Uster's SliverData NT interface, the exercise of extracting data becomes much simpler and more direct. This is the case because Uster now provides data elements in a Microsoft Access database, which is SQL/ODBC compliant. Once a physical connection has been established via an Ethernet connection, the data residing in the Access database can be transferred to an RDBMS server using some type of generic ODBC program as described earlier (again using ANSI-C for demonstration purposes). This generic ODBC program retrieves data by executing an SQL SELECT statement through an ODBC function call. The advantages to this method are obvious: (1) no proprietary commands need be known, and (2) nothing need be known about how or where the Access database stores the individual data elements. All that must be known to retrieve data are

1. The "short name" of the data elements
2. The table names in which they are located

The complexity of code needed to retrieve data in this way is drastically reduced. The generic ODBC software has been mentioned previously, and the following section details the software architecture.

15.5.19 Application of Common Database

The existence of a common database in a spinning facility will provide data in a readily accessible form for many applications that can transform this data into information and aid companies in accomplishing the ultimate goal of CIM. End users would ideally like to have available methods such as graphical user interfaces, statistical techniques and windowing mechanisms, and other visual techniques that allow access to the data being sought.

15.5.20 Flexible Applications

When discussing the design of the database, it has been discussed that the design depends on its intended use. This does not mean that the design in any way dictates the type of applications that could be applied, only that they must be within the realm of the data available. The data elements were grouped into tables based on general relationships they have with each other, as well as how the elements relate to the spinning machinery. Therefore, applications are limited to extracting data only within the parameters of the data available. This is quite obvious. But where, when, to whom, and in what form the data is presented is left up to the textile spinners. The data structure is not intended to be construed as an all-inclusive canned package but instead is meant to be a tool used by spinners to access what data they desire and display it in any form that they deem useful.

Many of the applications discussed here are not new; in fact, most are available in some form from the proprietary monitoring systems of the machinery manufacturers. However,

(1) these systems are valid usually for only one manufacturer's equipment, and (2) the applications are limited in the way data is presented. The graphical user interfaces (GUIs) for these systems are much different so training on each one is necessary, and there is seldom any room for changing or developing the type and way the information is presented. From the data as presented in the database with no modification, spinners can track production rates (by machine, position, section, or plant) or gather quality data on a machine running a certain style of yarn. A major application is to provide data from the common data location to statistical process control packages.

15.6 Summary

This chapter discusses the importance of CAD/CAM for the textile industry. The design procedure and the CAM usage in garment industries are explained in detail.

References

1. http://softwaresolutions.fibre2fashion.com/special-feature/designing-solutions/.
2. http://www.indiantextilejournal.com/articles/FAdetails.asp?id=4058.
3. https://textileapex.blogspot.in/2014/07/computers-use-in-textile-apparel.html.
4. http://www.fibre2fashion.com/industry-article/2433/technology-and-its-impacts-?page=18.
5. https://www.scribd.com/document/300024515/Importance-of-Computer-in-Field-of-Fashion-Designing-in-Apparel-Industrial.
6. http://ijoes.vidyapublications.com/paper/Vol13/06-Vol13.pdf.
7. http://edutechwiki.unige.ch/en/Computerized_embroidery.
8. https://www.textronic.com/design-dobby.html.
9. https://www.textronic.com/pdf/Design_Jacquard.pdf.
10. https://www.textronic.com/design-jacquard.html.
11. http://www.brstudio.com/news/textronics-design-jacquard-10-11-weave-maker-crack.html.
12. https://www.scribd.com/presentation/79684617/Fabric-Design-Software.
13. http://www.indiantextilejournal.com/articles/FAdetails.asp?id=553.
14. http://www.techexchange.com/library/The%20Software%20Fundamentals%20of%20Fashion%20Design%20-%20Getting%20It%20Right%20The%20First%20Time!.pdf.
15. https://www.prepressure.com/library/file-formats/bitmap-versus-vector.
16. https://helpx.adobe.com/photoshop/using/file-formats.html.
17. http://guides.lib.umich.edu/c.php?g=282942&p=1885348.
18. http://www.techexchange.com/library/Zooming%20in%20on%20Texture%20Mapping.pdf.
19. http://www.academia.edu/21601537/Communications_network_standards_in_the_textile_industry.
20. https://www.iso.org/obp/ui/#iso:std:iso:16373:-1:ed-1:v1:en.
21. https://www.uster.com/de/about-uster/press-room/news/2017/07/06/the-common-language-of-textile-quality/.
22. http://www.indiantextilejournal.com/articles/FAdetails.asp?id=5445.

Index

Note: Page numbers followed by f and t refer to figures and tables respectively.